SMART HORTICULTURE PRODUCTION

Integrating Advanced Technologies

Innovations in Horticultural Science

SMART HORTICULTURE PRODUCTION

Integrating Advanced Technologies

Rajni Rajan, PhD
Duniya Ram Singh, PhD
Editors

AAP | APPLE ACADEMIC PRESS

First edition published 2026

Apple Academic Press Inc.
1265 Goldenrod Circle, NE,
Palm Bay, FL 32905 USA

760 Laurentian Drive, Unit 19,
Burlington, ON L7N 0A4, Canada

CRC Press
2385 NW Executive Center Drive,
Suite 320, Boca Raton FL 33431

4 Park Square, Milton Park,
Abingdon, Oxon, OX14 4RN, UK

Apple Academic Press exclusively co-publishes with CRC Press, an imprint of Taylor & Francis Group, LLC

Library and Archives Canada Cataloguing in Publication

CIP data on file with Canada Library and Archives

Library of Congress Cataloging-in-Publication Data

CIP data on file with US Library of Congress

ISBN: 978-1-998511-66-2 (hbk)
ISBN: 978-1-998511-67-9 (ebk)

DOI: 10.1201/9781998511679

Innovations in Horticultural Science

Editor-in-Chief:

Dr. Mohammed Wasim Siddiqui, Assistant Professor-cum-Scientist
Bihar Agricultural University | www.bausabour.ac.in
Department of Food Science and Post-Harvest Technology
Sabour | Bhagalpur | Bihar | P. O. Box 813210 | INDIA
Contacts: (91) 9835502897
Email: wasim_serene@yahoo.com | wasim@appleacademicpress.com

The horticulture sector is considered as the most dynamic and sustainable segment of agriculture all over the world. It covers pre- and postharvest management of a wide spectrum of crops, including fruits and nuts, vegetables (including potatoes), flowering and aromatic plants, tuber crops, mushrooms, spices, plantation crops, edible bamboos etc. Shifting food pattern in wake of increasing income and health awareness of the populace has transformed horticulture into a vibrant commercial venture for the farming community all over the world.

It is a well-established fact that horticulture is one of the best options for improving the productivity of land, ensuring nutritional security for mankind and for sustaining the livelihood of the farming community worldwide. The world's populace is projected to be 9 billion by the year 2030, and the largest increase will be confined to the developing countries, where chronic food shortages and malnutrition already persist. This projected increase of population will certainly reduce the per capita availability of natural resources and may hinder the equilibrium and sustainability of agricultural systems due to overexploitation of natural resources, which will ultimately lead to more poverty, starvation, malnutrition, and higher food prices. The judicious utilization of natural resources is thus needed and must be addressed immediately.

Climate change is emerging as a major threat to the agriculture throughout the world as well. Surface temperatures of the earth have risen significantly over the past century, and the impact is most significant on agriculture. The rise in temperature enhances the rate of respiration, reduces cropping periods, advances ripening, and hastens crop maturity, which adversely affects crop productivity. Several climatic extremes such as droughts, floods, tropical cyclones, heavy precipitation events, hot extremes, and heat waves cause a negative impact on agriculture and are mainly caused and triggered by climate change.

In order to optimize the use of resources, hi-tech interventions like precision farming, which comprises temporal and spatial management of resources in horticulture, is essentially required. Infusion of technology for an efficient utilization of resources is intended for deriving higher crop productivity per unit of inputs. This would be possible only through deployment of modern hi-tech applications and precision farming methods. For improvement in crop production and returns to farmers, these technologies have to be widely spread and adopted. Considering the above-mentioned challenges of horticulturist and their expected role in ensuring food and nutritional security to mankind, a compilation of hi-tech cultivation techniques and postharvest management of horticultural crops is needed.

This book series, Innovations in Horticultural Science, is designed to address the need for advance knowledge for horticulture researchers and students. Moreover, the major advancements and developments in this subject area to be covered in this series would be beneficial to mankind.

Topics of interest include:

1. Importance of horticultural crops for livelihood
2. Dynamics in sustainable horticulture production
3. Precision horticulture for sustainability
4. Protected horticulture for sustainability
5. Classification of fruit, vegetables, flowers, and other horticultural crops
6. Nursery and orchard management
7. Propagation of horticultural crops
8. Rootstocks in fruit and vegetable production
9. Growth and development of horticultural crops
10. Horticultural plant physiology
11. Role of plant growth regulator in horticultural production
12. Nutrient and irrigation management
13. Fertigation in fruit and vegetables crops
14. High-density planting of fruit crops
15. Training and pruning of plants

16. Pollination management in horticultural crops
17. Organic crop production
18. Pest management dynamics for sustainable horticulture
19. Physiological disorders and their management
20. Biotic and abiotic stress management of fruit crops
21. Postharvest management of horticultural crops
22. Marketing strategies for horticultural crops
23. Climate change and sustainable horticulture
24. Molecular markers in horticultural science
25. Conventional and modern breeding approaches for quality improvement
26. Mushroom, bamboo, spices, medicinal, and plantation crop production

BOOKS IN THE SERIES

- **Spices: Agrotechniques for Quality Produce**
  Amit Baran Sharangi, PhD, S. Datta, PhD, and Prahlad Deb, PhD
- **Sustainable Horticulture, Volume 1: Diversity, Production, and Crop Improvement**
  Editors: Debashis Mandal, PhD, Amritesh C. Shukla, PhD, and Mohammed Wasim Siddiqui, PhD
- **Sustainable Horticulture, Volume 2: Food, Health, and Nutrition**
  Editors: Debashis Mandal, PhD, Amritesh C. Shukla, PhD, and Mohammed Wasim Siddiqui, PhD
- **Underexploited Spice Crops: Present Status, Agrotechnology, and Future Research Directions**
  Amit Baran Sharangi, PhD, Pemba H. Bhutia, Akkabathula Chandini Raj, and Majjiga Sreenivas
- **The Vegetable Pathosystem: Ecology, Disease Mechanism, and Management**
  Editors: Mohammad Ansar, PhD, and Abhijeet Ghatak, PhD
- **Advances in Pest Management in Commercial Flowers**
  Editors: Suprakash Pal, PhD, and Akshay Kumar Chakravarthy, PhD
- **Diseases of Fruits and Vegetable Crops: Recent Management Approaches**
  Editors: Gireesh Chand, PhD, Md. Nadeem Akhtar, and Santosh Kumar
- **Management of Insect Pests in Vegetable Crops: Concepts and Approaches**
  Editors: Ramanuj Vishwakarma, PhD, and Ranjeet Kumar, PhD

- **Temperate Fruits: Production, Processing, and Marketing**
  Editors: Debashis Mandal, PhD, Ursula Wermund, PhD,
  Lop Phavaphutanon, PhD, and Regina Cronje
- **Diseases of Horticultural Crops: Diagnosis and Management,**
  Volume 1: Fruit Crops
  Editors: J. N. Srivastava, PhD, and A. K. Singh, PhD
- **Diseases of Horticultural Crops: Diagnosis and Management, Volume 2: Vegetable Crops**
  Editors: J. N. Srivastava, PhD, and A. K. Singh, PhD
- **Diseases of of Horticultural Crops: Diagnosis and Management, Volume 3: Ornamental Plants and Spice Crops**
  Editors: J. N. Srivastava, PhD, and A. K. Singh, PhD
- **Diseases of Horticultural Crops: Diagnosis and Management, Volume 4: Important Plantation Crops, Medicinal Crops, and Mushrooms**
  Editors: J. N. Srivastava, PhD, and A. K. Singh, PhD
- **Biotic Stress Management in Tomato**
  Editors: Shashank Shekhar Solankey, PhD, and Md. Shamim, PhD
- **Medicinal Plants: Bioprospecting and Pharmacognosy**
  Editors: Amit Baran Sharangi, PhD, and K. V. Peter, PhD
- **Tropical and Subtropical Fruit Crops: Production, Processing, and Marketing**
  Editors: Debashis Mandal, PhD, Ursula Wermund, PhD,
  Lop Phavaphutanon, and Regina Cronje
- **Minor Fruits: Nutritional Composition, Bioactive Potential, and Their Food Applications**
  Editors: Sajad Ahmad Sofi, PhD, Khalid Muzaffar, PhD,
  Basharat Nabi Dar, PhD, and Shabir Ahmad Mir, PhD
- **Smart Horticulture Production: Integrating Advanced Technologies**
  Editors: Rajni Rajan, PhD, and Duniya Ram Singh, PhD

About the Editors

Rajni Rajan, PhD
Subject Matter Specialist (Climate Change), Bankers Institute of Rural Development, Lucknow, Uttar Pradesh, India

Rajni Rajan, PhD, is currently working as a Subject Matter Specialist (Climate Change) in the Bankers Institute of Rural Development, Lucknow, UP, India. Her research work focuses on fruit breeding, biotechnology, tissue culture, abiotic stress management, and mutation breeding in fruit crops. She explores the area of genomics, metabolomics, genome editing, and the implication of artificial intelligence in fruit crops. She qualified in JRF-NTS, SRF-NTS, and ARSB NET in Fruit Science. She also received a best oral and poster presentation awards for her papers at various international and national conferences. She has published 21 research papers, 10 book chapters, and several popular articles in reputed journals and magazines. She serves on the editorial boards of multiple reputable publications. She has also published two patents and was awarded one design patent by the Ministry of Commerce and Industry, Government of India. She is a lifelong member of many reputable organizations as well. She received an outstanding MSc Thesis award at the International Conference on Food Security through Agriculture & Allied Sciences, Kathmandu, Nepal.

Duniya Ram Singh, PhD
Vice Chancellor, Bihar Agricultural University (BAU), Sabour, Bihar, India

Duniya Ram Singh, PhD, currently serves as the Vice Chancellor of Bihar Agricultural University (BAU), Sabour, Bihar, India. Prior to this, he held the esteemed position of Vice Chancellor at Chandra Shekhar Azad University of Agriculture and Technology (CSAU&T), Kanpur, Uttar Pradesh, where he spearheaded its accreditation as the first NAAC-accredited State Agricultural University in the country. Additionally, he has served as the Vice Chancellor in charge of Chhatrapati Shahu Ji Maharaj University, Kanpur, and Sardar Vallabhbhai Patel University of Agriculture and Technology, Meerut, Uttar Pradesh.

Dr. Singh's tenure as the Director of ICAR-National Research Centre for Orchids in Sikkim and his 14-year career at ICAR-Central Island Agricultural Research Institute, Port Blair, as a Senior Scientist, Principal Scientist, Division Head, and Acting Director, have been marked by significant contributions to agricultural research and development. His academic journey began as an Assistant Professor at Acharya N. G. Ranga Agricultural University, Hyderabad, now known as Professor Jayashankar Telangana State Agricultural University. A prolific scholar, Dr. Singh has published 250 research papers in prestigious national and international journals, authored 19 books, 43 bulletins, and eight policy papers. His work has earned him 32 awards from national and international bodies, reflecting his profound impact on agricultural sciences. His research is extensively cited across multiple disciplines, underscoring his influence in the field.

Under his visionary leadership, CSAU&T, Kanpur, and BAU Sabour achieved notable milestones, including more than 10 patents and the release of several crop varieties each from the Central Varietal Release Committee (CVRC) and State Variety Release Committee (SVRC). He pioneered the concept of malnutrition-eradicated villages, exemplified by the establishment of India's first bio-fortified village, Anupur in Kanpur Dehat, which gained recognition from policymakers at various levels. His innovative approach during the COVID-19 pandemic included organizing a highly successful national webinar, appreciated by the Hon'ble Prime Minister of India. An environmental advocate, Dr. Singh initiated the planting of 60,000 plantlets, earning numerous green energy awards. He also transformed the CSAU&T Central Library into a 24/7 resource center and established a state-of-the-art Student Placement and Career Counseling Cell.

In the aftermath of the 2004 Tsunami, Dr. Singh played a pivotal role in domesticating the wild plant Noni (*Morinda citrifolia*), leading to the release of four commercial varieties and increasing its cultivation to over 100 acres. His work in promoting Noni as a major crop in the Andaman and Nicobar Islands, alongside coconut and areca nut, has been groundbreaking. He also developed a standardized tissue culture protocol for dragon fruits and introduced three new orchid varieties: Namo, Sardar, and Deendayal. His efforts have reduced India's dependency on foreign orchids and curbed biopiracy in the Northeastern states.

Dr. Singh's dedication to academic excellence is also demonstrated by his unique achievement of conducting two convocation ceremonies in a single day at Kanpur. His career stands as a testament to his commitment to advancing agricultural science and education, leaving an indelible mark on the institutions he has served and the broader agricultural community.

Contents

Contributors

Ali Abdullahi
Department of Agricultural Economics and Extension, Bayero University, PMB, Kano Nigeria

Ambuj
Indian Institute of Technology Kharagpur, Kharagpur, West Bengal, India

Mohd Talha Anees
Department of Geology, Faculty of Science, Universiti Malaya, Kuala Lumpur, Malaysia

Khalangre Atul
Department of Horticulture, Lovely Professional University, Phagwara, Punjab, India

Arindam Chakraborty
Department of Geology, Faculty of Science, Universiti Malaya, Kuala Lumpur, Malaysia

Rakesh Kumar Gupta
Department of Agricultural and Food Engineering Indian Institute of Technology, Kharagpur, West Bengal, India

K. Jaisuriyan
Department of Agronomy, School of Agriculture, Lovely Professional University, Phagwara, Punjab, India

Sapna Jarial
Department of Agricultural Economics and Extension, School of Agriculture, Lovely Professional University, Phagwara, Punjab, India

Pratap Kadam
National Referral Laboratory, ICAR-National Research Centre for Grapes, Pune, Maharashtra, India

Harjinder Kaur
Department of Horticulture, Lovely Professional University, Phagwara, Punjab, India

Rupesh Kaushik
Department of Horticulture, School of Agriculture, Lovely Professional University, Phagwara, Punjab, India

Pankaj Kumar
Department of Horticulture, School of Agriculture, Lovely Professional University, Phagwara, Punjab, India

Prasann Kumar
Department of Agronomy, School of Agriculture, Lovely Professional University, Phagwara, Punjab, India

Sushil Kumar
ICAR-Central Potato Research Institute, Regional Station, Jalandhar, Punjab, India

Rajendra Machavaram
Indian Institute of Technology Kharagpur, Kharagpur, West Bengal, India

Anis Mirza
National Referral Laboratory, ICAR-National Research Centre for Grapes, Pune, Maharashtra, India

P. K. Mishra
Acharya Narendra Deva Agriculture University and Technology, Kumarganj, Ayodhya, Uttar Pradesh, India

Sourabh Mishra
Hohal University, China, Xi Kang Lu, Gulou, Nanjing, Jiangsu, China

Harsh Nagar
Indian Institute of Technology Kharagpur, Kharagpur, West Bengal, India

S. K. Pandey
Acharya Narendra Deva Agriculture University and Technology, Kumarganj, Ayodhya, Uttar Pradesh, India

T. Pandiaraj
Acharya Narendra Deva Agriculture University and Technology, Kumarganj, Ayodhya, Uttar Pradesh, India

Amirhossein Pirmoradi
Department of Agricultural Extension, Faculty of Agriculture, Razi University, Kermanshah, Iran

T. P. Ahamed Shabeer
National Referral Laboratory, ICAR-National Research Centre for Grapes, Pune, India

Hrutik Shevkar
Department of Agriculture-Food Science and Technology (DISTAL), University of Bologna, Bologna, Italy

Gurpreet Singh
Department of Horticulture, Lovely Professional University, Phagwara, Punjab, India

Prem Prakash Srivastav
Department of Agricultural and Food Engineering Indian Institute of Technology, Kharagpur, West Bengal, India

Gajendra Singh
Indian Institute of Technology Kharagpur, Kharagpur, West Bengal, India

Harpinder Singh
School of Mechanical Engineering, Lovely Professional University, Phagwara, Punjab, India

Jatinder Singh
Department of Horticulture, School of Agriculture, Lovely Professional University, Phagwara, Punjab, India

Smita Rani Swain
Department of Agricultural and Food Engineering Indian Institute of Technology, Kharagpur, West Bengal, India

Bharat Taynath
National Referral Laboratory, ICAR-National Research Centre for Grapes, Pune, Maharashtra, India

Vandana Thakur
School of Agriculture, Department of Horticulture, Lovely Professional University, Phagwara, Punjab, India

Hina Upadhyay
Department of Agronomy, School of Agriculture, Lovely professional University, Phagwara, Punjab, India

Shashank Verma
Acharya Narendra Deva Agriculture University and Technology, Kumarganj, Ayodhya, Uttar Pradesh, India

Abbreviations

ACO	ant colony optimization
AGB	above ground biomass
AGV	automated guided vehicle
AI	artificial intelligence
ALNS	adaptive large neighborhood search
AMS	automated milking systems
An	antracol
ANN	artificial neural networks
ARIMA	autoregressive integrated moving average
ASCs	agricultural supply chains
BPNN	backpropagation neural network
CAE	convolutional autoencoder
CCD	charge-coupled device
CEA	controlled environment agriculture
CNN	convolutional neural network
CRISPR	Clustered Regularly Interspaced Short Palindromic Repeats
CWSI	Crop Water Stress Index
DCNN	deep convolutional neural networks
DGPS	differential global positioning system
DL	deep learning
DT	decision trees
ET	evapotranspiration
FAO	Food and Agriculture Organization
GANs	generative adversarial networks
GIS	geographic information systems
GM	genetically modified
GO	graphene oxide
GP	genomic prediction
GPS	global positioning system
GSM	global system for mobile
HDP	high-density planting
ICT	information and communications technology
INM	Integrated Nutrient Management
IoT	Internet of Things
IPM	integrated pest management
KNN	k-nearest neighbors

LAI	leaf area index
LCA	life cycle assessment
LGT	lane-guided transport
LIDAR	light detection and ranging
LSTM	long short-term memory
MAS	marker-assisted selection
ML	machine learning
MLR	multiple linear regression
nano-HAP	nanohydroxyapatite
NDSI	normalization different spectral index
NDVI	Normalized Difference Vegetation Index
NFs	nanofertilizers
NIRS	near infrared spectroscopy
NN	neuronal network
NP	nanoparticle
NSGA II	nondominated sorting genetic algorithm II
PAR	photosynthetically active radiation
PCA	principal component analysis
PLS	partial least squares
PLSR	partial least squares regression
PNN	probabilistic neural network
PPP	public–private partnership
ResNet	residual network
RFID	radio-frequency identification
RFR	random forest regression
RGB	red-green-blue
RH	relative humidity
RMSE	root mean square error
RNN	recurrent neural network
SDG	sustainable development goal
SIS	smart irrigation system
SND	smart nutrient delivery
SNPs	single nucleotide polymorphisms
SSC	soluble solids content
SSRs	simple sequence repeats
SVM	support vector machines
SVR	support vector regression
TA	titratable acid
TALENs	transcription activator-like effector nucleases
TIR	thermal infrared radiation
TP	total phenol
TSEB	two-source energy balance model

t-SNE	t-distributed stochastic neighbor embedding
UAVs	unmanned aerial vehicles
UGVs	unmanned ground vehicles
USGS	United States Geological Survey
VF	vertical farming
VGG	visual geometry group
VI	vegetation indices
VNIRS	visible and near-infrared spectroscopy
VRT	variable rate technology
WSM	warehouse management systems
WSN	wireless sensor networks
XAI	explainable AI
YOLO	you only look once
ZALDH	zinc/aluminum-layered double hydroxide
ZFNs	zinc finger nucleases

Preface

Welcome to *Smart Horticulture Production: Integrating Advanced Technologies*. In the vast landscape of agriculture, horticulture stands as a beacon of innovation and sustainability. With its focus on the cultivation of fruits, vegetables, ornamental plants, and more, horticulture plays a vital role in providing nutritious food, enhancing landscapes, and contributing to economic prosperity worldwide. As we stand at the intersection of technological advancement and agricultural practices, the concept of Smart Horticulture Production emerges as a guiding principle, promising efficiency, precision, and environmental stewardship.

This book, *Smart Horticulture Production: Integrating Advanced Technologies* is a comprehensive exploration of the transformative potential embedded within modern horticultural practices. Through a series of meticulously crafted chapters, we delve into the various facets of smart horticulture, elucidating its principles, methodologies, and practical applications. Each chapter serves as a gateway to a realm where traditional farming meets cutting-edge technology, where age-old wisdom converges with the latest innovations, and where sustainability is not just a goal but a way of life. The journey begins with an exploration of the foundational concepts of smart horticulture, providing readers with a solid understanding of its underlying principles and objectives. From there, we venture into the realm of sensor technologies and data analytics, unraveling the intricacies of monitoring and managing horticultural systems with unprecedented precision. Through discussions on automation, robotics, and artificial intelligence, we witness firsthand how technology is revolutionizing every aspect of horticultural production, from planting and irrigation to pest management and harvesting.

But smart horticulture production is not merely about technology; it is also about sustainability and resilience in the face of evolving environmental challenges. In chapters dedicated to climate-smart practices, resource optimization, and organic cultivation methods, we explore how smart horticulture serves as a bulwark against climate change and resource depletion while fostering biodiversity and ecosystem health. Moreover, this book does not exist in isolation.

It is a testament to the collaborative efforts of researchers, practitioners, and stakeholders who are dedicated to advancing the frontiers of horticulture. Through case studies, expert insights, and real-world examples, we aim to bridge the gap between theory and practice, empowering readers to embrace smart horticulture as a transformative force in their own agricultural endeavors.

As we embark on this journey together, I invite you to open your minds to the infinite possibilities that smart horticulture production offers. Whether you are a seasoned farmer, a budding horticulturist, or simply a curious enthusiast, there is something within these pages to inspire, educate, and ignite your passion for sustainable agriculture.

So, let us embark on this odyssey into the realm of smart horticulture production, where innovation meets tradition, and where the seeds of change blossom into a greener, smarter future.

This book brings together a collective wealth of knowledge, research, and innovation in the realm of temperate nut crops, offering valuable insights for horticulturists, scientists, growers, and nut enthusiasts.

With reverence for nature's bounty and gratitude for your companionship on this journey,

—Rajni Rajan, PhD
Subject Matter Specialist (Climate Change),
Bankers Institute of Rural Development, Lucknow, UP, India

D. R. Singh, PhD
Bihar Agricultural University, Sabour, Bhagalpur, India

CHAPTER 1

Introduction, Importance, Concept, and Application of Smart Farming in Horticulture

HARSH NAGAR*, AMBUJ, and RAJENDRA MACHAVARAM

Indian Institute of Technology Kharagpur, Kharagpur, West Bengal, India

*Corresponding author

ABSTRACT

The agricultural sector is undergoing a significant transformation through Smart Farming, driven by advancements in information and communication technologies (ICTs). This chapter delves into the evolving relationship between technology and horticulture, beginning with a historical overview of traditional farming practices and the growing need for innovation to address global challenges. It introduces the concept of Smart Farming, its components, and its integration into modern horticulture. The role of technologies such as the Internet of Things (IoT), data analytics, and precision agriculture is explored, illustrating how they enhance productivity, sustainability, and resource efficiency. Key innovations—including sensors, automation, robotics, and drones—are examined for their role in modernizing horticultural practices. The chapter also highlights practical applications such as crop monitoring, precision irrigation, pest and disease control, and supply chain optimization. Real-world case studies provide evidence of the tangible benefits and transformative potential of Smart Farming. Furthermore, the discussion addresses barriers to adoption and offers strategies to overcome them. Looking ahead, the chapter explores

Smart Horticulture Production: Integrating Advanced Technologies.
Rajni Rajan and Duniya Ram Singh (Eds.)

DOI: 10.1201/9781998511679-1

anticipated technological advancements and emphasizes the critical role of policy and regulation in supporting smart horticulture initiatives. Overall, this chapter serves as a comprehensive guide for researchers, practitioners, and policymakers interested in the future of technology-driven horticulture.

1.1 INTRODUCTION

Smart farming, which is also referred to as precision agriculture or smart agriculture, is an innovative method of farm management that improves agricultural sustainability and productivity through the use of cutting-edge technologies. It is a response to the increasing demands for agricultural food due to the growing world population, as well as the need to address the challenges posed by climate change and the depletion of natural resources (Ecer et al., 2023; Awan et al., 2021; Chergui & Kechadi, 2022). Utilizing information and communication technologies, including big data, cloud computing, and the Internet of Things (IoT), smart farming is characterized by its ability to monitor the field environment, analyze crop growth, and provide farmers with decision-making information (Awan et al., 2021). Integration of renewable energy sources, including solar and wind power, is another component of this strategy designed to reduce the environmental impact of agricultural operations (Ecer et al., 2023). Smart farming encompasses three main categories of technology: acquisition, evaluation, and analysis of data, as well as application of precision. These technologies are used to collect field-related data, analyze it to provide insights to farmers, and apply agricultural practices at variable rates, such as irrigation and pesticide application, to optimize productivity while protecting the environment. The adoption of smart farming is not limited to large-scale farms; smallholder farmers can also benefit from these technologies, which can help them ameliorate expenses, enhance the quality of their agricultural products, and augment crop yields.

However, the use of certain technologies, such as drones and aerial vehicles for crop monitoring, may be more suitable for large-scale farms due to their scale and margins (Chergui & Kechadi, 2022). In summary, smart farming represents a comprehensive methodology in the agricultural sector that exploits cutting-edge technologies to tackle the interconnected issues of climate change, environmental sustainability, and food security. It presents a viable trajectory toward enhancing the efficiency and resilience of the agricultural sector.

The evolution of farming practices in horticulture field, particularly in the context of sustainable agriculture, has been a topic of increasing interest in the scientific and policy communities (Rudel et al., 2016). Land highlights the call for sustainable intensification of horticulture, which entails the implementation of agricultural methods that not only enhance crop productivity but also supply essential ecosystem services for flora and fauna (Poláková, 2018). The aforementioned techniques—mulching, manuring, minimum tillage, and crop rotations—all contribute to enhanced nutrient cycling, diminished soil erosion, and increased soil biodiversity. A comparative analysis of crop rotation practices in conventional and organic agriculture. In one study (Heider et al., 2023), an examination revealed that organic farming induces variations in land use when compared to conventional farming. Specifically, crop rotations are 15% longer, which contributes to greater diversity and a more uniform distribution of crop species. The heightened intricacy associated with organic crop rotations is expected to improve the provision of ecosystem services to agroecosystems. The transition toward sustainable agriculture is not devoid of obstacles. The review by sustainability discusses the evolving nature of the Good Agricultural and Environmental Conditions principles, which are aimed at encouraging environmentally friendly farming methods. However, the review suggests that the shift toward a focus on resilience and risk within the context of rural farmland has displaced earlier sustainability approaches.

1.2 HISTORICAL PERSPECTIVE OF HORTICULTURE

1.2.1 MILESTONES IN HORTICULTURAL PRACTICES

Horticultural practices have evolved over centuries, marked by significant milestones that have shaped the way we cultivate plants for food, ornamental purposes, and environmental sustainability. Some notable milestones in horticultural practices include the following:

- **Domestication of plants:** One of the earliest milestones was the transition from wild plant harvesting to intentional cultivation. This marked the beginning of agriculture and horticulture as humans started selecting and cultivating plants for specific traits (Fuller, 2007; Purugganan & Fuller, 2009).

- **Development of crop rotation:** Ancient agricultural societies, such as those in Mesopotamia, adopted crop rotation to maintain soil fertility. This practice involves rotating crops in a systematic manner to improve soil structure and prevent the depletion of nutrients (Kirkegaard et al., 2008).
- **Introduction of greenhouses:** The use of greenhouses dates back to the Roman Empire, where they were used to grow exotic plants. However, it was during the 17th century in Europe that the concept of controlled environment agriculture (CEA) gained prominence (Barelli et al., 2008).
- **Discovery of plant nutrition:** In the 19th century, scientists like Justus von Liebig made groundbreaking discoveries about plant nutrition, highlighting the essential role of certain elements (nitrogen, phosphorus, and potassium) in plant growth. This laid the foundation for modern fertilizer use.
- **Invention of refrigeration:** The development of refrigeration technology in the 19th century allowed for the storage and transportation of perishable horticultural products over longer distances. This facilitated the growth of the horticultural industry on a global scale.
- **Advancements in plant breeding:** The 20th century saw significant advancements in plant-breeding techniques, including the discovery of hybridization and the development of high-yielding varieties. This led to increased crop productivity and improved disease resistance.
- **Introduction of integrated pest management (IPM):** With growing awareness of environmental sustainability, the concept of integrated pest management (IPM) emerged in the mid-20th century. IPM integrates biological, cultural, and chemical control methods to manage pests in a more environmentally friendly manner.
- **Precision agriculture and technology integration:** The late 20th and early 21st centuries witnessed the integration of technology into horticulture. Precision agriculture, which utilizes data-driven technologies like global positioning system (GPS), sensors, and

drones, allows farmers to optimize resource use and increase efficiency.

- **Organic farming movement:** The organic farming movement gained momentum in the late 20th century, emphasizing sustainable and environmentally friendly practices. Organic horticulture avoids synthetic pesticides and fertilizers, promoting soil health and biodiversity.
- **Genetic engineering and biotechnology:** Recent decades have seen breakthroughs in genetic engineering and biotechnology, enabling the development of genetically modified (GM) crops with enhanced traits such as pest resistance, drought tolerance, and improved nutritional content.

These milestones represent key developments in horticultural practices, reflecting a continuous process of innovation and adaptation to meet the changing needs of agriculture and society.

1.2.2 TRADITIONAL APPROACHES IN HORTICULTURE

Traditional approaches in horticulture refer to the conventional and time-tested methods used in the cultivation, management, and improvement of plants for food, ornamental, medicinal, and other purposes. Figure 1.1 illustrates how traditional agricultural practices in India can serve as strategic approaches to enhance food security while supporting environmental sustainability. These approaches have been developed and refined over centuries and have formed the foundation of horticultural practices.

While modern technologies and scientific advancements have introduced new techniques, traditional approaches still play a crucial role in many horticultural practices. Here are some key aspects of traditional approaches in horticulture.

1.2.1.1 PROPAGATION TECHNIQUES

- **Seed propagation:** Traditional horticulture often involves growing plants from seeds. This method is used for a wide range of crops and allows for the preservation of desirable traits through natural reproduction (Simmonds, 1997).

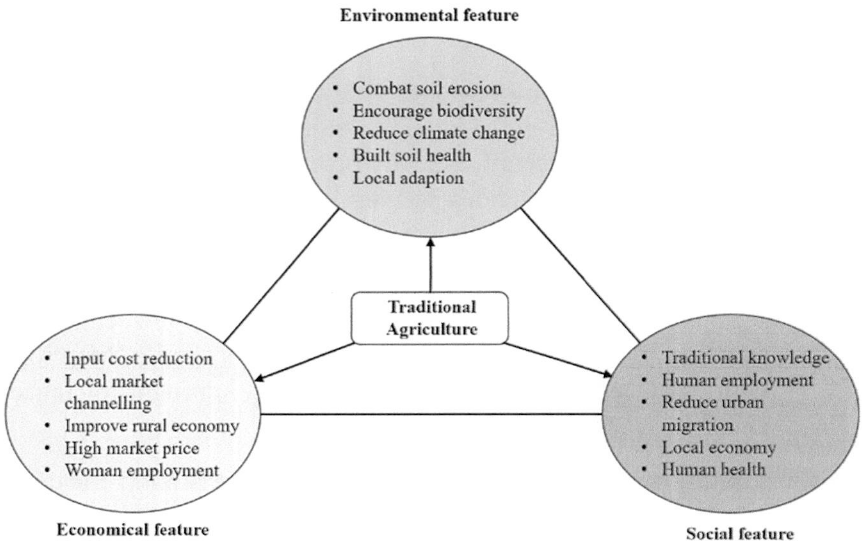

FIGURE 1.1 An examination of traditional agricultural practices in India as a strategy to promote food security and environmental sustainability.

- **Cutting propagation:** Taking cuttings from parent plants and allowing them to develop roots is a common traditional method, particularly for plants that are easily propagated this way, such as many fruit trees and ornamental shrubs (Singh et al., 2018).

1.2.1.2 SOIL MANAGEMENT

- **Crop rotation:** Traditional farmers rotate crops in specific sequences to maintain soil fertility and prevent the buildup of pests and diseases associated with specific crops (Bullock, 1992).
- **Composting:** The use of organic matter, such as compost, for soil enrichment is a traditional practice. Composting helps improve soil structure, water retention, and nutrient content.

1.2.1.3 WATER MANAGEMENT

- **Irrigation techniques:** Traditional methods of irrigation, such as furrow irrigation or drip irrigation, have been employed for centuries to ensure proper water supply to crops.

- **Rainwater harvesting:** Collecting and using rainwater is a traditional method in regions where water scarcity is an issue.

1.2.1.4 PEST AND DISEASE CONTROL

- **Biological controls:** Traditional horticulturists often use natural predators or parasites to control pests instead of relying solely on chemical pesticides.
- **Crop diversity:** Growing a variety of crops in proximity can disrupt pest and disease life cycles, thereby decreasing reliance on chemical interventions.

1.2.1.5 CULTURAL PRACTICES

- **Pruning and training:** Traditional horticulturists use pruning and training techniques to shape plants, improve air circulation, and enhance sunlight exposure.
- **Thinning:** Thinning of fruits or flowers is a traditional method to ensure proper development and quality.

1.2.1.6 SELECTION AND BREEDING

- **Selection of varieties:** Traditional farmers often select and save seeds from plants that exhibit desirable traits, leading to the development of locally adapted varieties.
- **Open pollination:** Allowing plants to be pollinated naturally, without controlled breeding, is a traditional approach that helps maintain genetic diversity.

While traditional approaches form the basis of horticulture, it is important to note that they are often complemented or enhanced by modern scientific methods to improve efficiency, productivity, and sustainability in agriculture. Integrated approaches that combine traditional wisdom with innovative technologies are increasingly common in contemporary horticultural practices.

1.3 NEED FOR INNOVATION IN HORTICULTURE

1.3.1 CHALLENGES FACED BY HORTICULTURISTS

Horticulturists grapple with a myriad of challenges, many stemming from environmental shifts due to climate change. Temperature variability and extreme weather events can disrupt plant growth, affecting flowering times and crop yields. Pests and diseases present ongoing threats, with the emergence of new species and the development of resistance to pesticides posing continuous challenges for effective control measures.

Water scarcity remains a critical concern, as horticulturists face competition for limited water resources, necessitating efficient water management practices. Globalization introduces complexities, with small-scale horticulturists struggling to access global markets and navigate trade regulations. Fluctuations in market dynamics and prices, influenced by consumer preferences and currency exchange rates, further impact profitability (Fereres et al., 2011).

Land-use changes, such as urbanization and degradation, reduce available space and contribute to soil quality issues. Energy and input costs can escalate, affecting the economic viability of horticultural operations. Labor shortages, especially for seasonal work, pose additional challenges, impacting crop management and productivity.

Biotic and abiotic stresses, including soil degradation and air pollution, compound the challenges faced by horticulturists. Technological adoption, particularly in developing regions, remains uneven, hindering access to modern agricultural tools. The overarching concern for sustainability underscores the need for balancing productivity with environmental conservation (Gull et al., 2019).

Addressing these challenges requires a holistic approach, combining traditional knowledge with innovative technologies, research advancements, and supportive policies. Collaboration between researchers, policymakers, and practitioners is essential to develop resilient and sustainable horticultural systems that can navigate the complexities of the modern agricultural landscape.

1.3.2 GLOBAL TRENDS AND DEMANDS IN AGRICULTURE

Global agriculture is experiencing transformative trends and demands, with a strong push toward sustainability at the forefront. Increasing awareness of environmental impact has led to a surge in sustainable agricultural practices aimed at conserving resources and promoting ecological balance. Precision agriculture, incorporating technologies like GPS guidance and drones, is gaining prominence for its ability to enhance efficiency and optimize resource utilization (Hazell & Wood, 2008).

Digital agriculture, characterized by the integration of data analytics, machine learning, and IoT devices, is revolutionizing farming practices and supply chain management. Climate change challenges have intensified the demand for climate-resilient crops, prompting research and development efforts to create varieties capable of withstanding extreme weather conditions, pests, and diseases (Klerkx et al., 2019).

The shift toward alternative proteins, including plant-based options and lab-grown meat, reflects changing consumer preferences driven by environmental and health concerns. Vertical farming and CEA are on the rise, offering year-round production in controlled conditions, particularly beneficial in urban areas.

Biotechnology and genetic engineering play a crucial role in crop improvement, with GM organisms and gene-editing technologies contributing to enhanced traits, disease resistance, and increased yields. Blockchain technology is being increasingly adopted in the agricultural supply chain, providing transparency and traceability from farm to consumer (Lock & Miles, 1993).

Water management practices, such as improved irrigation techniques and increased water-use efficiency, are vital in addressing water scarcity challenges. A set of regenerative agriculture principles that prioritize soil health, biodiversity, and carbon sequestration are gaining traction for their potential to create resilient and sustainable farming systems (Singh et al., 2001).

Rural development and social sustainability are gaining recognition, with initiatives supporting fair trade, smallholder farmers, and rural livelihoods. Consumer awareness and a desire for transparency in the food supply chain are influencing purchasing decisions, emphasizing the need for accountability and ethical practices in agriculture.

While these trends were significant, it is important to note that the agricultural landscape is dynamic, and new developments may have emerged since then. Ongoing global events, such as the COVID-19 pandemic, also have the potential to impact and shape future trends in agriculture. For the latest information, it is recommended to refer to recent sources and reports on global agricultural trends.

1.4 CONCEPT OF SMART FARMING

1.4.1 DEFINITION AND COMPONENTS

Smart farming, which is alternatively referred to as precision agriculture, involves the integration of sophisticated technologies in order to enhance agricultural procedures. A suite of components, including IoT devices and sensors, are utilized to collect real-time data on parameters, such as soil moisture, temperature, humidity, and crop health. These data are analyzed by data analytics tools, which support data-driven decision-making and provide valuable insights.

GPS technology is pivotal, facilitating precise mapping and monitoring of fields. GPS-guided tractors and equipment enable accurate application of inputs, such as seeds, fertilizers, and pesticides. Automation and robotics play a key role in tasks like planting, weeding, and harvesting, reducing reliance on manual labor while improving efficiency (Hofmann-Wellenhof et al., 2012).

Variable rate technology (VRT) optimizes resource utilization by permitting the targeted application of inputs based on the particular requirements of distinct field areas. Unmanned aerial vehicles (UAVs) and drones that are equipped with sensors enable aerial monitoring of superior resolution, aiding in crop stress detection and health assessment. Farm management software platforms integrate diverse data sources, offering a comprehensive overview of farm performance (Clark & McGuckin, 1996).

Accurate weather forecasting is crucial, and smart farming incorporates real-time weather data to plan and mitigate the impact of adverse conditions. Livestock monitoring systems employ wearable devices and sensors for health, location, and behavior tracking, facilitating early disease detection and comprehensive livestock management. Communication systems enable seamless connectivity between devices, machines, and farmers for real-time monitoring and control (Jain & Mallick, 2016).

The utilization of blockchain technology in smart farming is on the rise as a means to improve traceability and transparency throughout the agricultural supply chain. This is achieved by establishing immutable and secure ledgers of transactions and activities. The primary goal of smart farming is to optimize resource utilization, increase profitability, decrease environmental footprint, and enhance overall sustainability and efficiency in agricultural practices. The specific application of these technologies may vary based on agricultural type, local conditions, and specific crop or livestock considerations (Zheng et al., 2017). The smart farming framework in horticulture is visually outlined in Figure 1.2.

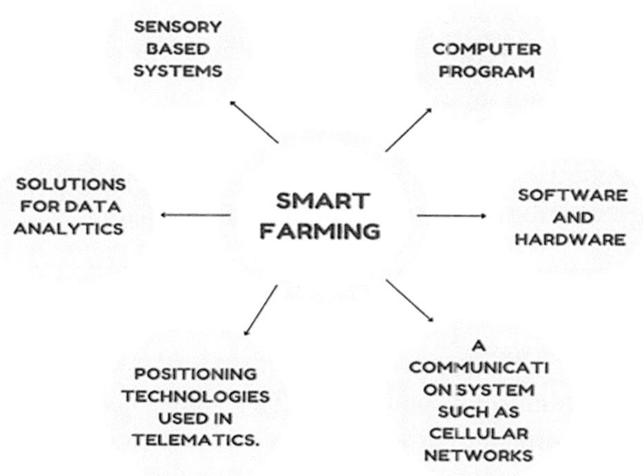

FIGURE 1.2 Pictorial representation of smart farming approach in horticulture.

1.4.2 INTEGRATION AND ROLE OF COMMUNICATION TECHNOLOGIES IN AGRICULTURE

1.4.2.1 DATA COLLECTION THROUGH SENSORS AND IOT DEVICES

The integration begins with the deployment of sensors and IoT devices in the field. These devices continuously collect real-time data on critical factors such as soil moisture, temperature, humidity, and crop health.

1.4.2.2 DATA ANALYTICS FOR INFORMED DECISION-MAKING

Advanced data analytics algorithms and tools are employed to process and examine the gathered data. Insights are generated from raw data at this stage, which assists farmers in making well-informed decisions regarding resource allocation, crop management, and overall farm operations.

1.4.2.3 GPS TECHNOLOGY FOR PRECISION FARMING

GPS technology is utilized for precise mapping and monitoring of fields. GPS-guided machinery enables accurate placement of inputs such as seeds, fertilizers, and pesticides, reducing waste and optimizing resource utilization.

1.4.2.4 AUTOMATION AND ROBOTICS

Instances where automation and robotics are indispensable include planting, weeding, and harvesting. The implementation of autonomous tractors and robotic systems results in enhanced productivity and accuracy through the reduction of manual labor requirements.

1.4.2.5 VARIABLE RATE TECHNOLOGY

The utilization of VRT permits the precise regulation of inputs in accordance with particular field conditions. This technology enables agricultural practitioners to tailor the utilization of resources, including fertilizers, to the specific requirements of various regions contained within a field.

1.4.2.6 AERIAL MONITORING WITH DRONES AND UAVS

UAVs and drones that are equipped with sensors acquire data and high-resolution images from above. Aerial monitoring plays a crucial role in evaluating the health of crops, identifying pests, and managing fields as a whole.

1.4.2.7 PLATFORMS FOR FARM MANAGEMENT SOFTWARE

Utilizing a centralized platform, farm management software integrates and administers data from a variety of sources. These platforms provide farmers with extensive information regarding their operations, facilitating the process of making informed decisions concerning planting, harvesting, and resource distribution.

1.4.2.8 WEATHER FORECASTING AND MONITORING

Real-time weather data is integrated into smart farming systems to enable farmers to monitor and respond to changing weather conditions. This assists in planning activities and mitigating risks associated with adverse weather events.

1.4.2.9 LIVESTOCK MONITORING SYSTEMS

In livestock farming, technology integration involves the use of wearable devices and sensors for monitoring the health, location, and behavior of animals. These data contribute to early disease detection and efficient herd management.

1.4.2.10 COMMUNICATION SYSTEMS

Communication systems facilitate connectivity between different components of the smart farming ecosystem. This connectivity enables real-time monitoring, control, and communication, allowing farmers to manage their operations remotely.

1.4.2.11 BLOCKCHAIN FOR TRANSPARENCY

The investigation of blockchain technology aims to improve traceability and transparency in the agricultural supply chain. By guaranteeing secure and unalterable records of transactions, it promotes accountability and instills confidence among stakeholders.

The overarching objective of technology integration in agriculture is to fundamentally transform conventional farming methods by imbuing them with enhanced precision, efficiency, and sustainability. By capitalizing on these sophisticated technologies, agricultural practitioners have the ability to augment output, mitigate ecological repercussions, and surmount the obstacles inherent in contemporary farming with greater efficacy.

1.5 IMPORTANCE OF SMART FARMING IN HORTICULTURE

1.5.1 ENHANCING PRODUCTIVITY AND EFFICIENCY

Smart farming enhances productivity and efficiency in horticulture by integrating precision agriculture techniques, data-driven decision-making, and advanced technologies. Precision planting, fertilization, and irrigation ensure optimal resource use, while real-time data from sensors and IoT devices enable informed decision-making about crop health and environmental conditions. Automation and robotics streamline tasks, reducing reliance on manual labor. VRT tailors input applications, minimizing waste and maximizing efficiency. Greenhouse and CEA create ideal growth conditions. Drones provide aerial monitoring for early detection of issues, and advanced irrigation management optimizes water use. Farm management software platforms centralize data for efficient planning and resource management. Quick responses to pest and disease threats and efficient supply chain management further contribute to increased productivity in horticulture.

- **Environmental stewardship:** Sustainable horticulture places a strong emphasis on responsible management of natural resources. This involves minimizing chemical inputs, reducing water usage, and preserving biodiversity to contribute to overall environmental stewardship. By prioritizing the health of ecosystems, sustainable practices aim to create a balance between cultivation and conservation (Bennett et al., 2018).
- **Water management:** Efficient water use is a fundamental aspect of sustainability in horticulture. Employing techniques such as drip irrigation, rainwater harvesting, and water recycling systems helps optimize water resources. These practices not only reduce water wastage but also contribute to the overall conservation of this vital resource in horticultural operations (Cosgrove & Loucks, 2015).
- **Soil health and conservation:** Sustainable horticulture places a strong emphasis on soil health through various practices. Techniques like cover cropping, crop rotation, and the incorporation of organic matter enhance soil fertility, structure, and water retention. By prioritizing soil health, sustainable horticulture minimizes the need for synthetic inputs and fosters a resilient and sustainable growing environment (Cárceles Rodríguez et al., 2022).
- **Integrated pest management:** To minimize the environmental impact of pest control, sustainable horticulture adopts integrated pest management approaches. This includes using biological controls, crop rotation, and resistant plant varieties. By integrating these practices, horticulturists aim to control pests while preserving the ecological balance of the surrounding environment (Kogan, 1998).

- **Organic farming practices:** Organic horticulture represents a commitment to sustainable practices by avoiding synthetic fertilizers and pesticides. Instead, natural alternatives are employed to enhance soil fertility and control pests. By embracing organic farming practices, horticulturists contribute to healthier ecosystems and reduce the environmental pollution associated with chemical inputs.
- **Energy efficiency:** Sustainable horticulture investigates methods of conserving energy in order to lessen its ecological impact. This consists of implementing energy-efficient irrigation systems, optimizing greenhouse designs for natural light, and utilizing renewable energy sources. By adopting these measures, horticulturists contribute to a more sustainable and energy-conscious approach to cultivation.
- **Waste reduction and recycling:** Promoting recycling and minimizing waste production are essential elements of sustainable horticulture. Agricultural by-product recycling, composting, and mulching are methods that aid in mitigating the environmental consequences linked to waste disposal. The implementation of these waste reduction strategies enhances the environmental responsibility and sustainability of the horticultural system.
- **Local and seasonal production:** By placing a greater emphasis on local and seasonal production, horticulture can reduce the environmental impact of long-distance transportation. In addition to promoting community connections and sustainability by reducing energy consumption and greenhouse gas emissions, this practice supports local economies.
- **Agroforestry and agroecology:** Integrating trees and shrubs into horticultural systems, known as agroforestry, and adopting agroecological principles are key components of sustainable horticulture. These practices enhance biodiversity, improve soil structure, and contribute to the creation of resilient ecosystems. By embracing these approaches, horticulturists promote sustainability and regenerative practices (Kaltoft, 1999).
- **Certification and standards:** Various certification programs and standards, such as organic certifications, play a crucial role

in guiding sustainable horticultural practices. Adhering to these standards ensures that production methods meet specific environmental and ethical criteria, providing consumers with assurance of environmentally responsible cultivation practices.

1.5.2 ECONOMIC AND ENVIRONMENTAL BENEFITS

Smart farming in horticulture offers a dual advantage by delivering economic benefits and promoting environmental sustainability. Through precision agriculture techniques, such as controlled environments and data-driven decision-making, farmers can achieve resource efficiency and cost savings. This is particularly evident in the optimized use of water, fertilizers, and pesticides, resulting in reduced waste and increased operational efficiency. The adoption of automation and robotics further contributes to labor savings, addressing challenges related to labor shortages and enhancing overall operational efficiency. VRT allows for targeted input applications, minimizing wastage and positively impacting both the economic bottom line and environmental footprint.

Increased productivity and yield are fundamental economic benefits derived from smart farming practices. Technologies like precision planting and monitoring systems contribute to higher crop yields, translating into economic gains for farmers. Quality improvement in produce, facilitated by controlled environments and monitoring, not only commands better prices in the market but also expands market access. The data-driven decision-making enabled by real-time information enhances the effectiveness of farming practices, leading to positive economic outcomes.

Environmental benefits arise from the water conservation initiatives embedded in smart farming technologies. Advanced irrigation management and soil moisture sensors help optimize water use, reducing water wastage and addressing environmental concerns associated with excessive water extraction. Moreover, the reduced environmental impact of horticultural activities, such as minimized chemical runoff and greenhouse gas emissions, aligns with sustainability goals and promotes environmentally friendly agricultural practices.

Smart farming practices in horticulture contribute to biodiversity preservation and long-term sustainability. By minimizing the negative impact on surrounding ecosystems, these practices help maintain a balanced and resilient agricultural environment. The combination of economic benefits, reduced environmental impact, and a focus on sustainability ensures that smart farming in horticulture is not only economically viable but also environmentally responsible, paving the way for a resilient and sustainable agricultural future.

1.6 TECHNOLOGICAL FOUNDATIONS

1.6.1 IOT IN HORTICULTURE

The IoT in horticulture refers to the integration of connected devices, sensors, and technologies that use the internet to collect, transmit, and receive data in the cultivation and management of plants. In this context, IoT applications are designed to enhance various aspects of horticultural practices, providing real-time information and enabling more precise and efficient decision-making (Ravi et al., 2015). Figure 1.3 presents an overview of the smart farming system implemented in modern agriculture. Here are key components and applications of IoT in horticulture:

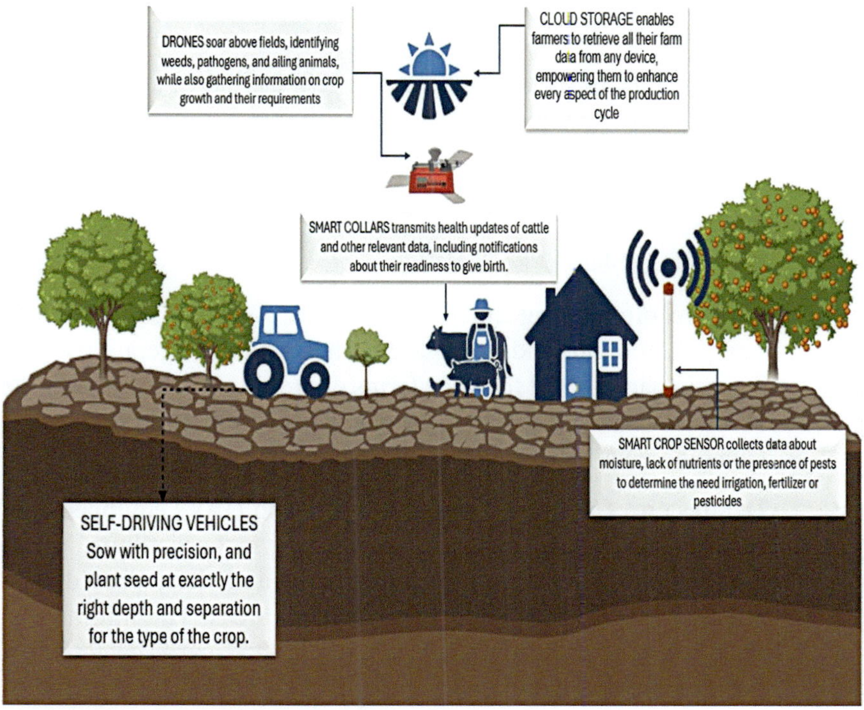

FIGURE 1.3 Smart farming system in agriculture.

- **Sensor Networks:** IoT in horticulture relies on sensor networks deployed in fields, greenhouses, or orchards. These sensors measure various environmental parameters such as soil moisture, temperature, humidity, light intensity, and nutrient levels. The data collected by these sensors provide valuable insights into the growing conditions (Rabka et al., 2022).
- **Data collection and monitoring:** IoT devices continuously collect data from the sensors, creating a network of interconnected information. These data are then transmitted to a central system, where it can be monitored in real-time. This real-time monitoring allows farmers to observe changes in environmental conditions and plant health promptly.
- **Precision agriculture:** Precision agriculture in horticulture is greatly enhanced by IoT. The real-time data from sensors helps farmers make informed decisions about the precise application of water, fertilizers, and pesticides. This targeted approach optimizes resource use, reduces waste, and improves overall efficiency (López et al., 2010).
- **Automation and control systems:** IoT facilitates automation in horticultural operations. Automated systems can be controlled and monitored remotely through IoT devices. This includes the automation of irrigation systems, climate control in greenhouses, and the operation of robotic devices for tasks like planting and harvesting (Shafi et al., 2019).
- **Remote monitoring and management:** With IoT, farmers can remotely monitor and manage their horticultural operations. This is particularly useful for large farms or greenhouse setups. Remote access to data and control systems allows farmers to respond to changing conditions promptly and efficiently.
- **Drones and UAVs:** IoT-enabled drones and UAVs equipped with sensors are utilized for aerial monitoring in horticulture. These devices capture high-resolution images and data, providing a comprehensive view of large agricultural areas. The information collected aids in crop health assessment and early detection of potential issues (Khort et al., 2020; Khadse, 2021).

- **Decision support systems:** The data generated by IoT in horticulture feed into decision support systems. These systems use analytics and algorithms to process the data and provide actionable insights. Farmers can use these insights to make informed decisions about crop management, resource allocation, and pest control.
- **Supply chain management:** IoT technology is applied in tracking and monitoring the supply chain in horticulture. This includes monitoring the transportation and storage conditions of harvested produce. By ensuring optimal conditions throughout the supply chain, IoT helps maintain the quality of the produce from farm to market (Robbins, 2018).
- **Smart greenhouses:** Greenhouses equipped with IoT technology can automatically control variables like temperature, humidity, and light based on real-time data. This precision control creates an optimal environment for plant growth, leading to improved yield and quality.
- **Predictive analytics:** IoT data, when combined with advanced analytics, enables predictive modeling. Predictive analytics can forecast trends, diseases, or optimal harvest times. This foresight allows farmers to plan and implement proactive measures, minimizing risks and maximizing productivity.

In summary, IoT in horticulture revolutionizes traditional farming practices by providing real-time data, automation, and decision support. It contributes to precision agriculture, resource optimization, and improved overall efficiency, ultimately enhancing the productivity and sustainability of horticultural operations.

1.6.2 DATA ANALYTICS AND DECISION SUPPORT SYSTEMS

Data analytics and decision support systems play a crucial role in horticulture by leveraging technology to process, analyze, and interpret the vast amounts of data generated in farming operations. These systems enable farmers to make informed decisions, optimize resource allocation, and enhance overall efficiency in horticultural practices. Here is an explanation of how data analytics and decision support systems are utilized in horticulture:

- **Data collection from sensors and IoT devices:** In modern horticulture, various sensors and IoT devices are deployed to collect data on environmental conditions, soil health, and crop performance. These devices continuously generate a stream of real-time data, forming the basis for informed decision-making.
- **Data preprocessing and integration:** Raw data from sensors often require preprocessing to clean and organize it for analysis. Decision support systems integrate data from diverse sources, including weather stations, soil sensors, and monitoring devices, ensuring a comprehensive dataset.
- **Data analytics techniques:** Data analytics involves employing statistical techniques, machine learning algorithms, and other analytical methods to uncover patterns, trends, and relationships within the collected data. These analyze extract meaningful insights that can inform decisions related to crop management, resource utilization, and pest control.
- **Predictive analytics for crop planning:** Predictive analytics uses historical data and modeling to forecast future trends. In horticulture, this can aid in predicting optimal planting times, estimating harvest yields, and identifying potential pest or disease outbreaks. Farmers can use these predictions for proactive planning and risk mitigation.
- **Monitoring crop health and stress:** Data analytics tools can process imagery and sensor data to monitor crop health and detect signs of stress. Early detection of issues such as nutrient deficiencies or pest infestations allows for timely interventions, minimizing crop losses and ensuring healthy yields.
- **VRT optimization:** Decision support systems optimize the implementation of VRT. By analyzing data on soil conditions and variability, these systems guide farmers in adjusting the rates of inputs such as fertilizers and pesticides. This targeted approach maximizes resource use efficiency.
- **Optimizing irrigation practices:** Data analytics help optimize irrigation practices by analyzing soil moisture data and

weather conditions. Decision support systems can recommend precise irrigation schedules, ensuring that plants receive the right amount of water at the right time. This minimizes water wastage and supports sustainable water management.

- **Resource allocation and efficiency:** Decision support systems aid in the efficient allocation of resources such as labor, water, and fertilizers. By analyzing data on resource usage and crop requirements, farmers can make informed decisions to optimize resource allocation, leading to increased efficiency and cost-effectiveness.
- **Integration with farm management software:** Data analytics and decision support systems are often integrated with farm management software platforms. This integration provides a centralized hub for farmers to access comprehensive insights into their horticultural operations, facilitating holistic decision-making.
- **Real-time monitoring and alerts:** Decision support systems can provide real-time monitoring and alerts based on predefined thresholds. If sensor data indicate deviations from optimal conditions or the presence of potential threats, farmers receive timely alerts, enabling them to take corrective actions promptly.

In summary, data analytics and decision support systems in horticulture empower farmers with actionable insights derived from real-time data. These systems contribute to precision agriculture, resource optimization, and informed decision-making, ultimately enhancing the productivity, sustainability, and economic viability of horticultural operations.

1.6.3 PRECISION AGRICULTURE TECHNIQUES

Precision agriculture techniques in horticulture involve the use of advanced technologies and data-driven approaches to optimize various aspects of crop management. These techniques aim to enhance efficiency, resource utilization, and overall productivity in horticultural practices. Here is an explanation of key precision agriculture techniques in horticulture:

- **GPS technology:** The use of GPS is essential for precision agriculture. It facilitates precise monitoring and mapping of horticultural fields. Utilizing GPS-guided machinery, including harvesters and tractors, guarantees accurate input application and navigation, thereby minimizing redundancies and maximizing resource utilization.

- **Variable rate technology:** Variable return to bed (VRT) systems enable the tailored application of inputs, including fertilizers, pesticides, and irrigation, to distinct zones within a given field. Fertilizer utilization can be optimized and environmental impact reduced by adapting input rates to the variability of soil and crop conditions.
- **Sensor technologies:** In order to gather up-to-the-minute information regarding environmental conditions, soil moisture, temperature, and crop health, horticultural fields employ a variety of sensors. The continuous monitoring provided by sensor technologies empowers agricultural practitioners to make well-informed and timely decisions pertaining to irrigation, pest management, and fertilization.
- **Drones and UAVs:** Horticulture employs drones that are equipped with sensors and cameras to conduct aerial monitoring. By capturing data and images at a high resolution, they enable farmers to evaluate the health of their crops, detect diseases, and pinpoint potential problem areas. Aerial perspective facilitates the process of making precise decisions.
- **Automated machinery and robotics:** Automated machinery and robotics play a key role in precision agriculture. Autonomous tractors, robotic harvesters, and other automated systems perform tasks with high precision. This reduces reliance on manual labor, ensures consistent operations, and contributes to overall efficiency.
- **Smart irrigation systems:** Intelligent irrigation systems optimize the timing and quantity of water applied to crops by utilizing sensor data and weather information. By conserving water resources and ensuring that plants receive the appropriate amount of water at the appropriate time, these systems prevent both under- and overirrigation.
- **Greenhouse and CEA:** Precision agriculture encompasses controlled environment and greenhouse setups as well. By utilizing

technology, these systems effectively regulate environmental variables including temperature, humidity, and light, thereby establishing ideal conditions that promote plant development and maximize overall output.

- **Data analytics and decision support systems:** Decision support systems and data analytics are utilized to process the enormous quantities of data produced by precision agriculture technologies. By conducting analyses on data pertaining to soil conditions, crop health, and environmental factors, these systems furnish decision-makers with actionable insights.
- **Smart farm management software:** Platforms for farm management software consolidate information from an assortment of precision agriculture technologies. These platforms provide an extensive synopsis of horticultural activities, facilitating the formulation of decisions pertaining to resource distribution, crop scheduling, and risk mitigation.
- **Real-time monitoring and alerts:** Real-time monitoring capabilities provided by precision agriculture technologies enable farmers to receive immediate alerts in response to deviations from optimal conditions or the presence of potential threats. This allows for prompt interventions and adjustments to maintain crop health and productivity.

In summary, precision agriculture techniques in horticulture leverage cutting-edge technologies to optimize every aspect of crop management. These techniques enhance resource efficiency, reduce environmental impact, and contribute to sustainable and economically viable horticultural practices.

1.7 KEY TECHNOLOGIES IN SMART HORTICULTURE

1.7.1 SENSOR TECHNOLOGIES

Horticulture sensor technologies encompass the implementation of diverse sensor systems to gather instantaneous data pertaining to soil properties, environmental circumstances, and the well-being of plants within agricultural environments. The sensors in question perform an essential function by delivering accurate data and ongoing surveillance, which empowers farmers to make well-informed choices. Here is an explanation of key sensor technologies used in horticulture:

- **Soil moisture sensors:** Soil moisture sensors quantify the soil's moisture content. The provision of real-time data on soil moisture levels by these sensors facilitates the optimization of irrigation practices for farmers. This promotes water conservation by ensuring that plants receive the appropriate quantity of water, thereby averting problems associated with inadequate or excessive irrigation.
- **Temperature and humidity sensors:** Temperature and humidity sensors monitor the atmospheric conditions in horticultural environments. These data are crucial for understanding the climate within greenhouses, ensuring optimal growing conditions, and preventing extremes that may adversely affect plant growth.
- **Light sensors:** Light sensors measure the intensity and duration of light in horticultural settings. This information is vital for understanding the light requirements of plants, especially in controlled environments like greenhouses. Light sensors help optimize lighting conditions to promote healthy plant growth.
- **Nutrient sensors:** Sensors for nutrients measure the concentrations of vital nutrients in the soil. Through ongoing monitoring of nutrient concentrations, agricultural practitioners are able to make well-informed judgments regarding fertilization methodologies. By employing this focused strategy, it is possible to guarantee that plants obtain the essential nutrients required for maximum development while reducing the need for excessive application.
- **Gas and air quality sensors:** Gas and air quality sensors measure parameters such as carbon dioxide (CO_2) levels and air composition. In closed environments like greenhouses, these sensors help maintain optimal air quality for plant growth. Monitoring CO_2 levels is particularly important for enhancing photosynthesis.
- **Weather stations:** Weather stations integrate multiple sensors to monitor various weather-related parameters, including temperature, humidity, wind speed, and precipitation. This comprehensive

data helps farmers anticipate and respond to changes in weather conditions, allowing for proactive measures to protect crops.

- **Leaf wetness sensors:** Leaf wetness sensors detect the presence and duration of moisture on plant surfaces. This information is valuable for disease management, as prolonged leaf wetness can create conditions conducive to certain plant diseases. Farmers can use these data to implement preventive measures.
- **Pest and disease monitoring sensors:** Specialized sensors can detect signs of pests and diseases in crops. These sensors may identify specific volatile compounds emitted by plants under stress. Early detection enables timely interventions, reducing the impact of pests and diseases on crop health.
- **Image sensors and cameras:** Image sensors and cameras capture visual data, allowing for visual monitoring of crop health. Drones equipped with cameras provide aerial imagery for large-scale assessments. This visual information aids in assessing plant vigor, identifying nutrient deficiencies, and detecting visual cues related to pest and disease infestations.
- **Vibration sensors:** Vibration sensors can be used to monitor mechanical aspects of machinery or infrastructure in horticulture. For example, they may be employed to detect irregularities in irrigation systems or machinery, ensuring timely maintenance to prevent disruptions in operations.

By integrating these sensor technologies, farmers can create a comprehensive data ecosystem that enables precision agriculture in horticulture. The continuous monitoring and analysis of these data empower farmers to make data-driven decisions, optimize resource use, and enhance overall crop management practices.

1.7.2 AUTOMATION AND ROBOTICS

Horticulture automation and robotics encompass the implementation of sophisticated technologies to mechanize and optimize a multitude of agricultural tasks. These technologies are crucial in horticulture for increasing overall productivity, decreasing labor requirements, and enhancing efficiency. Here is an explanation of key aspects related to automation and robotics in horticulture:

- **Planting automation:** Automated planting systems use robotic devices to precisely place seeds or seedlings in the soil. This ensures consistent spacing and optimal planting depth, contributing to uniform plant growth. Planting automation reduces the need for manual labor and accelerates the planting process.
- **Weeding robots:** Weeding robots are designed to identify and remove weeds in horticultural fields. These robots may utilize sensors and computer vision to distinguish between crops and weeds, enabling targeted weed removal. This automation minimizes the reliance on herbicides and manual weeding efforts.
- **Harvesting Automation:** Automated harvesting systems use robotics to harvest fruits, vegetables, or other crops. These systems are often equipped with sensors and computer vision to identify ripe produce and execute precise harvesting maneuvers. Automated harvesting reduces labor costs and ensures efficient and timely harvesting.
- **Pruning and trimming robots:** Pruning and trimming robots are designed to perform precision pruning and trimming of plants. These robots can be programmed to follow specific patterns and guidelines for shaping plants, ensuring uniform growth. Automated pruning contributes to maintaining plant health and optimizing yields.
- **Packaging and sorting automation:** Automation is applied to packaging and sorting processes in horticulture. Robotic systems can handle the sorting, grading, and packaging of harvested produce based on predefined criteria such as size, weight, and quality. This enhances the efficiency of postharvest operations.
- **Autonomous vehicles and tractors:** Autonomous vehicles and tractors equipped with GPS and sensor technologies can perform various tasks in horticultural fields. These vehicles can navigate autonomously, execute precision planting, and carry out other field operations without direct human intervention. This automation improves operational efficiency.

- **Greenhouse automation:** Greenhouse automation involves the integration of robotic systems to manage environmental conditions within greenhouses. Automated systems control factors such as temperature, humidity, and ventilation, ensuring optimal conditions for plant growth. This precision control enhances greenhouse productivity.
- **Precision irrigation systems:** Automated irrigation systems regulate irrigation in response to detected soil moisture levels via the use of sensors. On the basis of real-time data, these systems can be programmed to deliver precise quantities of water to various areas of the field. Automated irrigation enhances water-use efficiency and promotes healthy plant growth.
- **Labor-saving devices:** Various robotic devices and automation technologies are designed to save labor in horticulture. This includes robotic carts for transporting harvested produce, automated equipment for mulching or applying protective covers, and other labor-saving devices that contribute to operational efficiency.
- **Data-driven decision-making:** Automation and robotics in horticulture generate vast amounts of data. These data can be analyzed to derive insights for decision-making. Farmers can use information from automated systems to optimize crop management, resource allocation, and overall farm operations.

In summary, automation and robotics in horticulture introduce a new era of efficiency and precision. By automating labor-intensive tasks and incorporating robotics into various stages of the agricultural process, farmers can achieve higher productivity, reduce operational costs, and promote sustainable and technologically advanced horticultural practices.

1.7.3 DRONES AND REMOTE SENSING

The term "drones and remote sensing in horticulture" pertains to the utilization of UAVs, more colloquially recognized as drones, that are outfitted with an assortment of sensors and imaging technologies in order to gather data and oversee horticultural crops. These technologies enable farmers to make well-informed decisions by providing valuable insights into crop health, growth patterns, and environmental conditions. Here is an explanation of the key aspects of drones and remote sensing in horticulture:

- **Aerial imaging and surveillance:** Aerial imagery of horticultural fields is acquired by drones by means of high-resolution sensors and cameras. The aforementioned imagery affords farmers an aerial perspective of the crops, facilitating the surveillance of general crop well-being, detection of possible complications, and evaluation of crop distribution across the field.
- **Multispectral and hyperspectral imaging:** Drones can carry multispectral and hyperspectral cameras that capture imagery beyond the visible spectrum. This allows for the identification of specific wavelengths associated with plant health, nutrient content, and stress factors. Multispectral and hyperspectral imaging provide detailed insights into the physiological condition of crops.
- **Thermal imaging:** Thermal cameras on drones detect variations in surface temperatures. Thermal imaging is particularly useful for identifying variations in plant health, water stress, and the presence of diseases. It helps farmers assess the effectiveness of irrigation practices and detect anomalies not visible in standard imagery.
- **NDVI (Normalized Difference Vegetation Index) Mapping:** NDVI mapping is a remote sensing technique that uses near-infrared and red light to assess plant health and vigor. Drones equipped with NDVI sensors provide farmers with a quantitative measure of vegetation health. This information aids in identifying areas of the field that may require attention or adjustments in management practices.
- **Crop health monitoring:** Drones facilitate real-time monitoring of crop health by capturing frequent and up-to-date aerial images. This allows farmers to detect changes in plant conditions, such as nutrient deficiencies, pest infestations, or diseases, enabling timely interventions to maintain crop health.
- **Precision agriculture applications:** Drones support precision agriculture by providing detailed spatial information about

the field. This information can be utilized by farmers to apply inputs like fertilizers, pesticides, and water at variable rates. The implementation of precision agriculture techniques maximizes the utilization of resources and contributes to enhanced operational effectiveness.

- **3D mapping and modeling:** Light detection and ranging or photogrammetry-enabled drones are capable of generating three-dimensional maps and models of horticultural fields. This spatial information aids in assessing terrain variations, drainage patterns, and overall topography, providing valuable insights for land management.
- **Early detection of issues:** Drones enable early detection of crop issues by regularly monitoring the field. Rapid identification of potential problems, such as pest outbreaks or water stress, allows farmers to take immediate corrective actions, minimizing the impact on crop yield.
- **Disease and pest monitoring:** Drones assist in monitoring for signs of diseases and pest infestations. High-resolution imagery captured by drones can reveal subtle visual cues indicative of these issues. Early detection supports targeted interventions and reduces the need for widespread pesticide application.

In summary, drones and remote sensing technologies in horticulture revolutionize crop monitoring and management. By providing real-time, high-resolution data, these tools empower farmers to make precise and informed decisions, ultimately contributing to improved crop health, resource optimization, and increased productivity.

1.8 APPLICATIONS OF SMART FARMING IN HORTICULTURE

1.8.1 CROP MONITORING AND MANAGEMENT

Crop monitoring and management in horticulture is a comprehensive approach that encompasses various facets of cultivation. Continuous environmental monitoring, including temperature, humidity, and light intensity, ensures optimal growing conditions in greenhouses or open fields. Soil health is assessed through tools such as moisture sensors and nutrient analyzers, guiding farmers in precise irrigation and fertilization practices. Observing and tracking crop growth stages is crucial for timely interventions and resource allocation, facilitated by visual assessments and growth models. Pest and disease monitoring involves methods ranging from visual inspections to advanced technologies like drones, enabling early detection and targeted management strategies. Water and nutrient management are essential components, with precision irrigation and fertilization practices ensuring efficient resource use and preventing stress factors. Remote sensing technologies, such as drones and satellites, provide high-resolution imagery for comprehensive field assessments, aiding in crop health monitoring and stress identification. Integrated pest management strategies promote a balanced approach to pest control, minimizing reliance on chemical inputs. Precision agriculture techniques, including GPS and automated machinery, contribute to resource optimization. Decision support systems integrate data from various sources, enhancing data-driven decision-making for improved crop management practices. Postharvest monitoring, involving technologies like cold storage monitoring and traceability systems, ensures the preservation of harvested produce and efficient supply chain management. Overall, this holistic approach optimizes growing conditions, prevents issues, and maximizes yields in horticulture.

1.8.2 PRECISION IRRIGATION

By applying water to crops in a targeted and efficient manner, precision irrigation in horticulture maximizes the utilization of this vital resource. By employing technological advancements including automated irrigation systems, weather data, soil moisture sensors, and soil moisture sensors, precision irrigation guarantees that plants are supplied with an optimal quantity of water in the correct locations and timing within a given field. By customizing irrigation techniques to suit particular crop and soil circumstances, agricultural practitioners can reduce water wastage, avert water stress, and improve overall water utilization efficiency. By adopting this methodology, horticulture not only preserves water but also promotes enhanced crop well-being, augmented crop production, and sustainable agricultural methods.

1.8.3 DISEASE AND PEST CONTROL

Disease and pest control in horticulture involves the systematic management of threats to plant health, including pathogens and harmful insects. This process employs various strategies such as cultural practices, biological control, and judicious use of pesticides. IPM is commonly employed, emphasizing a holistic approach that combines biological control agents, habitat manipulation, and targeted use of chemical controls when necessary. Regular monitoring for signs of diseases and pests, early detection, and strategic interventions are essential elements of effective control. By implementing these practices, farmers can mitigate the impact of pests and diseases, minimize reliance on chemical inputs, and promote sustainable and environmentally friendly horticultural practices.

1.8.4 SUPPLY CHAIN MANAGEMENT

Supply chain management within the horticulture industry pertains to the methodical organization and enhancement of the procedures entailed in the manufacturing process, distribution, and consumption of horticultural products, from farm to consumer. It encompasses the entire journey of horticultural goods, including planting, harvesting, processing, packaging, transportation, storage, and retailing. Effective supply chain management in horticulture involves strategic planning to ensure the timely and efficient flow of products while minimizing waste and costs. Key aspects include logistics, quality control, traceability, and coordination among various stakeholders such as farmers, processors, distributors, and retailers. By streamlining these processes, supply chain management in horticulture aims to deliver fresh, high-quality produce to consumers, reduce postharvest losses, and enhance overall market efficiency.

1.9 CASE STUDIES AND SUCCESS STORIES

1.9.1 REAL-WORLD EXAMPLES OF SMART HORTICULTURE IMPLEMENTATION

Intelligent horticulture implementations optimize agricultural practices by utilizing cutting-edge technologies to increase output, resource efficiency, and sustainability. Here are some real-world examples:

- **Greenhouse automation in the Netherlands:** Dutch greenhouse growers extensively use smart technologies for climate control. Automated systems regulate temperature, humidity, and light, optimizing conditions for plant growth. Sensors monitor and adjust parameters in real time, ensuring precise control and resource efficiency.
- **Precision irrigation in Israel:** Israel, known for its expertise in agricultural innovation, employs precision irrigation techniques in horticulture. Frequently incorporating automation and sensors, drip irrigation systems deliver water directly to the plant roots. This maximizes water use efficiency, a crucial factor in water-scarce regions.
- **Smart pest management in California vineyards:** Vineyards in California utilize smart pest management systems. Automated traps equipped with sensors monitor pest populations, and data analytics help predict potential outbreaks. This enables growers to implement targeted interventions, reducing the need for widespread pesticide use.
- **Vertical farming in Japan:** Japan, with limited arable land, has embraced vertical farming for efficient space utilization. Indoor vertical farms use controlled environments with LED lighting, hydroponic systems, and automated climate control. This allows year-round cultivation of crops with optimal resource efficiency.
- **Drone-based crop monitoring in Australia:** Australian farmers employ drones for crop monitoring in vast agricultural landscapes. Drones equipped with sensors capture high-resolution imagery, providing insights into crop health, pest infestations, and overall field conditions. This technology aids in timely decision-making and targeted interventions.
- **Smart greenhouses in China:** China utilizes smart greenhouse technologies to enhance horticultural practices. These greenhouses integrate sensors, automation, and data analytics to optimize conditions for plant growth. The systems monitor factors such as temperature, humidity, and nutrient levels, ensuring high-quality produce.
- **Blockchain traceability in European berry farms:** Berry farms in Europe implement blockchain technology for supply

chain traceability. Each stage of the supply chain is recorded on a transparent and secure blockchain, allowing consumers to trace the origin and quality of berries. This enhances transparency and trust in the produce.

- **Automated harvesting robots in the United States:** In the United States, robotics companies have developed automated harvesting robots for fruits and vegetables. These robots use computer vision and robotic arms to identify and pick ripe produce, reducing labor costs and addressing challenges related to seasonal labor shortages.
- **Smart orchard management in New Zealand:** New Zealand orchards utilize smart management systems for fruit production. Sensors measure soil moisture, weather conditions, and plant health, providing real-time data to farmers. This information guides irrigation schedules, pest control, and overall orchard management.
- **IoT-enabled precision agriculture in India:** Indian farmers are adopting IoT-enabled precision agriculture for various crops. Soil sensors, weather stations, and mobile apps provide farmers with real-time data and recommendations for irrigation, fertilization, and pest control. This technology empowers farmers to make informed decisions and improve crop yields.

These real-world examples showcase the diverse applications of smart horticulture practices globally, highlighting the versatility and effectiveness of technology-driven approaches in enhancing agricultural productivity and sustainability.

1.10 SUMMARY, DISCUSSIONS, AND FUTURE PERSPECTIVES

This book delves into the multifaceted realm of smart farming applications in horticulture, unraveling a tapestry of technological innovations reshaping traditional agricultural practices. The exploration encompasses precision agriculture techniques, automated systems, and data-driven decision-making, underscoring their transformative impact on resource efficiency, crop management, and sustainability. Real-world examples from global horticultural practices, including Dutch greenhouses, Israeli precision irrigation, Californian smart pest management, and Japanese vertical farming, illuminate the diverse applications of smart technologies. The integration of IoT, sensors, drones, and automation emerges as a driving force behind enhanced productivity, reduced environmental impact, and resilient farming systems. As the book culminates, the synthesis of insights emphasizes the economic viability, adaptability, and collaborative nature of smart farming in horticulture, heralding a new era of precision and efficiency in agricultural landscapes.

The discussion unfolds a narrative of technological synergies and their impact on horticulture. Precision irrigation emerges as a cornerstone, optimizing water use and mitigating environmental stress. Automated systems, including robotics and smart pest management, exemplify the shift toward labor-saving practices and sustainable pest control. The utilization of data analytics and decision support systems is crucial, as they empower farmers to make well-informed decisions by incorporating real-time data. The global perspective offered by real-world examples accentuates the versatility and adaptability of these technologies across diverse agricultural landscapes. Supply chain optimization, traceability systems, and collaborative efforts further underscore the holistic nature of smart farming in horticulture. The interplay of these elements positions smart farming not merely as a technological trend but as a fundamental shift toward precision, sustainability, and resilience in horticultural practices.

Looking ahead, the future perspectives outlined in this book envision a continued evolution of smart farming in horticulture. Anticipated advancements include the integration of emerging technologies such as AI and blockchain for even greater precision and transparency. The expansion of data analytics capabilities, coupled with the proliferation of sensor technologies, is expected to refine decision-making processes further. The collaborative spirit among stakeholders is poised to deepen, fostering a culture of knowledge exchange and innovation. As smart farming becomes more accessible, efforts toward education and skill-building will likely be pivotal in empowering farmers to harness the full potential of these technologies. The ongoing quest for economic viability and sustainability will drive continuous refinement and adaptation, ensuring that smart farming remains at the forefront of the agricultural revolution, contributing to global food security and environmental stewardship.

KEYWORDS

- **smart farming**
- **horticulture technology**
- **precision agriculture**
- **agricultural innovation**
- **sustainable farming practices**
- **Internet of Things in agriculture**
- **data-driven agriculture**
- **crop management technologies**
- **environmental sustainability in agriculture**
- **collaborative agriculture**

REFERENCES

Awan, S., Ahmed, S., Ullah, F., Nawaz, A., Khan, A., Uddin, M.I., Alharbi, A., Alosaimi, W., & Alyami, H. (2021). IoT with BlockChain: A futuristic approach in agriculture and food supply chain. In: S. Khan, Ed., *Wireless Communications and Mobile Computing*. Hindawi Limited. http://doi.org/10.1155/2021/5580179.

Barelli, L., Bidini, G., Gallorini, F., & Servili, S. (2008). Hydrogen production through sorption-enhanced steam methane reforming and membrane technology: A review. *Energy*, 33, 554–570. https://doi.org/10.1016/J.ENERGY.2007.10.018.

Bennett, N.J., Whitty, T.S., Finkbeiner, E., Pittman, J., Bassett, H., Gelcich, S., & Allison, E.H. (2018). Environmental stewardship: A conceptual review and analytical framework. *Environmental Management*, 61, 597–614.

Bullock, D.G. (1992). Crop rotation. *Critical Reviews in Plant Sciences*, 11(4), 309–326.

Cárceles Rodríguez, B., Durán-Zuazo, V.H., Soriano Rodríguez, M., García-Tejero, I.F., Gálvez Ruiz, B., & Cuadros Tavira, S. (2022). Conservation agriculture as a sustainable system for soil health: A review. *Soil Systems*, 6(4), 87.

hChergui, N., & Kechadi, M.T. (2022). Data analytics for crop management: A big data view. *Journal of Big Data*, 9, 123. http://doi.org/10.1186/s40537-022-00668-2.

Clark, R.L., & McGuckin, R.L. (1996). Variable rate application technology: An overview. In: *Proceedings of the Third International Conference on Precision Agriculture*. Madison, WI, USA: American Society of Agronomy, Crop Science Society of America, Soil Science Society of America, pp. 855–862.

Cosgrove, W.J., & Loucks, D.P. (2015). Water management: Current and future challenges and research directions. *Water Resources Research*, 51(6), 4823–4839.

Ecer, F., Ögel, İ. Y., Krishankumar, R., & Tirkolaee, E.B. (2023). The q-rung fuzzy LOPCOW-VIKOR model to assess the role of unmanned aerial vehicles for precision agriculture realization in the Agri-Food 4.0 era. *Artificial Intelligence Review,* 1–34. http://doi.org/10.1007/s10462-023-10476-6.

Fereres, E., Orgaz, F., & Gonzalez-Dugo, V. (2011). Reflections on food security under water scarcity. *Journal of Experimental Botany*, 62(12), 4079–4086.

Fuller, D. (2007). Contrasting patterns in crop domestication and domestication rates: Recent archaeobotanical insights from the old world. *Annals of Botany*, 100, 903–924. https://doi.org/10.1093/aob/mcm048.

Gull, A., Lone, A.A., & Wani, N.U.I. (2019). Biotic and abiotic stresses in plants. In: *Abiotic and Biotic Stress in Plants*, pp. 1–19.

Hazell, P., & Wood, S. (2008). Drivers of change in global agriculture. *Philosophical Transactions of the Royal Society B: Biological Sciences*, 363(1491), 495–515.

Heider, K., Rodriguez Lopez, J.M., Bischoff, A., Balbo, A.L., & Scheffran, J. (2023). Toward climate-resilient and biodiverse agriculture in the Mediterranean region: Experiences and perceptions of farmers engaged in sustainable food production. *Organic Agriculture*, 13, 513–529. http://doi.org/10.1007/s13165-023-00444-3

Hofmann-Wellenhof, B., Lichtenegger, H., & Collins, J. (2012). *Global Positioning System: Theory and Practice*. Springer Science & Business Media.

Jain, G., & Mallick, B. (2016). A review on weather forecasting techniques. International *Journal of Advanced Research in Computer and Communication Engineering,* 5(12), 177–180.

Kaltoft, P. (1999). Values about nature in organic farming practice and knowledge. *Sociologia Ruralis*, 39(1), 39–53.

Khadse, K. (2021). To study applications of agricultural drones in irrigation and agriculture. *Bioscience Biotechnology Research Communications*, 14(9), 81–86.

Khort, D., Kutyrev, A., Smirnov, I., & Voronkov, I. (2020). Automated system for designing and management of agricultural technologies in horticulture. In: *2020 IEEE International Conference on Problems of Infocommunications. Science and Technology (PIC S&T)*. IEEE, pp. 1–6.

Klerkx, L., Jakku, E., & Labarthe, P. (2019). A review of social science on digital agriculture, smart farming and agriculture 4.0: New contributions and a future research agenda. *NJAS-Wageningen Journal of Life Sciences*, 90, 100315.

Kirkegaard, J., Christen, O., Krupinsky, J., & Layzell, D. (2008). Break crop benefits in temperate wheat production. *Field Crops Research*, 107, 185–195. https://doi.org/10.1016/J.FCR.2008.02.010.

Kogan, M. (1998). Integrated pest management: Historical perspectives and contemporary developments. *Annual Review of Entomology*, 43(1), 243–270.

Lock, R., & Miles, C. (1993). Biotechnology and genetic engineering: Students' knowledge and attitudes. *Journal of Biological Education,* 27(4), 267–272.

López, J. A., Soto, F., Iborra, A., Sánchez, P., & Suardíaz, J. (2010). Design and implementation of a wireless sensor network for precision horticulture. In: *Sensor Applications, Experimentation, and Logistics: First International Conference, SENSAPPEAL 2009*, September 25, 2009, Revised Selected Papers 1. Athens, Greece: Springer Berlin Heidelberg, pp. 27–42.

Poláková, J. (2018). Sustainability—Risk—Resilience: How Does the case of the good agricultural and environmental conditions measure up? *Sustainability, 10*(5), 1614. http://doi.org/10.3390/su10051614

Purugganan, M., & Fuller, D. (2009). The nature of selection during plant domestication. *Nature,* 457, 843–848. https://doi.org/10.1038/nature07895.

Rabka, M., Mariyanayagam, D., & Shukla, P. (2022). IoT-based horticulture monitoring system. In: *Intelligent Sustainable Systems: Selected Papers of WorldS4 2021.* Singapore: Springer, Vol. 2, pp. 765–774.

Ravi, P., Barrios, E., Bayala, J., Diby, L., Donovan, J., Gyau, A., Graudal, Lars., Jamnadass, R., Kahia, J.W., Kehlenbeck, K., Kindt, R., Kouame, C., McMullin, S., Van Noordwijk, M., Shepherd, K.D., Sinclair, F., Vaast, P., Vågen, T.-G. & Xu, J. (2015). Agroforestry: Realizing the promise of an agroecological approach. In: *Agroecology for Food Security and Nutrition, Proceedings of the FAO International Symposium*, 18–19 September 2014, Rome, Italy: Food and Agriculture Organization of the United Nations (FAO), pp. 201–224.

Robbins, J.A. (2018). Small Unmanned Aircraft Systems (sUAS) an emerging technology for horticulture. *Horticultural Reviews,* 45, 33–71.

Rudel, T., Kwon, O.-J., Paul, B., Boval, M., Rao, I., Burbano, D., McGroddy, M., Lerner, A.M., White, D., Cuchillo, M., Luna, M., & Peters, M. (2016). Do smallholder, mixed crop-livestock livelihoods encourage sustainable agricultural practices? A meta-analysis. *Land, 5*(1), 6. http://doi.org/10.3390/land5010006

Shafi, U., Mumtaz, R., García-Nieto, J., Hassan, S.A., Zaidi, S.A. R., & Iqbal, N. (2019). Precision agriculture techniques and practices: From considerations to applications. *Sensors*, 19(17), 3796.

Simmonds, N.W. (1997). A review of potato propagation by means of seed, as distinct from clonal propagation by tubers. *Potato Research*, 40(2), 191–214.

Singh, K.B., Gajri, P.R., & Arora, V.K. (2001). Modelling the effects of soil and water management practices on the water balance and performance of rice. *Agricultural Water Management*, 49(2), 77–95.

Singh, K.K., Krishan, C., & Singh, K. (2018). Propagation of citrus species through cutting: A review. *Journal of Medicinal Plants Studies*, 6(1), 167–172.

Zheng, Z., Xie, S., Dai, H., Chen, X., & Wang, H. (2017). An overview of blockchain technology: Architecture, consensus, and future trends. In: *2017 IEEE International Congress on Big Data* (BigData Congress). IEEE, pp. 557–564.

CHAPTER 2

Artificial Intelligence System and Horticultural Production: A New Paradigm for High Productivity

RAKESH KUMAR GUPTA*, SMITA RANI SWAIN, and PREM PRAKASH SRIVASTAV

Department of Agricultural and Food Engineering, Indian Institute of Technology, Kharagpur, West Bengal, India

*Corresponding author

ABSTRACT

The integration of artificial intelligence (AI) systems into horticultural production represents a transformative paradigm, offering the potential to revolutionize traditional farming methods and enhance productivity on a significant scale. This chapter explores the multifaceted impacts and innovations that AI brings to horticulture. AI's role in horticulture begins with data-driven insights, where machine learning algorithms analyze vast datasets encompassing climate conditions, soil health, crop characteristics, and historical yield data. This deep analysis enables the development of predictive models that can forecast crop yields with unprecedented accuracy. By understanding the intricate relationships between various variables, farmers gain valuable insights into optimal resource allocation, leading to precision farming practices. The implementation of AI extends to automated processes, such as robotic harvesting and sorting technologies, transforming labor-intensive tasks and increasing operational efficiency. By leveraging technologies like blockchain and IoT devices, AI ensures the traceability of horticultural products, instilling confidence in

Smart Horticulture Production: Integrating Advanced Technologies.
Rajni Rajan and Duniya Ram Singh (Eds.)

DOI: 10.1201/9781998511679-2

consumers regarding safety, sustainability, and ethical sourcing practices. This chapter encapsulates the essence of AI as a catalyst for a new era in horticultural production, where technology converges with traditional farming wisdom, paving the way for high productivity, sustainability, and a redefined agricultural landscape..

2.1 INTRODUCTION TO AI IN HORTICULTURAL PRODUCTION

The agricultural landscape is undergoing a profound transformation with the integration of artificial intelligence (AI), particularly in the intricate domain of horticultural production. This introduction aims to provide a comprehensive overview of the fundamental concepts, significance, and transformative potential of AI in the context of horticulture. As the global demand for food continues to escalate in the face of a burgeoning population and environmental challenges, the incorporation of AI emerges as a promising solution to revolutionize traditional agricultural practices and usher in a new era of productivity and sustainability. The advent of AI technologies represents a paradigm shift in horticulture, offering novel approaches to address longstanding challenges. Unlike conventional agricultural methods that rely heavily on manual labor and experience-driven decision-making, AI introduces a data-driven, intelligent system capable of analyzing vast amounts of information to optimize various aspects of horticultural processes (Cembrowska-Lech et al., 2023; Javaid et al., 2023). From crop monitoring to resource management, the infusion of AI has the potential to enhance precision, efficiency, and overall productivity in horticultural practices.

The significance of AI in horticultural production lies in its capacity to revolutionize the way we cultivate, manage, and harvest crops. With the world facing the daunting task of feeding an ever-expanding population, the need for sustainable and high-yield agricultural practices has never been more pressing. AI, through its ability to process and interpret complex datasets, offers a means to achieve this balance by providing insights that lead to informed decision-making. This not only ensures optimal resource utilization but also contributes to the development of resilient and adaptive agricultural systems. At the core of AI's integration into horticulture are several key concepts and technological foundations that drive its functionality.

Machine learning, a subset of AI, enables systems to learn from data, recognize patterns, and make predictions (Ghahramani, 2015). In the context of horticulture, machine learning algorithms can analyze historical and real-time data to predict crop yields, identify potential diseases, and optimize irrigation schedules (Akhter & Sofi, 2022). Additionally, computer vision, a technology that enables machines to interpret visual information, plays a crucial role in tasks such as crop monitoring and automated harvesting. The synergy of these technologies, coupled with advancements in sensor networks and data analytics, forms the backbone of AI applications in horticultural production.

While the integration of AI in horticulture presents unprecedented opportunities, it also brings forth a set of challenges that demand careful consideration. Technical barriers, ethical concerns, and the need for extensive data infrastructure are among the challenges that must be addressed to ensure the responsible and equitable adoption of AI in agriculture (Alexander et al., 2024). However, these challenges are accompanied by a wealth of opportunities, including the potential for increased yields, resource efficiency, and sustainable agricultural practices. The journey toward harnessing the full potential of AI in horticultural production requires a balanced approach that acknowledges and mitigates challenges while capitalizing on the transformative opportunities that lie ahead. In the chapters that follow, we will delve deeper into specific applications of AI in horticulture, exploring its role in crop monitoring, precision agriculture, automated harvesting, and more. By understanding the nuances of AI integration in each facet of horticultural production, we can unravel the full spectrum of benefits and challenges associated with this technological revolution, paving the way for a future where agriculture is not only productive but also sustainable and resilient in the face of evolving global challenges.

2.2 FUNDAMENTALS OF HORTICULTURAL AI SYSTEMS

The fundamentals of horticultural AI systems encompass a broad spectrum of concepts and technologies that revolutionize traditional horticultural practices. AI brings a data-driven and intelligent approach to the cultivation and management of crops, offering unprecedented precision and efficiency (Javaid et al., 2023; Jung et al., 2021). In this section, we will explore the key components that constitute the foundation of horticultural AI systems. Machine learning is a pivotal aspect of AI systems in horticulture. This subset of AI empowers machines to learn from data, recognize patterns, and make predictions without explicit programming. In horticulture, machine learning algorithms analyze vast datasets, encompassing information about soil conditions, weather patterns, crop health, and historical yield data (Ayoub Shaikh et al., 2022). By identifying patterns and correlations within these data, machine learning models can predict optimal planting times, recommend irrigation schedules, and even anticipate potential pest outbreaks.

Computer vision, another crucial element, allows AI systems to interpret visual information. In horticulture, computer vision is employed for crop monitoring, disease detection, and automated harvesting (Fuentes-Peñailillo et al., 2023; Tian et al., 2020). High-resolution cameras and sensors capture images of crops, and AI algorithms analyze these images to assess crop health, identify signs of diseases or nutrient deficiencies, and monitor growth patterns. The real-time insights provided by computer vision enable farmers to take proactive measures to address issues, ultimately enhancing crop productivity. The integration of sensors and Internet of Things (IoT) devices is fundamental to creating a networked and interconnected AI ecosystem in horticulture. Soil moisture sensors, temperature gauges, and other IoT devices continuously collect real-time data from the field. This information is then fed into AI systems, allowing farmers to make informed decisions about irrigation, nutrient application, and overall crop management. The synergy between sensors and AI contributes to a more responsive and adaptive agricultural system. Data analytics plays a crucial role in extracting meaningful insights from the vast amounts of data generated in horticulture. AI systems leverage advanced analytics tools to process, analyze, and visualize data, providing farmers with actionable information (Atheeq et al., 2023; Patil et al., 2023; Subudhi et al., 2023). Through predictive analytics, farmers can anticipate crop yields, identify trends, and optimize resource allocation. This data-driven decision-making not only enhances productivity but also supports more sustainable and resource-efficient horticultural practices.

Precision agriculture involves the targeted management of resources to optimize crop yields. AI systems play a pivotal role in precision agriculture by precisely tailoring interventions based on real-time data (Patil et al., 2023; Khan & Shorna, 2023). This includes precision planting, variable rate irrigation, and targeted pest control. The result is a more efficient use of resources, reduced environmental impact, and increased overall productivity. The use of autonomous vehicles and robots in horticulture further exemplifies the application of AI. AI-driven robots can perform tasks such as harvesting, weeding, and pruning with precision and efficiency (Balaska et al., 2023). These technologies not only alleviate the labor-intensive nature of certain horticultural tasks but also contribute to consistent and high-quality crop production.

2.3 AI-DRIVEN CROP MONITORING AND MANAGEMENT

AI-driven crop monitoring and management represent a transformative paradigm in agriculture, offering farmers unprecedented insights and tools to optimize crop health, resource allocation, and overall productivity. This section explores the key components and applications of AI in crop monitoring and management, illustrating how advanced technologies are reshaping traditional farming practices. AI-enhanced crop monitoring often begins with the utilization of remote sensing technologies and satellite imagery (Madasamy, 2023). Satellites capture high-resolution images of agricultural fields, providing a wealth of data on crop conditions. AI algorithms analyze these images, detecting subtle variations in crop health, identifying potential issues such as diseases or nutrient deficiencies, and offering a comprehensive overview of the entire field (Venkatasaichandrakanth & Iyapparaja, 2023). This approach allows for early intervention and targeted management strategies. The integration of AI in crop monitoring transforms agriculture into a data-driven industry. AI systems analyze vast datasets from various sources, including sensors, satellite imagery, and weather stations (Ashraf & Akanbi, 2023). This analysis produces actionable insights that inform decision-making processes. Farmers can rely on real-time data to make informed choices about planting, harvesting, and resource allocation, leading to improved overall efficiency and productivity. AI's adaptability is a key asset in crop management. As environmental conditions change, AI systems can dynamically adjust recommendations for crop care. This adaptability is particularly valuable in the face of climate variability, allowing farmers to respond to unforeseen challenges and optimize crop management strategies for resilience.

2.4 PRECISION AGRICULTURE WITH AI

Precision agriculture, coupled with AI, is reshaping the landscape of horticulture by introducing advanced technologies that enhance decision-making, optimize resource utilization, and promote sustainable practices (Karunathilake et al., 2023). This section explores the specific applications and benefits of AI-driven precision agriculture in horticulture production. AI facilitates automated crop monitoring through the analysis of high-resolution images captured by drones or other imaging devices. These images provide detailed insights into crop health, growth patterns, and the presence of pests or diseases. By employing machine learning algorithms, farmers can detect issues early, enabling timely intervention and minimizing the use of pesticides (Attri et al., 2023). The concept of variable rate technology is applied in horticulture through AI-driven systems. By analyzing soil data, historical performance, and crop requirements, AI generates prescription maps for the variable application of inputs such as fertilizers and pesticides (Javaid et al., 2023). This targeted application minimizes waste, reduces environmental impact, and optimizes crop nutrition. AI-powered autonomous machinery, including robotic harvesters and precision planters, is increasingly employed in horticulture. These machines can navigate fields, identify and harvest ripe fruits or vegetables, and perform other tasks with precision. The use of AI-driven robotics not only streamlines labor-intensive processes but also ensures accuracy in handling delicate crops (Mankar et al., 2023). AI optimizes planting and seeding processes in horticulture by analyzing data on soil conditions, climate, and crop types. Machine learning models assist in determining the optimal planting density and patterns, ensuring efficient use of space and resources (Yoosefzadeh Najafabadi et al., 2023). This precision contributes to improved crop uniformity and overall yield.

2.5 SMART IRRIGATION SYSTEMS

In the realm of horticultural production, the adoption of smart irrigation systems represents a significant leap forward, transforming traditional irrigation practices into technologically advanced and data-driven processes (Prakash et al., 2023). This comprehensive exploration delves into the fundamental concepts, key components, applications, and the broader impact of smart irrigation systems in the context of horticulture. From optimizing water usage to enhancing crop health, these systems are at the forefront of sustainable agriculture, providing a blueprint for the future of efficient and precision irrigation. Smart irrigation systems leverage cutting-edge technologies to intelligently manage the application of water in agricultural fields (Abioye et al., 2020). At their core, these systems integrate sensors, data analytics, and automation to monitor, control, and optimize irrigation processes. Unlike conventional irrigation methods that often rely on fixed schedules or manual assessments, smart irrigation adapts to real-time conditions, ensuring that crops receive the right amount of water at the right time (Bwambale et al., 2022; Gu et al., 2020). This adaptability is crucial for horticultural production, where different crops, soil types, and environmental factors demand tailored irrigation approaches. Water scarcity and the increasing demand for food necessitate a paradigm shift in agricultural practices. Horticulture, which encompasses the cultivation of fruits, vegetables, and ornamental plants, relies heavily on precise and efficient irrigation. Smart irrigation systems play a crucial role in addressing the challenges associated with conventional irrigation methods by leveraging technology to optimize water use, reduce waste, and enhance overall crop performance.

2.5.1 KEY COMPONENTS OF SMART IRRIGATION SYSTEMS

Sensor's Technology and IoT Connectivity: Sensor technologies are the backbone of smart irrigation systems. Soil moisture sensors, weather sensors, and humidity sensors provide real-time data about the environmental conditions and moisture levels in the soil (Pramanik et al., 2022). These data are instrumental in determining when and how much water should be applied to the crops. Sophisticated sensors can measure soil salinity, enabling even more precise irrigation strategies (Casterad et al., 2018). The IoT is a key enabler of smart irrigation. IoT devices, including sensors and actuators, are interconnected through a network, allowing seamless communication and data exchange. This connectivity enables remote monitoring and control of irrigation systems, providing farmers with the flexibility to manage irrigation operations from anywhere with internet access.

Data Analytics and AI: The integration of data analytics and AI enhances the capabilities of smart irrigation systems. AI algorithms analyze data from sensors, historical weather patterns, and crop requirements to make predictive decisions. Machine learning models can learn and adapt, optimizing irrigation schedules based on evolving conditions (Glória et al., 2021; Saggi & Jain, 2022). This intelligent approach ensures that water is applied precisely when and where it is needed, minimizing water wastage.

Automated Actuators and Valves: Automated actuators and valves are responsible for controlling the flow of water in response to the information provided by sensors and AI algorithms (Kirve et al., 2023). These components enable precise and timely adjustments to irrigation parameters, ensuring that each plant receives the optimal amount of water. Automation reduces the reliance on manual intervention, leading to more efficient water use.

2.6 AI-ENHANCED PEST AND DISEASE MANAGEMENT

AI-enhanced pest and disease management represents a revolutionary approach to agriculture, leveraging the power of AI to optimize and streamline the monitoring and control of pests and diseases in crops (Figure 2.1). Traditional methods of pest and disease management often rely on manual observation, which can be time-consuming and may not provide real-time insights. With AI, farmers can deploy sophisticated sensors, drones, and imaging technologies to collect vast amounts of data from fields, enabling early detection and precise identification of potential threats (Karunathilake et al., 2023). By analyzing historical data, current environmental conditions, and other relevant factors, AI algorithms can forecast the likelihood of pest infestations or disease outbreaks. This early warning system allows farmers to implement preventive measures, such as adjusting irrigation practices or deploying targeted pesticides before the issue escalates. Predictive modeling is particularly valuable in regions where specific pests or diseases exhibit cyclical patterns or are influenced by seasonal variations (Tonnang et al., n.d.). By understanding these patterns, farmers can optimize their pest management strategies, leading to more effective and sustainable agricultural practices. AI technologies enable the automation of pest identification and monitoring processes. Computer vision, a subset of AI, allows cameras and sensors to capture images of crops and identify pests or signs of diseases with high precision (Pakeerathan, 2023). This automation significantly reduces the time and effort required for manual monitoring, especially in large-scale agricultural operations. Automated pest identification systems can distinguish between beneficial insects and harmful pests, facilitating targeted interventions. Additionally, these systems can monitor the progression of diseases, enabling early detection and swift action to mitigate the impact on crops. Despite this, AI contributes to the development and implementation of Integrated Pest Management (IPM) strategies. IPM is an approach that combines various pest control methods, emphasizing biological control, cultural practices, and the judicious use of pesticides when necessary. AI technologies enhance the effectiveness of IPM by providing real-time data and insights into the factors influencing pest dynamics (Batz et al., 2023). For example, AI algorithms can analyze data on natural predators, weather conditions, and crop health to optimize the timing of beneficial insect releases. This level of precision ensures that biological control methods align with the specific needs of the crop and the prevailing environmental conditions.

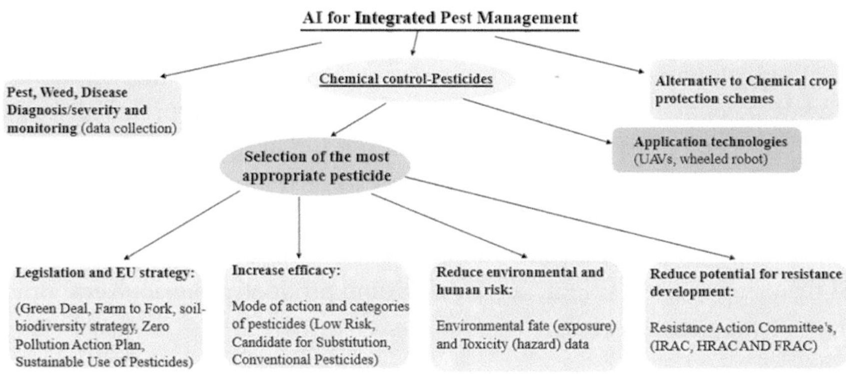

FIGURE 2.1 A flowchart of the crop protection system with AI and robotics techniques.

Source: Reprinted from Balaska et al. (2023). Copyright © 2023 by the authors. Licensee MDPI, Basel, Switzerland. https://creativecommons.org/licenses/by/4.0/

2.7 PREDICTIVE ANALYTICS FOR CROP YIELD

Predictive analytics for horticulture crop yield using AI systems is revolutionizing traditional farming practices by harnessing the power of advanced technologies to make accurate predictions and informed decisions. This approach relies on the integration of diverse datasets, including climate conditions, soil health indicators, and historical yield data. Machine learning algorithms, a core component of AI, analyze these data to identify patterns and relationships, enabling the creation of predictive models (Sarker, 2021). Historical data analysis serves as the foundation for training these models, allowing them to understand the complex interactions between various variables that influence crop yields (Manjunath & Palayyan, 2023). Feature selection and engineering further refine the models by identifying the most relevant factors and creating new variables to capture specific nuances of the horticultural system. The integration of remote sensing technologies, such as satellites and drones, provides real-time and high-resolution data, enhancing the precision of predictive analytics.

The applications and benefits of predictive analytics in horticulture are vast. Farmers can engage in precision farming by adjusting irrigation, nutrient applications, and other management practices based on real-time predictions (Adeyemi et al., 2017). This not only optimizes resource allocation but also contributes to sustainable and efficient agricultural practices. Predictive analytics supports risk management by allowing farmers to anticipate and mitigate the impact of adverse weather events, diseases, or pest outbreaks. Additionally, it aids in decision-making for crop planning, resource optimization, and overall farm efficiency. However, challenges such as data quality, model interpretability, and environmental variability need to be addressed. Ongoing advancements, including the integration of the IoT, explainable AI (XAI), and collaboration platforms, are shaping the future of predictive analytics for horticulture crop yield. This transformative approach holds the promise of fostering more resilient, sustainable, and productive agricultural practices in the years to come.

2.8 AUTOMATED HARVESTING AND SORTING TECHNOLOGIES

The agricultural industry faces a major challenge from the growing world population, which is expected to reach 9.7 billion in 2050 and 10.4 billion in 2100. As a result, there is an increasing need to modify agricultural practices to account for population expansion. The labor-intensive and ineffective nature of traditional hand harvesting techniques such as sickles, scythes, and knives were used to manually cut crops close to the ground (Kaur et al., 2023). Automated harvesting and sorting technologies, empowered by AI systems, are revolutionizing the agricultural landscape by enhancing efficiency, reducing labor costs, and improving overall productivity.

These technologies leverage cutting-edge AI algorithms and robotics to streamline the harvesting and sorting processes in various crops (Kaur et al., 2023). One of the key advancements is the integration of computer vision, a subset of AI and deep learning (DL) (Figure 2.2), which enables machines to visually analyze and comprehend the characteristics of crops, facilitating precise identification and selective harvesting. The basic architecture of DL-based artificial neural network for fruit detection and recognition is shown in Figure 2.3.

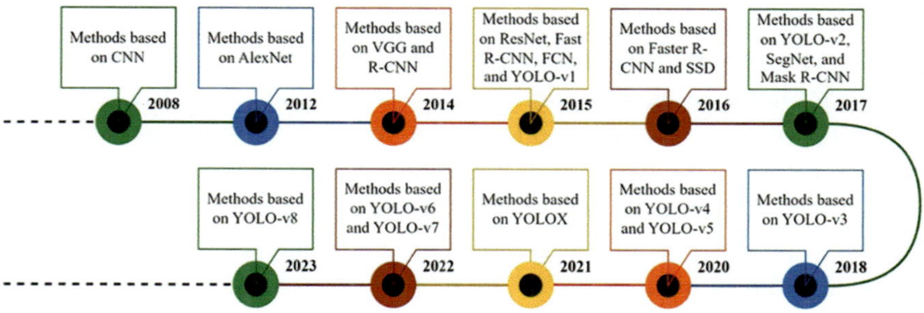

FIGURE 2.2 Main research processes of fruit detection and recognition methods based on DL.

Source: Reprinted from Xiao et al. (2023). Copyright © 2023 by the authors. Licensee MDPI, Basel, Switzerland. https://creativecommons.org/licenses/by/4.0/

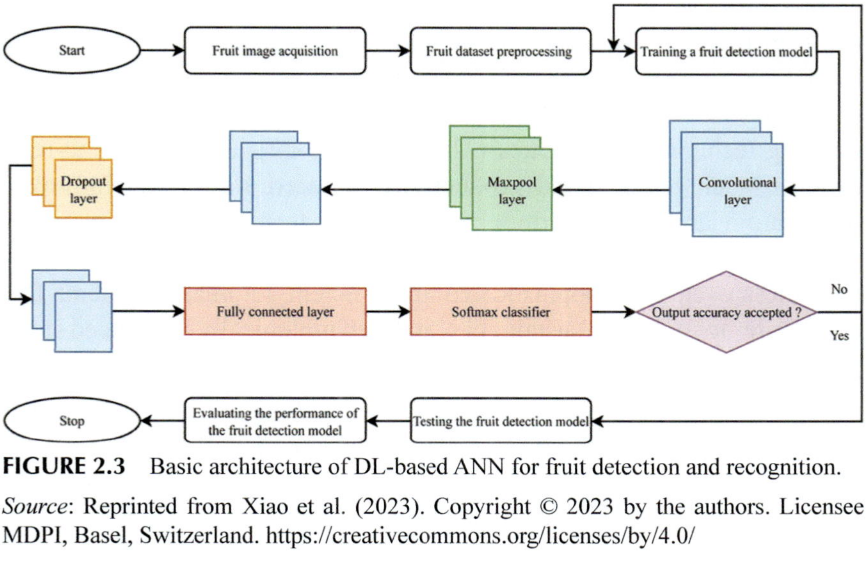

FIGURE 2.3 Basic architecture of DL-based ANN for fruit detection and recognition.

Source: Reprinted from Xiao et al. (2023). Copyright © 2023 by the authors. Licensee MDPI, Basel, Switzerland. https://creativecommons.org/licenses/by/4.0/

In automated harvesting, AI-equipped robotic systems can distinguish between ripe and unripe fruits or vegetables with remarkable accuracy. These systems use cameras and sensors to capture detailed images of the crops, and machine learning algorithms process this information to make real-time decisions on whether a particular item is ready for harvest (Bazame et al., 2021). This targeted harvesting not only minimizes waste but also ensures that crops are picked at their peak ripeness, contributing to improved quality and taste.

Sorting technologies, augmented by AI, play a pivotal role in optimizing post-harvest processes. AI algorithms can analyze the size, shape, color, and defects of harvested produce, enabling automated sorting systems to categorize items based on predefined criteria (Shandookh et al., 2022). This level of precision ensures that only the highest-quality produce reaches the market, reducing the likelihood of rejected batches and minimizing losses for farmers. Additionally, automated sorting enhances the speed and scalability of postharvest operations, allowing for quicker delivery of fresh produce to consumers.

The implementation of AI in automated harvesting and sorting technologies is not only enhancing efficiency but also addressing labor challenges in the agriculture sector. As the demand for manual labor in harvesting and sorting diminishes, farmers can overcome labor shortages and reduce their dependency on seasonal workers. This technological shift promotes sustainability by creating a more reliable and continuous workflow, especially in regions where labor availability is a persistent concern. While the integration of AI in automated harvesting and sorting is promising, challenges remain, including the initial investment costs and the need for ongoing technological updates (Zhou et al., 2022). However, long-term benefits, such as increased productivity, reduced waste, and improved economic viability for farmers, make these technologies increasingly attractive. As AI continues to evolve, the future of agriculture is likely to witness even more sophisticated and efficient automated systems that contribute to a more sustainable and technologically advanced food production ecosystem.

2.9 FARM-TO-TABLE TRACEABILITY WITH AI

Farm-to-table traceability, facilitated by AI, is transforming the horticulture industry by providing consumers with unprecedented transparency into the journey of their produce from the farm to their tables. This innovative approach employs AI technologies to track and trace the entire lifecycle of horticultural products, ensuring quality, safety, and sustainability. At the core of farm-to-table traceability is the use of AI for data collection, integration, and analysis. Through the implementation of sensors, radio-frequency identification tags, and other IoT devices, detailed information about each stage of production is captured. These data include crucial factors such as soil conditions, weather patterns, cultivation practices, and pesticide usage. Machine learning algorithms process this vast amount of data, identifying patterns and correlations that contribute to a comprehensive understanding of the product's journey.

The application of AI in traceability significantly enhances food safety. In the event of a contamination or outbreak, AI can swiftly trace the origin of affected produce, allowing for targeted recalls and minimizing the impact on consumers (Misra et al., 2022). This level of precision not only safeguards public health but also protects the reputation of farmers and producers. Additionally, consumers are increasingly concerned about the environmental and ethical aspects of their food choices. AI-driven traceability enables them to make informed decisions by providing insights into the sustainability practices of the farms, including water usage, carbon footprint, and ethical labor practices (Namkhah et al., 2023). Supply chain optimization is another crucial aspect of farm-to-table traceability with AI. By leveraging predictive analytics, the technology can forecast crop yields, predict harvest times, and optimize transportation routes. This not only reduces food waste but also ensures that products reach consumers at the peak of freshness. AI algorithms can dynamically adjust supply chain processes based on real-time data, improving overall efficiency and reducing the environmental impact associated with transportation and storage (Ghahremani Nahr et al., 2021).

The implementation of blockchain technology often complements AI in farm-to-table traceability. Blockchain creates an immutable and transparent ledger that records every transaction and movement of produce along the supply chain. AI algorithms can then analyze this blockchain data, providing an additional layer of verification and enabling consumers to access a tamper-proof record of the product's journey. This combination of AI and blockchain establishes a robust and trustworthy traceability system. From a consumer perspective, farm-to-table traceability with AI fosters a deeper connection with the food they consume. Mobile apps and QR codes on product packaging allow consumers to scan and access detailed information about the specific journey of the produce they are buying. This transparency not only builds trust but also aligns with the growing demand for ethically sourced and locally produced food. Despite these advantages, challenges exist in the widespread adoption of farm-to-table traceability with AI, including the need for standardized data formats, interoperability between different systems, and the upfront costs of implementing these technologies. However, as the technology matures and consumer awareness increases, the benefits of enhanced traceability, improved food safety, and sustainable practices are likely to drive the continued integration of AI in horticulture supply chains.

2.10 ECONOMIC AND ENVIRONMENTAL IMPACTS OF AI ADOPTION

The adoption of AI in horticulture production has profound economic and environmental implications, reshaping traditional farming practices and paving the way for a more sustainable and efficient agricultural future. Economically, AI technologies offer the promise of increased productivity and efficiency. Precision farming, enabled by AI, allows for optimized resource utilization, minimizing waste and reducing operational costs. Automated processes, such as robotic harvesting and sorting, streamline labor-intensive tasks, addressing concerns related to labor shortages and contributing to overall cost-effectiveness in horticulture production. In addition to improving efficiency, AI adoption in horticulture production has far-reaching environmental benefits. Precision agriculture, facilitated by AI-driven analytics, enables farmers to precisely tailor inputs such as water, fertilizers, and pesticides (Balaska et al., 2023). This targeted approach minimizes environmental impact by reducing excess chemical usage and mitigating the runoff of harmful substances into surrounding ecosystems. Furthermore, AI technologies contribute to sustainable pest and disease management, allowing for early detection and targeted interventions that minimize the need for broad-spectrum pesticides, protecting beneficial organisms and promoting biodiversity. The optimized supply chain management facilitated by AI not only enhances economic efficiency but also reduces the environmental footprint of horticulture production. AI-driven logistics and distribution systems enable streamlined transportation routes, reducing greenhouse gas emissions associated with food transport. Moreover, the reduction of food waste through predictive analytics ensures that fewer resources are expended on producing uneaten food, contributing to a more sustainable and circular agricultural economy. The adoption of AI in horticulture production supports responsible water management through data-driven irrigation systems. AI algorithms can analyze soil moisture levels and weather patterns, enabling precise irrigation scheduling those conserves water resources. This not only addresses concerns related to water scarcity but also contributes to sustainable farming practices. While the economic and environmental impacts of AI adoption in horticulture production are promising, challenges exist, including the initial investment costs, the need for technological infrastructure, and considerations regarding data privacy. However, long-term benefits, such as increased yields, resource efficiency, and a reduced environmental footprint, position AI as a transformative force in creating a more resilient, economically viable, and environmentally sustainable horticulture sector. Policymakers, industry stakeholders, and farmers need to collaborate to ensure that the integration of AI in agriculture aligns with sustainability goals and ethical considerations. As technologies continue to advance, the ongoing evolution of AI in horticulture production will play a pivotal role in addressing the complex challenges facing global food systems.

2.11 CHALLENGES IN IMPLEMENTING AI IN HORTICULTURE PRODUCTION

- The integration of AI into horticulture production faces several challenges that need careful consideration for successful implementation. One major obstacle is the quality and accessibility of data. Agricultural environments often lack standardized and comprehensive datasets, making it challenging to train accurate AI models. Efforts are needed to improve data collection infrastructure and promote collaborative data-sharing initiatives among stakeholders to overcome this hurdle.
- Another significant challenge is the initial cost of implementing AI technologies on farms. The upfront expenses associated with acquiring AI-enabled equipment, sensors, and systems can be a barrier for many farmers, especially those operating on limited budgets. Finding ways to make AI technologies more financially

accessible, potentially through government subsidies or private-sector partnerships, is crucial to encourage broader adoption among a diverse range of agricultural practitioners.

- Technology integration and compatibility present additional challenges. Many existing farming practices may not seamlessly align with new AI systems, leading to difficulties in incorporating these technologies into established workflows. Ensuring that AI solutions are user-friendly, compatible with existing infrastructure, and accompanied by adequate training and support mechanisms can facilitate a smoother integration process.
- Lastly, the skills and education gap pose a significant challenge. Farmers and agricultural workers may lack the technical expertise needed to operate and maintain AI systems effectively. Bridging this knowledge gap requires investment in education and training programs tailored to the specific needs of the agricultural community, empowering them to leverage AI tools effectively for improved horticulture production.

2.12 FUTURE PROSPECTS AND CONCLUSION

The future prospects of AI in horticulture production hold exciting possibilities, with ongoing innovations poised to transform the industry. Several key trends and developments are likely to shape the future landscape of AI in horticulture:

Advancements in Precision Agriculture: Precision agriculture, driven by AI technologies, will continue to evolve. This includes the development of more sophisticated sensors, drones, and robotics for precise monitoring and targeted interventions. AI-driven precision agriculture will enable farmers to optimize resource usage, enhance crop management, and minimize environmental impact.

Integration of IoT and Edge Computing: The integration of the IoT and edge computing will play a crucial role in enhancing real-time data processing on the farm. IoT devices, such as sensors and actuators, will generate vast amounts of data, and edge computing will enable on-site processing. This integration will improve the responsiveness and efficiency of AI systems in horticulture.

XAI: As AI systems become more complex, there will be a growing emphasis on XAI. Providing clear and interpretable explanations for AI-generated insights and decisions will be essential for gaining the trust of farmers and stakeholders. XAI ensures transparency, allowing users to understand and validate the reasoning behind AI recommendations.

Climate-Resilient Agriculture: AI will play a pivotal role in helping agriculture become more resilient to climate change. By integrating climate models with AI-driven analytics, farmers can anticipate and adapt to changing weather patterns. This will involve developing AI models that can predict the impact of climate change on crop yields and recommend adaptive strategies.

Personalized Crop Management: AI will enable more personalized and adaptive crop management strategies. Machine learning algorithms can analyze data on individual plant characteristics, soil conditions, and environmental factors to tailor cultivation practices for optimal growth. This personalized approach can maximize yields while minimizing resource inputs.

Blockchain for Supply Chain Transparency: The use of blockchain technology, coupled with AI, will further enhance supply chain transparency. Blockchain creates an immutable and transparent ledger, and AI can analyze data from this ledger to provide insights into the entire journey of horticultural products. This ensures traceability, quality assurance, and ethical sourcing, meeting the increasing consumer demand for transparency.

AI-Enabled Crop Breeding and Genetic Engineering: AI will accelerate advancements in crop breeding and genetic engineering. Machine learning algorithms can analyze vast genomic datasets to identify desirable traits and predict the performance of new crop varieties. This can expedite the development of crops with improved yields, resilience, and nutritional content.

Collaborative Platforms and Knowledge Sharing: Collaborative platforms that facilitate data sharing and knowledge exchange among farmers, researchers, and industry stakeholders will become more prevalent. These platforms will create a networked ecosystem where insights, best practices, and AI models can be shared, fostering a community-driven approach to innovation in horticulture production.

In conclusion, the future of AI in horticulture production is marked by continuous innovation and the integration of advanced technologies. These developments have the potential to revolutionize the way crops are grown, monitored, and managed, contributing to more sustainable, efficient, and resilient horticulture practices. As these trends unfold, collaboration between the agricultural community, technology developers, and policymakers will be essential to harness the full potential of AI in shaping the future of horticulture.

ACKNOWLEDGMENTS

The authors thank the Indian Institute of Technology Kharagpur for their assistance in this study. The Ministry of Education (formerly the Ministry of Human Resource Development), Government of India, is also appreciated by the authors for providing an institutional research assistantship.

KEYWORDS

- **artificial intelligence in horticulture**
- **high productivity**
- **precision farming**
- **automated harvesting**
- **deep learning**

REFERENCES

Abioye, E.A., Abidin, M.S.Z., Mahmud, M.S.A., Buyamin, S., Ishak, M.H.I., Rahman, M.K.I.A., Otuoze, A.O., Onotu, P., & Ramli, M.S.A. (2020). A review on monitoring and advanced control strategies for precision irrigation. *Computers and Electronics in Agriculture,* 173, 105441. https://doi.org/10.1016/j.compag.2020.105441

Adeyemi, O., Grove, I., Peets, S., & Norton, T. (2017). Advanced monitoring and management systems for improving sustainability in precision irrigation. *Sustainability,* 9(3), 353. https://doi.org/10.3390/su9030353

Akhter, R., & Sofi, S.A. (2022). Precision agriculture using IoT data analytics and machine learning. In: *Journal of King Saud University—Computer and Information Sciences.* Elsevier. Vol. 34, Issue 8, pp. 5602–5618. https://doi.org/10.1016/j.jksuci.2021.05.013

Alexander, C.S., Yarborough, M., & Smith, A. (2024). Who is responsible for 'responsible AI'?: Navigating challenges to build trust in AI agriculture and food system technology. *Precision Agriculture,* 25, 146–185. https://doi.org/10.1007/s11119-023-10063-3

Ashraf, H., & Akanbi, M.T. (2023). Sustainable agriculture in the digital age: Crop management and yield forecasting with IoT, Cloud, and AI. *Tensorgate Journal of Sustainable Technology and Infrastructure for Developing Countries,* 6(1), 64–71.

Atheeq, L.K., Venkatesh, R., Rajashekharappa, M.T., & Kumar, S.J. (2023). Data analytics and agricultural transformation: A potential disrupter. In: *Agricultural Sector in India: Accelerating Growth and Enhancing Competitiveness*. Taylor & Francis, pp. 321–343. https://doi.org/10.4324/9781003434672-25

Attri, I., Awasthi, L.K., Sharma, T.P., & Rathee, P. (2023). A review of deep learning techniques used in agriculture. *Ecological Informatics,* 77, 102217. https://doi.org/10.1016/j.ecoinf.2023.102217

Ayoub Shaikh, T., Rasool, T., & Rasheed Lone, F. (2022). Towards leveraging the role of machine learning and artificial intelligence in precision agriculture and smart farming. In: *Computers and Electronics in Agriculture*. Elsevier, Vol. 198, p. 107119. https://doi.org/10.1016/j.compag.2022.107119

Patil, B.D., Gupta, S., Sheikh, A.I., Kumar, S.S., Kumari, L., Raj, G.B. (2023). IoT and big data integration for real-time agricultural monitoring. *Journal of Advanced Zoology,* 44(S2), 3079–3089.

Balaska, V., Adamidou, Z., Vryzas, Z., & Gasteratos, A. (2023). Sustainable crop protection via robotics and artificial intelligence solutions. *Machines,* 11(8), 774. https://doi.org/10.3390/machines11080774

Batz, P., Will, T., Thiel, S., Ziesche, T.M., & Joachim, C. (2023). From identification to forecasting: The potential of image recognition and artificial intelligence for aphid pest monitoring. *Frontiers in Plant Science,* 14, 1150748. https://doi.org/10.3389/fpls.2023.1150748

Bazame, H.C., Molin, J.P., Althoff, D., & Martello, M. (2021). Detection, classification, and mapping of coffee fruits during harvest with computer vision. *Computers and Electronics in Agriculture,* 183, 106066. https://doi.org/10.1016/j.compag.2021.106066

Bwambale, E., Abagale, F.K., & Anornu, G.K. (2022). Smart irrigation monitoring and control strategies for improving water use efficiency in precision agriculture: A review. In: *Agricultural Water Management*. Elsevier, Vol. 260, p. 107324. https://doi.org/10.1016/j.agwat.2021.107324

Casterad, M.A., Herrero, J., Betrán, J.A., & Ritchie, G. (2018). Sensor-based assessment of soil salinity during the first years of transition from flood to sprinkler irrigation. *Sensors (Switzerland)*, 18(2), 616. https://doi.org/10.3390/s18020616

Cembrowska-Lech, D., Krzemińska, A., Miller, T., Nowakowska, A., Adamski, C., Radaczyńska, M., Mikiciuk, G., & Mikiciuk, M. (2023). An integrated multi-omics and artificial intelligence framework for advance plant phenotyping in horticulture. *Biology*, 12(10), 1298. https://doi.org/10.3390/biology12101298

Fuentes-Peñailillo, F., Carrasco Silva, G., Pérez Guzmán, R., Burgos, I., & Ewertz, F. (2023). Automating seedling counts in horticulture using computer vision and AI. *Horticulturae*, 9(10), 1134. https://doi.org/10.3390/horticulturae9101134

Ghahramani, Z. (2015). Probabilistic machine learning and artificial intelligence. *Nature*, 521(7553), 452–459. https://doi.org/10.1038/nature14541

Ghahremani Nahr, J., Nozari, H., & Sadeghi, M.E. (2021). Green supply chain based on artificial intelligence of things (AIoT). *International Journal of Innovation in Management, Economics and Social Sciences*, 1(2), 56–63. https://doi.org/10.52547/ijimes.1.2.56

Glória, A., Cardoso, J., & Sebastião, P. (2021). Sustainable irrigation system for farming supported by machine learning and real-time sensor data. *Sensors*, 21(9), 3079. https://doi.org/10.3390/s21093079

Gu, Z., Qi, Z., Burghate, R., Yuan, S., Jiao, X., & Xu, J. (2020). Irrigation scheduling approaches and applications: A review. *Journal of Irrigation and Drainage Engineering*, 146(6), 04020007. https://doi.org/10.1061/(ASCE)IR.1943-4774.0001464

Javaid, M., Haleem, A., Khan, I.H., & Suman, R. (2023). Understanding the potential applications of artificial intelligence in agriculture sector. *Advanced Agrochem,* 2(1), 15–30. https://doi.org/10.1016/j.aac.2022.10.001

Jung, J., Maeda, M., Chang, A., Bhandari, M., Ashapure, A., & Landivar-Bowles, J. (2021). The potential of remote sensing and artificial intelligence as tools to improve the resilience of agriculture production systems. *Current Opinion in Biotechnology,* 70, 15–22. https://doi.org/10.1016/j.copbio.2020.09.003

Karunathilake, E.M.B.M., Le, A.T., Heo, S., Chung, Y.S., & Mansoor, S. (2023). The path to smart farming: Innovations and opportunities in precision agriculture. In: *Agriculture (Switzerland)*. Multidisciplinary Digital Publishing Institute, Vol. 13, Issue 8, p. 1593. https://doi.org/10.3390/agriculture13081593

Kaur, B., Mansi, Dimri, S., Singh, J., Mishra, S., Chauhan, N., Kukreti, T., Sharma, B., Singh, S.P., Arora, S., Uniyal, D., Agrawal, Y., Akhtar, S., Rather, M.A., Naik, B., Kumar, V., Gupta, A.K., Rustagi, S., & Preet, M.S. (2023). Insights into the harvesting tools and equipment's for horticultural crops: From then to now. *Journal of Agriculture and Food Research,* 14, 100814. https://doi.org/10.1016/j.jafr.2023.100814

Khan, I., & Shorna, S.A. (2023). Cloud-based IoT solutions for enhanced agricultural sustainability and efficiency. *AI, IoT and the Fourth Industrial Revolution Review,* 13(7), 18–26. https://scicadence.com/index.php/AI-IoT-REVIEW/article/view/3

Kirve, M., Deshmukh, O., Kamalekar, P., Fulsunge, K., Kale, S., & Mehta, S. (2023). Smart water supply in irrigation using raspberry Pi. *SSRN Electronic Journal.* https://doi.org/10.2139/ssrn.4640034

Madasamy, S. (2023). Analysis of long-range wireless area network enabled smart agriculture for sustainable food production. *Asian Journal of Research in Computer Science,* 16(3), 243–253. https://doi.org/10.9734/ajrcos/2023/v16i3359

Manjunath, M.C., & Palayyan, B.P. (2023). An efficient crop yield prediction framework using hybrid machine learning model. *Revue d'Intelligence Artificielle,* 37(4), 1157–1167.

Mankar, S., Patil, A., Suthar, R., & Kshirsagar, V. (2023). Modern agribot seed sowing with enhanced features. *International Research Journal of Modernization in Engineering Technology and Science,* 11, 2507–2510.

Misra, N.N., Dixit, Y., Al-Mallahi, A., Bhullar, M.S., Upadhyay, R., & Martynenko, A. (2022). IoT, big data, and artificial intelligence in agriculture and food industry. *IEEE Internet of Things Journal,* 9(9), 6305–6324. https://doi.org/10.1109/JIOT.2020.2998584

Namkhah, Z., Fatemi, S.F., Mansoori, A., Nosratabadi, S., Ghayour-Mobarhan, M., & Sobhani, S.R. (2023). Advancing sustainability in the food and nutrition system:

A review of artificial intelligence applications. *Frontiers in Nutrition,* 10. https://doi.org/10.3389/fnut.2023.1295241

Pakeerathan, K. (2023). Image processing: A smart technology for early detection of crop pests and diseases. In: *Smart Agriculture for Developing Nations*. Singapore: Springer, pp. 135–150. https://doi.org/10.1007/978-981-19-8738-0_10

Prakash, C., Singh, L.P., Gupta, A., & Lohan, S.K. (2023). Advancements in smart farming: A comprehensive review of IoT, wireless communication, sensors, and hardware for agricultural automation. In: *Sensors and Actuators A: Physical*. Elsevier, Vol. 362, p. 114605. https://doi.org/10.1016/j.sna.2023.114605

Pramanik, M., Khanna, M., Singh, M., Singh, D.K., Sudhishri, S., Bhatia, A., & Ranjan, R. (2022). Automation of soil moisture sensor-based basin irrigation system. *Smart Agricultural Technology,* 2, 100032. https://doi.org/10.1016/j.atech.2021.100032

Saggi, M.K., & Jain, S. (2022). A survey towards decision support system on smart irrigation scheduling using machine learning approaches. *Archives of Computational Methods in Engineering,* 29(6), 4455–4478. https://doi.org/10.1007/s11831-022-09746-3

Sarker, I.H. (2021). Machine learning: Algorithms, real-world applications and research directions. *SN Computer Science,* 2(3), 160. https://doi.org/10.1007/s42979-021-00592-x

Shandookh, R.A., Salman, T.M., & Miry, A.H. (2022). Crop sorting using image segmentation and artificial intelligence: A review. In: *Proceedings—CSCTIT 2022: 5th College of Science International Conference on Recent Trends in Information Technology*, pp. 118–124. https://doi.org/10.1109/CSCTIT56299.2022.10145693

Subudhi, S., Dabhade, R.G., Shastri, R., Gundu, V., Vignesh, G.D., & Chaturvedi, A. (2023). Empowering sustainable farming practices with AI-enabled interactive visualization of hyperspectral imaging data. *Measurement: Sensors,* 30, 100935. https://doi.org/10.1016/j.measen.2023.100935

Tian, H., Wang, T., Liu, Y., Qiao, X., & Li, Y. (2020). Computer vision technology in agricultural automation—A review. *Information Processing in Agriculture,* 7(1), 1–19. https://doi.org/10.1016/j.inpa.2019.09.006

Tonnang, H., Sokame, B.M., & Villena, O. (n.d.). A system dynamics model examines how temperature influences the phenology of insect pests within crop ecosystems. *SSRN Electronic Journal.* https://doi.org/10.2139/SSRN.4601577

Venkatasaichandrakanth, P., & Iyapparaja, M. (2023). Review on pest detection and classification in agricultural environments using image-based deep learning models and its challenges. *Optical Memory and Neural Networks,* 32(4), 295–309. https://doi.org/10.3103/S1060992X23040112

Yoosefzadeh Najafabadi, M., Hesami, M., & Eskandari, M. (2023). Machine learning-assisted approaches in modernized plant breeding programs. *Genes,* 14(4), 777. https://doi.org/10.3390/genes14040777

Zhou, H., Wang, X., Au, W., Kang, H., & Chen, C. (2022). Intelligent robots for fruit harvesting: Recent developments and future challenges. *Precision Agriculture,* 23(5), 1856–1907. https://doi.org/10.1007/s11119-022-09913-3

CHAPTER 3

Use of Artificial Intelligence in Modern Agricultural Practices

MOHD TALHA ANEES,[1] ARINDAM CHAKRABORTY,[1] and HINA UPADHYAY[2*]

[1]*Department of Geology, Faculty of Science, Universiti Malaya, Kuala Lumpur, Malaysia*

[2]*Department of Agronomy, School of Agriculture, Lovely Professional University, Phagwara, Punjab, India*

*Corresponding author

ABSTRACT

Agriculture, grappling with outdated methods amid surging global demands and rapid population growth, is undergoing a transformative shift driven by innovative automated solutions. The integration of Internet of Things (IoT), big data, artificial intelligence (AI), and machine learning (ML) is shaping the landscape of "smart agriculture." This study aims to comprehensively explore the applications of AI in diverse agricultural domains. The major objectives encompass examining the overview and pivotal role of AI in agriculture, as well as identifying specific applications within various agricultural sectors. To fulfill these goals, a literature review was conducted across different agricultural fields where AI and related methods have been employed. This study delves into AI applications in crop monitoring, predictive analysis for farming precision, smart agriculture, crop yield prediction, weed and pest control, and supply chain optimization. Notably, in crop management, AI and ML are instrumental in refining pesticide use through precision agriculture, accounting for

Smart Horticulture Production: Integrating Advanced Technologies.
Rajni Rajan and Duniya Ram Singh (Eds.)

DOI: 10.1201/9781998511679-3

timing, location, and crop specifics. ML's ability to accurately detect crop quality features contributes to improved pricing and reduced waste. Water management reaps benefits from ML's estimation of evapotranspiration, optimizing irrigation efficiency. In summary, technological advancements are steering agriculture toward sustainable and intelligent practices, effectively addressing the challenges posed by labor scarcity, evolving regulations, and population growth.

3.1 INTRODUCTION

Agriculture, being among the world's most ancient and vital industries, faces the challenge of meeting the escalating demand for food and employment spurred by the rapid growth of the global population. Traditional farming methods are insufficient to satisfy these burgeoning needs and concurrently provide job opportunities for billions of people worldwide. Due to a scarcity of labor, rigorous regulations, growing global population, and a declining agricultural workforce, innovative automated techniques are being implemented to transform the farming practices.

Technological progress encompassing artificial intelligence (AI), machine learning (ML), Internet of Things (IoT), and big data are permeating nearly every sector, and agriculture is no exception. Initiatives and studies are actively pursuing the enhancement of both the quality and quantity of agricultural goods, aiming to imbue them with connectivity and intelligence through the paradigm of "smart farming." Addressing the challenges of labor scarcity, evolving regulations, and population growth necessitates a paradigm shift in the farming practices (Dharmaraj & Vijayanand, 2018).

In the realm of crop management, AI and ML play pivotal roles. Traditional approaches involve open-air or greenhouse pesticide spraying to enhance yield, but precision agriculture management, incorporating ML, allows agrichemicals to be applied judiciously based on factors such as timing, location, and the crops affected. Accurate detection and classification of crop quality features using ML contribute to higher product prices and reduced waste, as machines can uncover new traits significant to crop quality.

Water management is a critical aspect of agriculture, affecting agronomy, climate, and hydrology. ML-based applications contribute by estimating evapotranspiration daily, weekly, or monthly, enabling more efficient use of irrigation systems. Additionally, precise predictions of daily dew point temperature aid in anticipating weather phenomena, further enhancing evaporation and evapotranspiration estimation.

AI and ML models are increasingly embraced by farmers to boost productivity, with the food-tech sector reaping substantial benefits. Robots and sensors are deployed for crop management, data collection, and monitoring, underscoring the growing prospects of ML in digital agriculture. The ecological footprint of agriculture diminishes with the assistance of AI, facilitating informed decision-making, especially in areas such as weed management, optimal harvest timing, and monitoring the health of soil and crops (Smith, 2018).

The continuous evolution of AI, fueled by increment in cloud penetration and advancement in computational capabilities, has begun to revolutionize various sectors of the global economy, and agriculture is at the forefront of this transformation. AI and ML are proving to be invaluable tools for farmers, contributing to increased efficiency in decision-making processes related to agricultural and livestock production (Zha, 2020).

The burgeoning availability of data and the accessibility thereof are pivotal factors driving the agricultural sector's holding of AI technology. AI and ML-based surveillance systems provide insights for monitoring crops, detecting pests, and diagnosing soil issues, empowering farmers to make informed decisions. Agricultural robots are increasingly employed for labor-intensive tasks, offering cost savings and reducing the physical workload on human labor (Rodzalan et al., 2020).

In conclusion, the myriad applications of AI in agriculture spans from precision farming and crop management to decision-making processes that enhance productivity and sustainability. The integration of AI and ML technologies is reshaping traditional agricultural practices, presenting an opportunity for farmers to navigate the challenges posed by a growing global population and resource constraints. The transformative potential of AI in agriculture is a testament to the adaptability of technology in addressing the evolving needs of one of humanity's oldest and most vital industries.

This study was conducted to analyze the AI application in various fields of agriculture. The objectives of this study are (1) to examine the overview and necessities of AI in agriculture, and (2) to identify the AI applications in different sectors of agriculture. The overall structure of this study is shown in Figure 3.1. This study would be helpful for researchers to understand the recent development of AI usage in different agriculture sectors.

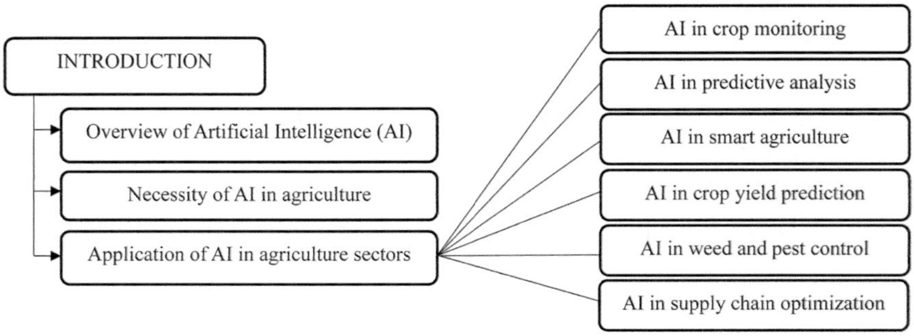

FIGURE 3.1 Structure of this study.

3.2 OVERVIEW OF ARTIFICIAL INTELLIGENCE

AI, is an interdisciplinary field dedicated for replicating human intelligence in machines, seeks to imbue robots with capabilities mirroring human thinking processes and behaviors, encompassing acquiring knowledge and addressing challenges. In the realm of agriculture, researchers and extension specialists are harnessing AI technology to address issues related to productivity, thereby revolutionizing traditional farming practices.

It has vast potential in agriculture, offering assistance to farmers in crucial decision-making processes. By aiding in appropriate crop types selection, crop production estimation, soil optimization and management of nutrients, pests and diseases management, and forecasting commodity prices, AI contributes to increased yields and enhanced efficiency in farming. Utilizing cutting-edge methods such as IoT, ML, deep learning (DL), image processing, robotics, artificial neural networks, AI, and wireless sensor networks (WSN) addresses the complex and dynamic nature of agricultural systems (Liu, 2020).

Real-time monitoring is a key feature facilitated by AI technologies, enabling farmers to receive continuous updates on various parameters such as temperature, weather conditions, soil condition, and water usage. This data-driven approach empowers farmers to make informed decisions, reducing losses while maximizing the crop yields. The development of smart farming practices is a testament to AI's transformative influence on agriculture, ushering in a new era of efficiency and sustainability (Ben & Hanana, 2021).

At the core of AI's functionality are ML and DL algorithms that learn from and analyze data to emulate human intelligence. These algorithms dynamically link input and output variables, providing predictive capabilities for a range of situations, from straightforward to complex. AI's ubiquitous presence in daily life, from facial recognition apps to self-driving vehicles, underscores its versatility and impact across various industries (Linaza et al., 2023). While AI has already yielded considerable productivity gains in other sectors, its transformative potential in agriculture is notable. Precision agriculture, enabled by AI, encompasses tasks such as pest control, crop selection, planting, harvesting, and crop rotation, harnessing the power of ML to process and utilize data effectively.

The foundational assumption of AI lies in the notion that human intellect can be codified, allowing machines to comprehend and execute a myriad of tasks. The primary objectives of AI include learning, logical thinking, and perception. As it continues to evolve, its impact extends across diverse industries, automating specific tasks and revolutionizing conventional practices. The potential of AI in agriculture is not merely confined to automation; it has the capacity to transform the agricultural landscape, simplifying tasks from the mundane to the complex. By aiding in decision-making, optimizing resource utilization, and enhancing its overall efficiency, this technology has the potential to contribute significantly to the advancement of agriculture, modernizing one of the most ancient industries for the digital era. As the world witnesses the increasing integration of intelligent machinery, the transformative power of AI in agriculture is growing, offering a future where technology is central to sustainability and productive farming practices (Shadrin et al., 2019).

3.3 NECESSITY OF ARTIFICIAL INTELLIGENCE IN AGRICULTURE

In the face of a growing global population and escalating demands for agricultural production, the labor-intensive nature of agriculture is pushing the need for automation to the forefront. AI emerges as a crucial ally for farmers, offering substantial assistance across different elements, technologies, and uses. Predictive analytics and sophisticated agricultural and crop management systems play a pivotal role in ensuring quality and availability of crops. By leveraging satellite imagery and meteorological data, businesses can assess land area, continuously monitor crop health, and make well-informed decisions (Jung et al., 2023).

Incorporating big data, AI, and ML technologies enables businesses to predict pricing trends, estimate crop production and yield, and detect pest and disease outbreaks. This wealth of information enables businesses to provide farmers with valuable insights, providing guidance on demand levels, selecting the most profitable crop varieties, and advising on pesticide usage, and future pricing patterns. As the complexity of modern agriculture grows, AI becomes a potent instrument, significantly mitigating resource and labor shortages. It is imperative for large corporations to invest in AI for agriculture, recognizing its potential to enhance productivity and efficiency across various industries. Industries such as agriculture, transportation, finance, and healthcare are already experiencing the advantages of AI applications, overcoming traditional challenges and fostering innovation (Skvortsov, 2020).

The global population is expanding rapidly, accompanied by accelerated urbanization and evolving consumer behavior. Farmers face increasing pressure to meet the rising demand for food, necessitating innovative strategies to boost output. However, challenges such as limited rich soil, climate variability, monocropping, and extensive pesticide application create significant obstacles for food producers. The implementation of AI in agriculture offers an intriguing opportunity to tackle these challenges, offering a means to control risks posed by pests, diseases, and environmental factors.

The agricultural industry, heavily dependent on natural forces, is under stress due to unpredictable weather patterns, labor shortages, and the continual need for increased yields. To meet future demands, the industry must undergo substantial scaling up, and farm efficiency needs to double. In this context, AI emerges as a key player in automating agriculture, providing innovative solutions to enhance productivity, reduce risks, and manage the complexities of modern farming. As this sector looks to the future, AI stands as a transformative force, offering the potential for sustainability and efficient agricultural practices (Lee et al., 2020).

Some applications of AI in agriculture are also provided in Table 3.1 (Javaid et al., 2023).

TABLE 3.1 Application of AI in Agriculture Sectors with Their Keywords

S. No.	Application	Keywords Related to the Application
1.	Prediction of weather	AI technology, agricultural conditions, crop health monitoring, unsupervised ML algorithms, and precision farming
2.	Allowing farmers to achieve efficient results	AI in agriculture, creative farming, cognitive IoT, data analysis, precision farming
3.	Help yield healthier crops	AI in agriculture, weather analysis, crop selection, predictive analytics, soil chemical analysis
4.	Predict plant diseases	AI in weed elimination, predictive analytics in agriculture, automated harvesting, real-time crop management, cognitive solutions for farming
5.	Self-driving tractors	Precision farming technologies, agricultural robotics, AI in farm labor solutions, data analytics in agriculture, predictive analytics for farming
6.	Improve decision making	AI in agricultural decision-making, precision farming technologies, automation in irrigation, sensor technology in agriculture, data analytics for farm efficiency
7.	Crop and soil monitoring	AI in crop and soil monitoring, precision agriculture applications, ML algorithms for data analysis, smart agriculture technologies, predictive pest control with AI
8.	Proper information on crop	AI in agricultural advisory, crop damage prevention, AI-enabled image recognition, IoT and AI-driven Solutions, robotic farming with AI
9.	Student studying regarding Agriculture	Agricultural education and AI, collaborative AI applications in agriculture, remote sensing and multispectral imagery, AI for crop health monitoring, soil health assessment with AI
10.	Recognize inefficient resource usage trends.	AI resource optimization, preventive livestock health monitoring, profitability enhancement through AI, sustainable agriculture with AI, diverse AI algorithms in agriculture
11.	Detect soil defects	AI soil analysis, predictive agriculture with AI, environmental impact monitoring, IoT and AI in farming, disaster management with AI
12.	Determining suitable areas for planting particular crops.	Drone-based crop planning, AI seed quality assessment, soil flaw detection with AI, AI in flora pattern analysis, weed control with computer vision
13.	Offer accurate instructions for managing water resources	AI in precision farming, ML for crop sustainability, robotics and AI in gardening, disease identification in plant roots, digital agriculture advantages

TABLE 3.1 *(Continued)*

S. No.	Application	Keywords Related to the Application
14.	Detect the smallest insects	AI in pest detection, climate-responsive AI farming, crop management with technology, AI for yield pattern analysis, real-time crop health monitoring
15.	Assist food supply chain	AI in agriculture revolution, economic efficiency with AI, risks of implementing AI, agricultural market structure, technology and sustainability in agriculture
16.	Enhance crop yield and productivity	AI in agricultural productivity, predictive analytics in agribusiness, soil monitoring and pest control, ML for crop performance analysis, leaf vein morphology in plant classification
17.	Proper guidance	Precision agriculture with AI, predictive analytics for farming, cost savings in agriculture, AI in plant disease detection, AI robots in farming fields
18.	Allocation of resources at the farm level	Technology in agricultural finance, real-time crop monitoring, AI for efficiency in agriculture, SAAS platforms with AI algorithms, democratization of IoT and AI
19.	Drive predictive analytics	AI in agricultural sowing, precision farming with IoT, predictive analytics in agriculture, technology for crop productivity, autonomous systems in farming
20.	Attain effective outcomes with reduced effort	Agricultural intelligence, smart farming with IoT, drone-enabled farming, AI in plant disease detection, image recognition in agriculture
21.	Assists with crop cultivation and harvesting	AI in crop cultivation, predictive analysis in agriculture, early problem detection, precision agriculture with AI, drone-based crop monitoring
22.	Increase farmer profitability	AI in pest detection, increased food yields, water-efficient agriculture, job creation with AI, AI in gardening and landscaping
23.	Improve agricultural efficiency	Sensor fusion in agriculture, AI in video analytics for field security, plant health monitoring with drones, AI advancements in agriculture, data-driven decision-making in farming
24.	Assists in addressing the issue of food scarcity.	AI in agriculture for food security, cognitive computing in crop recommendations, customized farming advice, AI considering market trends in farming, ML for informed decision-making in agriculture
25.	Create accurate and controlled methodologies.	Precision farming with AI, emerging technologies in agriculture, sustainable agriculture, AI-driven agricultural revolution, environmental impact of agricultural practices
26.	Detect anomalies and impurities	ML algorithms in food quality control, efficient food production, precision farming with AI, AI in food and beverage packaging, ensuring food security with AI

3.4 APPLICATION OF ARTIFICIAL INTELLIGENCE IN DIFFERENT AGRICULTURE SECTORS

3.4.1 AI IN CROP MONITORING

In crop monitoring, evaluating crucial biochemical and physical characteristics for field management, such as leaf area index (LAI), leaf nitrogen concentration (N), and aboveground biomass (AGB), has become crucial. AI plays a key role in overcoming challenges linked to traditional monitoring methods, supplying a cost-effective and noninvasive method with great precision.

Traditionally, field-based surveys and lab analyses were labor-intensive and impractical for large-scale use. Satellite remote sensing offered an alternative, estimating crop parameters such as N, AGB, and LAI. However, challenges such as limited spatial resolution, revisiting frequencies, soil effects, and atmospheric conditions restricted its applicability, particularly in dense and diverse canopies. To address these hurdles, several vegetation indices are used. The suggestion involves using three spectral bands, narrow spectral bands, or the spectral region known as the red-edge. Additionally, metrics derived from microwave radar and the integration of light detection and ranging (LiDAR) data, offering three-dimensional canopy structural information, have enhanced the precision of crop monitoring techniques. While LiDAR has been extensively utilized for tree-related studies, its application in crop studies faced limitations such as point density and canopy penetration capacity, coupled with higher costs (Wang et al., 2017).

The emergence of unmanned aerial vehicles (UAVs) has revolutionized crop monitoring by offering enhanced spatial and temporal resolution. UAVs equipped with RGB cameras provide operational simplicity and flexibility, offering detailed 3D canopy information through cost-efficient methods like structure from motion. Although UAV-RGB imagery may lack the canopy penetration capacity of LiDAR, it provides dense and accurate point clouds.

The successful application of combining satellite imagery and UAV data in forest/tree studies has demonstrated the potential for enhancing estimations of forest properties. Nevertheless, the integration of rich spectral information from satellite data with high-resolution canopy structural features from UAV imagery for crop monitoring remains largely unexplored in agricultural applications.

Maimaitijiang et al. (2020) discussed the use of ML methods such as random forest regression (RFR), support vector regression (SVR), and partial least squares regression (PLSR) in crop monitoring. PLSR stands as a widely utilized statistical technique in the realm of crop monitoring, specializing in the analysis of relationships among multiple variables. In agriculture, PLSR finds application in correlating spectral data, often sourced from remote sensing technologies, with specific crop characteristics. By establishing these connections, PLSR facilitates the prediction of crucial parameters such as nutrient levels, biomass, and overall crop health. SVR emerges as an ML algorithm tailor-made for regression tasks in the context of crop monitoring. Its prowess lies in adeptly managing nonlinear correlations among input variables, rendering it a suitable option for capturing the intricate dynamics of agricultural systems. SVR proves instrumental in predicting various crop attributes, including yield, drawing insights from a combination of factors such as weather conditions, soil properties, and historical data. Within the domain of crop monitoring, RFR stands out as a versatile ensemble learning technique. Recognized for its ability to navigate large datasets featuring diverse variables, RFR is employed to model relationships between different environmental factors and crop outcomes. Its strength lies in capturing intricate interactions and dependencies within the data, rendering it invaluable for predicting crop yields and identifying influential variables. The utilization of PLSR, SVR, and RFR exemplifies the sophisticated integration of statistical and ML techniques in the advancement of precise and data-driven crop monitoring practices.

3.4.2 USE OF AI IN PREDICTIVE ANALYSIS FOR FARMING PRECISION

The undulating prices of agricultural commodities wield a profound influence on a country's gross domestic product, creating a ripple effect that resonates emotionally and financially with farmers whose painstaking efforts may go unrewarded. The unpredictable nature of these price fluctuations necessitates a proactive approach, wherein predictive analytics emerges as a pivotal tool in the agriculture supply chain, enabling informed decisions to mitigate and manage the risks associated with such volatility.

The repercussions of diminished agricultural production, attributed to factors such as unstable climatic conditions and global warming, underline the significance of predictive analytics in addressing the challenges faced by the common man. Agricultural commodities, spanning crops, livestock, and dairy products, face price dynamics influenced by an array of factors such as climate patterns, government policies, crop diseases, and overall availability. The resultant price volatility presents a dual-edged sword, capable of either benefiting or detrimentally impacting stakeholders. Insight into these price fluctuations becomes instrumental for governmental bodies and organizations, allowing them to make strategic decisions to manage associated risks effectively.

Understanding the intricacies of price fluctuations opens avenues for governments to implement support measures, offering loans and insurance at reduced interest rates to shield farmers from financial instability. Farmers, armed with predictive insights, can take timely actions to enhance crop production and adapt to changing market conditions. In the financial markets, analytics has become a dominant force, enabling organizations to anticipate and respond to price fluctuations for the sake of economic stability. Price variability is inherent in agricultural markets, prompting the utilization of predictive analytics and data mining techniques to discern trends and seasonality within price data. The role of data mining in agriculture, specifically in predicting crop prices, has been explored through various models such as autoregressive integrated moving average (ARIMA), naive, exponential smoothing, and more. The application of these predictive models has proven successful in forecasting the prices of diverse agricultural commodities at both national and international levels.

Research findings emphasize the superiority of ML models over traditional statistical methods. Artificial neural networks (ANN) showcase their prowess by learning and adapting to the intricate fluctuations, outperforming statistical methods such as exponential smoothing and ARIMA in predicting the rice exports of Thailand. Recurrent neural network (RNN) models, particularly long short-term memory (LSTM), are lauded for their effectiveness in handling sequential time-series data, surpassing the performance of ARIMA models in predicting retail trends and wheat production in Pakistan (Sabu & Kumar, 2020).

Hybrid models, combining linear and nonlinear components, further enhance predictive accuracy. Notable examples include the fusion of ARIMA and Elman neural network to forecast hog and canola prices in Germany. As agriculture grapples with the challenges posed by price volatility, the adoption of ML models emerges as a transformative solution, offering a more nuanced understanding of market dynamics and empowering stakeholders to navigate the complexities of the agricultural landscape with greater foresight and resilience (Rather et al., 2015).

3.4.3 USE OF AI IN SMART AGRICULTURE

Water, a precious and scarce natural resource, is indispensable for life and agriculture. Its management demands careful planning, development, conservation, and, most importantly, sustainable usage. With escalating demands and finite resources, optimal management of water at the farm level becomes imperative. Balancing increasing agricultural needs with limited water availability necessitates efficient and sustainable practices to boost crop yields. The challenge lies in maximizing crop production under restricted water sources to meet the escalating demands of future food production. This requires a paradigm shift toward using limited water supplies judiciously, aiming to irrigate more areas with the same amount of water.

In recent years, the integration of AI systems and the IoT has emerged as a promising approach to precisely address irrigation challenges, primarily utilizing linear models. Past research endeavors have focused on irrigation control models employing real-time detection of root zone soil moisture deficit. These models utilize water balance data to formulate linear time-series models. Noteworthy among them is the application of the deep LSTM model in predicting soil moisture for a lychee orchard, showcasing the effectiveness of linear time-series models with sequential processing over time. Additionally, an intelligent smart irrigation decision support system has been proposed, incorporating both linear (PLSR) and layered (adaptive neuro-fuzzy inference system) models to estimate irrigation needs over a week (Tace et al., 2020).

While these models enhance irrigation control, their integration with data analysis techniques often results in slow and regionally limited sampling. The novel irrigation prediction approach for efficient management of intelligent automatic irrigation, encompasses four key steps. First, the installation of sensors (measuring soil moisture, temperature, and rainfall) forms the foundation. Second, the sensors are linked to an acquisition system, establishing a data acquisition framework. The third step involves utilizing the Node-RED platform, designed for streamlined supervision, storage, and notification. Finally, the collected data undergoes processing using multiple algorithms, including K-nearest neighbors (KNN), neural networks, support vector machines (SVMs), naive Bayes, and logistic regression. The results highlight the superior decision-making rate of the KNN algorithm, reaching an impressive rate of 98.3% (Zhang, 2016; Tace et al., 2020).

This innovative approach holds significant promise, particularly in large-scale irrigation systems or domestic applications. It not only eases the labor-intensive task of watering plants but also enhances water management, ensuring a more sustainable and efficient use of this precious resource. By seamlessly integrating technology with agriculture, this approach paves the way for a more sustainable and technologically advanced future in irrigation practices.

3.4.4 USE OF AI IN CROP YIELD PREDICTION

Extensive research has been driven by the global focus on crop yield prediction, revealing varied approaches harnessing advanced technologies. A particular study introduced a DL framework for forecasting crop yield, utilizing remote sensing data. The model, incorporating a convolutional neural network (CNN) with a Gaussian process component and dimensional reduction technique, aimed to predict crop yield across developing countries throughout the year. Demonstrated with a comprehensive dataset involving soybeans and encompassing sensing, soil, and climate data from the United States, the model's precision, as measured by root mean square error (RMSE), was notably enhanced, particularly with the Gaussian approach when compared to both LSTM and CNN models.

A notable undertaking integrated agronomic principles of crop modeling with ML to establish a foundation for predicting large-scale crop yields. Emphasizing accuracy, modularity, and reusability, this process utilized crop simulation results and diverse environmental data sourced from a database. Three ML algorithms—gradient boosting, SVR, and KNN—were employed to forecast yields for various crops across regions in the Netherlands, Germany, and France.

In a distinct approach, researchers proposed a novel multilevel ML model, fusing RNN and CNN architectures. This integration aimed to extract both spatial and temporal features for enhanced crop yield prediction. Their investigation focused on the US Corn Belt, utilizing time-series remote sensing data and soil property information as inputs. The study evaluated the model's performance from 2003 to 2015 at the county level, demonstrating the effectiveness of the proposed approach (You et al., 2017).

An investigative study delved into the synergy of crop modeling and ML to improve corn yield predictions in the US Corn Belt. By incorporating simulation crop model variables into ML models, the study achieved a notable decrease in yield prediction RMSE, emphasizing the inadequacy of weather information alone. The findings underscored the importance of integrating hydrological inputs into ML models for more accurate yield predictions (Shahhosseini et al., 2023). Further contributions extended the application of ML to predict corn hybrid yields based on genotype and environmental data. Leveraging a deep neural network-based solution, this work achieved remarkable accuracy in predicting average yield and standard deviation. Additionally, a study focused on predicting potato tuber yield using soil and crop properties collected through proximal sensing was conducted. SVR emerged as the top-performing model in this context (Khaki & Wang, 2019). Expanding the scope to a 43-year benchmark dataset, another study employed an ML-based RNN model to predict wheat crop yield in the northern region of India. Addressing the vanishing gradient problem, the study employed LSTM, showcasing the model's superiority over other techniques (Bali & Singla, 2023). Taking a unique direction, a recent study concentrated on African countries, utilizing an ML architecture on satellite image data to predict maize yields at the district level. Diverging from image-centric approaches, this study took a holistic approach by incorporating climate, chemical, and agricultural parameters, recognizing the pivotal impact of climate change on crop productivity (Kaneko et al., 2019). In essence, these diverse studies underscore the evolving landscape of crop yield prediction, emphasizing the integration of advanced technologies, ML algorithms, and comprehensive datasets to enhance precision and address the challenges faced by agricultural systems globally (Cedric et al., 2022).

3.4.5 USE OF AI IN WEED AND PEST CONTROL

Automated weed detection in agricultural fields commonly employs sensors to differentiate weeds from crops, soil, and plant residue. Weed detection methods are broadly categorized as contact or noncontact approaches, with the latter involving sensing conducted via satellites, airborne vehicles, ground-based vehicles, or robotics. Spectroscopy, known for its speed, plays a crucial role in distinguishing spectral differences between crop plants and weeds. Spatial attributes, vital for measurements, are emphasized, requiring incorporation into the analysis. These methods rely on the reflectance, absorbance, or emission of inherent signals or features from plants, detected by nonimaging or imaging sensors such as color imaging, hyperspectral imaging, and multispectral imaging. Incorporating these sensing methods with ML algorithms such as Bayesian classifiers and CNN has shown effectiveness in swiftly distinguishing between crops and weeds (Su, 2020).

Visible/near-infrared spectroscopy, which analyzes the interaction between the object and radiation intensity, has effectively differentiated between silver beet and maize plants. This discrimination was achieved using an SVM algorithm based on three characteristic variables. Another study was conducted on sugar beet and weeds by utilizing spectral reflectance. It is important to note that nonimaging spectroscopy provides spectral information but lacks spatial details of the entire object.

Color imaging has demonstrated efficiency in weed classification. With the help of color imaging, CNN-based classification systems can operate on small device hardware with a very low processing time. Hyperspectral imaging captures numerous narrow-band images, generating spectra for all pixels in an image. Hyperspectral imaging-based models have attained accurate classification rates in distinguishing nightshade weed from tomato plants.

Multispectral imaging, a successor to hyperspectral imaging, has a wide range of spectral reflectance. While achieving high accuracy for weed classification in soybean fields, it faces challenges in distinguishing weeds in heterogeneous environment such as weed with high densities. Chlorophyll fluorescence imaging captures the fluorescence emission of plant chlorophyll but faces challenges in distinguishing healthy crop plants from weeds due to identical fluorescence peaks.

These widely used sensors often rely on plant inherent features, such as chlorophyll and nitrogen, for weed detection. However, the effective implementation of these methods requires extensive learning of the characteristics of numerous weeds or crops to establish a robust ML model before undertaking plant identification operations. The ongoing challenge lies in streamlining models, reducing redundant information and enhancing the speed of online detection for efficient and accurate weed management in agriculture.

3.4.6 USE OF AI IN SUPPLY CHAIN OPTIMIZATION

Agricultural supply chains (ASCs) bear similarities to fast-moving consumer goods supply chains but exhibit distinctions in raw material procurement and final product delivery. These supply chains encompass diverse operations, including preproduction, production, storage, processing, retail, and distribution, all contributing to the journey of agricultural products from the field to end consumers. Inherently complex, ASCs involve multiple stakeholders such as farmers, processors, certification agencies, traders, retailers, distributors, and consumers, necessitating effective coordination across strategic, tactical, and operational levels (Sharma et al., 2020).

The global challenge of sustainable food production amid rapid industrialization and overpopulation underscores the need for transformative approaches in ASCs. Existing issues such as lack of industrialization, information asymmetry, inadequate management practices, poor traceability, and compliance with food safety norms demand urgent attention. Heightened awareness of social, economic, and environmental aspects by various stakeholders, including nongovernmental organizations, consumer organizations, and policymakers, has led to increased pressure for sustainable food production strategies. Aligning with the United Nations' 2030 Sustainable Development Goals, there is a recognized imperative for ASCs to shift drastically toward sustainability.

Emerging technologies contribute to a substantial volume of data, encompassing soil, weather, geospatial, irrigation, and livestock information, driving decision-making in different ASC phases. The application of the IoT platform, coupled with drones and sensors, emerges as a promising technology for data analytics, supporting agile field operations and intelligent assistance. Drones, in particular, exhibit significant potential for crop and livestock monitoring, while sensors and actuators enhance understanding of farming conditions, including weather, environmental factors, pest management, and disease detection.

ML algorithms play a pivotal role in optimizing ASC performance across various phases. Supervised, unsupervised, and reinforcement learning techniques contribute to specific objectives aimed at improving decision-making and overall efficiency. A feedback loop between ASC phases and ML algorithms facilitates the development of ML capabilities for extracting valuable information. ML-based analyses, integrated at different stages, enhance ASC performance economically, environmentally, and socially.

The utilization of ML algorithms in the preproduction phase enables timely rainfall forecasting, supporting effective water planning and resource management. ML-driven site-specific nutrient management in the production phase ensures accurate crop yield predictions and optimized resource use. In the processing phase, ML algorithms contribute to reduced greenhouse gas emissions, improved environmental quality, and minimized postharvest losses. ML applications in the distribution phase enhance social and economic sustainability by improving food safety, quality, demand forecasting, and economic savings.

In essence, the integration of ML algorithms across ASC phases serves to develop efficient and sustainable agricultural practices, addressing economic, environmental, and social dimensions of food production and distribution.

3.5 CONCLUSION

In summary, the integration of AI into agriculture marks a transformative shift toward precision, sustainability, and efficiency. AI's role in crop monitoring, driven by ML methods such as PLSR, SVR, and RFR, enhances predictions related to nutrient levels, biomass, and overall crop health. Predictive analysis for farming precision, powered by AI-driven ML models, provides invaluable insights into crop prices and yields, empowering decision-making for economic stability and sustainable practices.

In smart agriculture, AI's impact on water management, utilizing linear models and ML techniques, ensures efficient irrigation control. Crop yield prediction sees unprecedented advancements through AI-driven methodologies such as CNN, RNN, and SVR, emphasizing precision and adaptability. Weed and pest control benefit from AI and sensor technology fusion, streamlining detection processes.

Supply chain optimization, using AI technologies, IoT platforms, drones, and ML algorithms, contributes to economic, environmental, and social sustainability. The amalgamation of AI with various technologies signifies a promising trajectory toward sustainable, data-driven, and technologically advanced farming practices, addressing challenges and propelling agriculture into a future characterized by efficiency, precision, and resilience.

ACKNOWLEDGMENTS

MTA and AC are thankful to the Department of Geology, Faculty of Science, Universiti Malaya, Malaysia for publishing this work. HU is thankful to the School of Agriculture, Lovely Professional University, Punjab, India for publishing this work and also for providing necessary laboratory facilities.

KEYWORDS

- **Internet of Things**
- **artificial intelligence**
- **machine learning**
- **agriculture sectors**
- **crop monitoring**

REFERENCES

Bali, N., & Singla, A. (2023). ML based wheat crop yield prediction model in Punjab region of North India. *Applied Artificial Intelligence,* 35(35), 3304–3328.

Ben Ayed, R., & Hanana M. (2021). Artificial intelligence to improve the food and agriculture sector. *Journal of Food Quality* 2021(1), 5584754.

Cedric, L.S., Adoni, W.Y.H., Aworka, R., Zoueu, J.T., Mutombo, F.K., Krichen, M., & Kimpolo, C.L.M. (2022). Crops yield prediction based on ML models: Case of West African countries. *Smart Agricultural Technology,* 2, 300049.

Dharmaraj, V., & Vijayanand, C. (2018). Artificial intelligence (AI) in agriculture. *International Journal of Current Microbiology and Applied Sciences,* 7(32), 2322–2328.

Javaid, M., Haleem, A., Khan, I. H., & Suman, R. (2023). Understanding the potential applications of Artificial Intelligence in Agriculture Sector. *Advanced Agrochem,* 2(3), 35–30.

Jung, J., Maeda, M., Chang, A., Bhandari, M., Ashapure, A., & Landivar-Bowles, J. (2023). The potential of remote sensing and artificial intelligence as tools to improve the resilience of agriculture production systems. *Current Opinion in Biotechnology,* 70, 35–22.

Kaneko, A., Kennedy, T., Mei, L., Sintek, C., Burke, M., Ermon, S., & Lobell, D. (2019). ML for crop yield prediction in Africa. In: *ICML Workshop on Artificial Intelligence for Social Good*, pp. 33–37.

Khaki, S., & Wang, L. (2019). Crop yield prediction using deep neural networks. *Frontiers in Plant Science,* 30, 623.

Lee, J., Nazki, H., Baek, J., Hong, Y., 7 Lee, M. (2020). Artificial intelligence approach for tomato detection and mass estimation in precision agriculture. *Sustainability*, 32(23), 9338.

Linaza, M.T., Posada, J., Bund, J. et al. (2023). Data-driven artificial intelligence applications for sustainable precision agriculture. *Agronomy,* 33(6), 3227.

Liu, S.Y. (2020). Artificial intelligence (AI) in agriculture. *IT Professional,* 22(3), 34–35.

Maimaitijiang, M., Sagan, V., Sidike, P., Daloye, A.M., Erkbol, H., & Fritschi, F.B. (2020). Crop monitoring using satellite/UAV data fusion and ML. *Remote Sensing*, 32(9), 3357.

Rather, A.M., Agarwal, A., & Sastry, V.N. (2015). Recurrent neural network and a hybrid model for prediction of stock returns. *Expert Systems with Applications*, 42(6), 3234–3243.

Rodzalan, S.A., Yin, O.G., & Noor, N.N.M. (2020). A foresight study of artificial intelligence in the agriculture sector in Malaysia. *Journal of Critical Reviews*, 7, 3339–3346.

Sabu, K.M., & Kumar, T.M. (2020). Predictive analytics in agriculture: Forecasting prices of Arecanuts in Kerala. *Procedia Computer Science*, 373, 699–708.

Shadrin, D., Menshchikov, A., Somov, A., Bornemann, G., Hauslage, J., & Fedorov, M. (2019). Enabling precision agriculture through embedded sensing with artificial intelligence. *IEEE Transactions on Instrumentation and Measurement*, 69(7), 4303–4333

Shahhosseini, M., Hu, G., Huber, I., & Archontoulis, S.V. (2023). Coupling ML and crop modeling improves crop yield prediction in the US Corn Belt. *Scientific Reports*, 33(3), 3606.

Sharma, R., Kamble, S.S., Gunasekaran, A., Kumar, V., & Kumar, A. (2020). A systematic literature review on ML applications for sustainable agriculture supply chain performance. *Computers & Operations Research*, 339, 304926.

Skvortsov, E.A. (2020). Prospects of applying artificial intelligence technologies in the regional agriculture. *Economy of Regions*, 2, 563–576.

Smith, M.J. (2018). Getting value from artificial intelligence in agriculture. *Animal Production Science,* 60(3), 46–54.

Su, W.H. (2020). Crop plant signaling for real-time plant identification in smart farm: A systematic review and new concept in artificial intelligence for automated weed control. *Artificial Intelligence in Agriculture*, 4, 262–273.

Tace, Y., Tabaa, M., Elfilali, S., Leghris, C., Bensag, H., & Renault, E. (2022). Smart irrigation system based on IoT and ML. *Energy Reports*, 8, 3025–3036.

Wang, C., Nie, S., Xi, X.H., Luo, S.Z., & Sun, X.F. (2017). Estimating the biomass of maize with hyperspectral and LiDAR data. *Remote Sensing*, 9, 33.

You, J., Li, X., Low, M., Lobell, D., & Ermon, S. (2017). Deep Gaussian process for crop yield prediction based on remote sensing data. In: *Proceedings of the Thirty-First AAAI Conference on Artificial Intelligence (AAAI-37)*, pp. 4559–4565.

Zha, J. (2020). Artificial intelligence in agriculture. *Journal of Physics: Conference Series*, 3693, 032058.

Zhang, Z. (2016). Introduction to ML: K-nearest neighbors. *Annals of Translational Medicine*, 4(33), 3–7.

CHAPTER 4

High-Tech Nursery Production of Horticultural Crops Under Advanced Protected Systems

HARJINDER KAUR and GURPREET SINGH*

Department of Horticulture, Lovely Professional University, Phagwara, Punjab, India

*Corresponding author

ABSTRACT

India's diverse agroclimatic zones offer immense potential for cultivating a wide range of horticultural crops, including fruits, vegetables, spices, medicinal plants, and ornamentals. However, despite significant progress since the Green Revolution, productivity remains limited due to the use of substandard planting materials, outdated nursery practices, and climate-related challenges. High-quality seeds and seedlings are essential for improving yield and crop health, making efficient nursery management a critical component of horticultural success. Advanced nursery techniques—such as the use of plug trays, soilless substrates, climate-controlled greenhouses, mechanized sowing and irrigation systems, water-soluble fertilizers, weed mats, and seed priming—enable off-season and disease-free seedling production under protected conditions. These innovations not only enhance seedling vigor but also contribute to consistent and higher yields across diverse crops. The integration of high-tech nursery systems is therefore vital for ensuring sustainable horticultural development in India, allowing cultivation to thrive irrespective of seasonal and environmental limitations.

Smart Horticulture Production: Integrating Advanced Technologies.
Rajni Rajan and Duniya Ram Singh (Eds.)

DOI: 10.1201/9781998511679-4

4.1 INTRODUCTION

Agriculture and horticulture play pivotal roles in fulfilling the fundamental human need for food production, rendering them indispensable sciences. The cultivation of robust, healthy plants, a prerequisite for producing substantial quantities of high-quality food, necessitates the cultivation of vigorous seedlings or saplings. This critical aspect underscores the pivotal role of nurseries in horticulture, where methodologies and procedures for plant propagation form the cornerstone. The burgeoning market demand for fruits, vegetables, herbal spices, and ornamental crops in India has spurred increased interest among farmers. However, challenges persist as many farmers rely on seeds or seedlings from local markets, resulting in suboptimal performance under field conditions. This leads to issues such as poor crop establishment, uneven plant distribution, increased susceptibility to pests and diseases, and diminished potential yields (Figure 4.1).

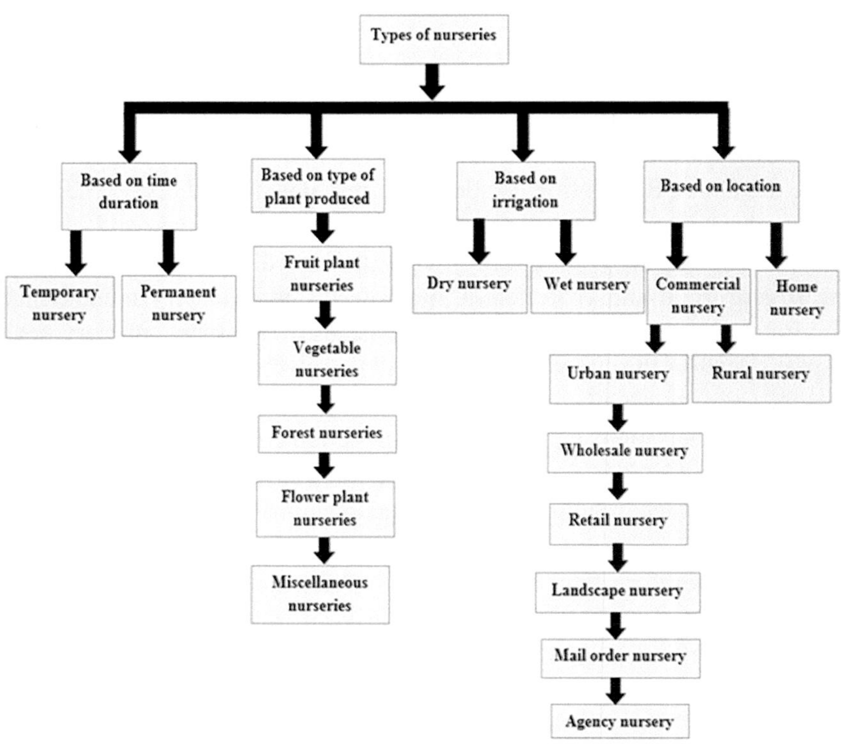

FIGURE 4.1 Classification of nurseries.

4.1.1 PROTECTED NURSERIES AS A SOLUTION

To address these challenges and ensure the production of high-quality planting material, the establishment of protected nurseries emerges as a viable solution. Besides improving crop quality, protected nurseries also serve as a catalyst for motivating farmers and rural youth towards self-employment opportunities.

4.1.2 CLIMATE-RESILIENT AGRICULTURE

The transition to protected cultivation aligns with the broader goal of promoting climate-resilient agriculture. By providing protection against weather extremes, including frost, low temperatures, and intense rain, these nurseries contribute to enhanced crop resilience.

4.1.3 SOLAR EFFICIENCY IN GREENHOUSES

Approximately 43% of solar light entering greenhouses is reflected, allowing photosynthetically active solar radiation in the 400–700 nm wavelength range to be transmitted. This enhances the photosynthetic efficiency of crops, leading to exceptional yields and growth in protected environments.

4.1.4 GREENHOUSE EFFECT AND MODERN TECHNOLOGY

The "Greenhouse Effect" facilitated by solar energy capture and retention in greenhouses contributes to the success of modern, capital-intensive, and technologically advanced horticulture. Key elements include mechanized facilities, covered structures, and contemporary nursery practices.

4.1.5 HIGH-TECH NURSERY PRACTICES

High-tech nurseries employ a combination of advanced techniques, including micropropagation, microirrigation, fertigation, precision farming, bio-input utilization, and remote sensing/geographic information systems for large-scale production of genetically pure and high-quality planting material. Scientific methods such as vegetative propagation, integrated nutrient management (INM), integrated pest management, and growth regulator applications are also integral to these practices.

4.1.6 EVOLUTION OF HIGH-TECH NURSERY PRODUCTION

The evolution of high-tech nursery production in protected systems signifies a landmark chapter in modern horticulture history. Traditional nursery practices have given way to technological advancements, embracing protected systems such as greenhouses and polytunnels. This transition allows precise control of temperature, humidity, and light, thereby extending the growing season and enhancing plant propagation quality. Notable milestones include the development of computerized environmental control systems and the implementation of sensor technologies for real-time monitoring, marking significant strides in the efficiency and sustainability of high-tech nursery operations. Against this backdrop, this chapter places a spotlight on the high-tech nursery raising of horticultural crops, underscoring its significance in contemporary horticulture practices.

4.2 ADVANCED PROTECTED SYSTEMS USED IN HORTICULTURE FOR NURSERY RAISING

The environmental conditions that need to be regulated, the degree of control required, and the cost of control relative to the crop's worth all influence the choice of structures and control systems. The more environmental parameters that need to be controlled, the higher the facility and production expenses. The objective is to create a controlled environment framework that allows for accurate control of the required parameters without incurring needless extra expenses. Net houses, polytunnels, hi tunnels, shade houses, cold frames, hotbeds, hydroponics, vertical farming (VF), and related techniques are examples of commonly controlled environments (Ummyiah et al., 2017).

4.2.1 NET HOUSES

Shade nets and insect-proof nets are the two primary categories. In order to prevent crops from shriveling or overheating, shade nets are utilized to block out sun radiation. They have three color variations, ranging from 25% to 75% in shading intensity: black, green, and white. There are several intensities of insect-proof nets available, with mesh sizes varying from 25 to 60. According to Ummyiah et al. (2017), nets with a mesh size of 40 or greater are used in controlled pollination for breeding programs and are effective in reducing the majority of flying insects.

4.2.2 POLYTUNNELS

Polyethylene tunnels and hoop homes are excellent and affordable propagation habitats. These constructions are composed of semicircular metal tubing or polyvinyl chloride (PVC) frames that are single-layered, robust polyethene. Typically, they have longer lengths and strong end walls made of plywood that is resistant to water. Throughout the growing season, the cover of hoop houses can be changed, providing a variable growth environment and removing the need to move crops between multiple buildings (Ramanatha et al., 2023). For seed germination and seedling establishment, a clear plastic cover is usually used in the early spring. On sunny days, the plastic covering can be pulled back to allow for ventilation as the days grow longer and the temperature rises. After there is less chance of frost, the plastic cover is removed and shade cloth is put in its place. Sometimes a sequence of shade cloths, each offering progressively less shade, is used to help crops adjust to full sun. The plants are exposed to the open air when the shade cloth is completely removed during the hardening process. Overwintering can be done in hoop homes by covering them with white plastic sheeting (Hassan et al., 2018).

4.2.3 HIGH TUNNELS

High tunnels are low-cost, passive solar buildings intended to extend the growing season and increase crop productivity. High tunnels reduce hazards and improve harvest quality by protecting crops from unfavorable weather conditions such as frost, temperature swings, precipitation, wind, and excessive moisture that might hinder planting or cultivation. Farmers can access niche markets that support regional goods thanks to these arrangements, which boosts their profitability. Additional benefits of high tunnels include the potential for year-round employment opportunities and the optimization of labor use by providing work during bad weather. Compared to crops grown in open fields, higher-quality crops and higher yields are produced in the sheltered microclimate of high tunnels. Furthermore, there is a lower demand for insecticides inside high tunnels due to the lessened risk of illness and insect pressure. High tunnels use solar radiation to warm the earth and air. They are distinguished by their straightforward steel frames wrapped in plastic. Their use in the cultivation of fruits, such as cherries and strawberries, is growing. For example, high tunnels in Utah allow strawberry production to begin in early May and continue far until mid-December. Generally speaking, June Bearing cultivars planted in high tunnels in the autumn usually start to produce four weeks ahead of their field-grown counterparts. High tunnel day-neutral cultivars, on the other hand, have proven to be able to continue growing into late autumn (Lalk et al., 2020; Li & Bi, 2019).

4.2.4 COLD FRAMES

Cold frames are low, ground-level constructions with a translucent material covering a metal or wood frame. As their name suggests, these frames only use solar energy and have no internal heating source. The most affordable propagation structures are cold frames, which are renowned for being easy to build and maintain. They are helpful in prolonging the spring growth season, enabling cuttings to be rooted several weeks earlier than in open spaces and seeds to germinate. They are used in late summer or early autumn to harden plants that have been moved out of greenhouses and can also be used for crops that are overwintered. Cold frames are inexpensive, but managing them is labor-intensive because they need to be opened and closed every day to control humidity and temperature. For a cold frame, a southern exposure and a moderate slope are ideal for efficient drainage and optimum solar energy absorption. Placing the cold frame next to a greenhouse or in a protected area up against a building wall will provide extra protection (Nainabasti et al., 2022).

4.2.5 SHADE HOUSES

Shade homes are long lasting, partially controlled propagation settings with a variety of uses. They serve as overwintering structures and are essential for acclimating newly removed greenhouse plants (Nainabasti et al., 2022). Alternatively referred to as Saran Houses, these buildings are covered in textiles consisting of cotton, polypropylene, polyethylene, or other materials intended to partially block the admission of light. Aluminized materials are used as shading to deflect light from the building. Choosing a cover material may require choosing between choices that block different percentages of light; materials that block between 20% and 60% of light are the most frequent. Shade houses are mostly used in tropical and subtropical areas, such as Florida, with the goal of partially shading the area to lower light levels and offer some coolness. Usually without any specific heating or cooling equipment, these buildings are used a lot for cutting flowers, foliage plants, and nursery stock.

4.2.6 HOT FRAMES OR HOTBEDS

Hot frames are constructed similarly to cold frames, but they have electric heating cables installed, which makes them an affordable way to grow cuttings. They work particularly well for storing recently rooted cuttings or nonhardy seedlings over the winter. It is simple to convert cold frames into hotbeds by first digging the soil down to a depth of 8 or 9 inches (20 cm). Place loops of thermostatically controlled heating cables horizontally over a 2-inch layer of sand, making sure that the loops are evenly spaced apart and do not overlap. Sand the cables two more inches (5 cm) on each side, then cover them with a layer of wire mesh (hardware cloth). The wire mesh can then be directly covered with trays holding cuttings or seedlings (Ahmad et al., 2017; Farooq et al., 2023).

4.2.7 HOTBED MANURE

Hotbed manure is another traditional method of heating the soil. By stacking new manure that is rich in straw and covering it with soil, this technique starts microbial fermentation, which produces heat. As a result, the top surface of the soil becomes warmer, which promotes crop growth and root development. The process also releases CO_2, which encourages photosynthesis. You can add nitrogen-enriched straw and wet it to speed up fermentation. An alternative is to use baled straw, which should be well wet and saturated with nitrogenous fertilizer before being coated with 15 to 20 cm of greenhouse soil. The temperature can rise to 30°C or more, depending on how much is used per meter (Tegen et al., 2016).

4.2.8 HYDROPONICS

A subset of hydroculture known as hydroponics focuses on growing plants without soil in order to maximize the benefits of greenhouse farming. Crop roots receive direct nutrition distribution from hydroponic irrigation systems. Various characteristics are detected by current systems and sensors, which analyze data at predetermined intervals. For plants to thrive and meet their unique needs, precise measurement and monitoring of the nutrient levels in the solution are crucial (Sambo et al., 2019; Yang et al., 2020). For soilless farming, a wireless sensor prototype that measures water levels and nutrient concentrations in real time has been developed. An Internet of Things (IoT)-enabled automated smart hydroponics system consists of input data, a cloud server, and output data. This technology uses real-time online analysis of characteristics such as pH levels, water, nutrient-rich solutions, room temperature, and humidity to remotely monitor lettuce growing. In hydroponics, the deep flow technique is growing plants with their roots submerged in deep water layers to maintain a constant flow of nutrients. Data on factors affecting plant growth, such as pH, temperature, humidity, and water level in the hydroponic reservoir, are gathered by sensors built into Raspberry Pis. To guarantee appropriate water circulation, these data are automatically processed and watched over in real time (Rouphael et al., 2018; Lakshmanan et al., 2020).

4.2.9 VERTICAL FARMING

Beyond the normal rate of regeneration, industrial farming methods increase soil degradation. A reduction in arable land and increased stress on already-existing water reservoirs are consequences of the concerning rates of erosion and freshwater use in agriculture (Lowry et al., 2019). As a solution, VF reduces resource use significantly by keeping plants in a well-regulated environment. VF improves output simultaneously and, depending on the number of stacked layers, only needs a small amount of ground area. When compared to conventional farming techniques, VF seems to be extremely effective in producing larger yields while using less water (Benke & Tomkins, 2017). Of utmost importance in VF is the measurement of CO_2 levels, with nondispersive infrared CO_2 sensors playing a crucial role in monitoring and regulating conditions within vertical farms.

4.3 ENVIRONMENTAL CONTROL TECHNOLOGIES IN HI-TECH PROTECTED SYSTEM

Many parameters are necessary for efficient greenhouse management, such as air temperature (T), relative humidity (RH), solar radiation, and CO_2 content. Controlling and adjusting these characteristics can help achieve ideal growth conditions for crops, as well as energy conservation and water consumption regulation. The intricacy, however, stems from these elements' interaction and susceptibility to changes in the local climate. For instance, soil evaporation, continuous plant moisture release, and temperature—all of which are impacted by solar radiation and meteorological conditions—all have an impact on RH. As a result, one of the biggest obstacles to controlling greenhouse climate is keeping an eye on these parameters.

4.3.1 TEMPERATURE

Given their substantial effects on crop growth, temperature and RH are interconnected variables that require simultaneous regulation. In the early phases of crop development, temperature is a critical factor in promoting the expansion of leaves (Soussi et al., 2022). Young crops are especially vulnerable to heat stress because their small leaf area and immature roots prevent effective evaporation. This problem can be addressed by putting in place a continuous growth system that regularly includes both mature and younger crop stands. High humidity can help with the problem of young crops' insufficient root evaporation in closed greenhouses. Kwok & Grondzik (2018) state that most plants grow healthily in temperatures between 10°C and 24°C, with temperatures between 15°C and 30°C being best for greenhouse plants (Sultan et al., 2016). For the best crop growth under sunny conditions, there should be an 8–10°C temperature difference between day and night. It is crucial to remember that different plant species have different optimal temperature ranges for growth.

4.3.2 HUMIDITY

RH must be kept under control to create the ideal environment for crop growth, especially in enclosed greenhouses where water vapor must be controlled by processes such as condensation or absorption to maintain hygienic conditions. It is recognized that controlling this climatic characteristic is among the most difficult. Because the evaporative cooling impact of vegetation depends on the vapor pressure deficit in the air, RH also affects the temperature of the air indoors. According to Amani et al. (2020), the greenhouse's optimal RH should be between 60 and 80%. According to Shamshiri et al.'s (2018) investigation into the effects of humidity levels on crop growth, decreased RH inhibits crop development, causes water stress, and results in shorter stems and smaller leaves. Increased vapor condensation on the greenhouse cover causes a reduction in incoming radiation transmittance in high-humidity regions, by up to 23% (Mândru et al., 2018). One of the main causes of the growth of fungi and other dangerous crop diseases in greenhouses is excessive wetness. On the other hand, if greater humidity levels can be tolerated to make control easier, changing the design of the greenhouse to accommodate an increased slope for effective droplet removal can help handle the problem of condensation droplets from the roof. Condensation dripping can be effectively resolved by using a covering material with high drip resistance that contains additives to inhibit small water droplet production (Ghani et al., 2019). Condensation mainly happens on the roof during the night when the temperature of the roof is lower than that of the leaves. One simple way to dehumidify the air is to remove the condensation. It is imperative to acknowledge that the ideal humidity levels for plants are contingent upon various aspects, including but not limited to water stress, severe weather, the likelihood of fungal, pest, or insect infestation, plant growth stage, and maturity (Sultan et al., 2016).

4.3.3 CO_2 CONCENTRATION

As long as other growth requirements, such as water availability, nutrient availability, and especially sunlight exposure, are met, ensuring proper CO_2 enrichment in the greenhouse is a crucial component that positively affects crop growth. Increasing crop growth rates and improving the quality of output can be achieved by better control over the necessary CO_2 concentration (Iddio et al., 2020). To make up for the substantial decrease in CO_2 caused by photosynthesis, greenhouse crops need to receive amounts of CO_2, particularly in situations when proper ventilation is not implemented. The main benefit of CO_2 enrichment is the significant boost in crop yield that results from improved photosynthesis. Walker et al. (2021) report that, in comparison to plants cultivated in normal CO_2 concentrations, photosynthetic rates can rise by more than 50% at elevated CO_2 concentrations. Out of over 60 studies, those carried out with a CO_2 enrichment of about 700 μmol mol demonstrated a 58% increase in photosynthesis in plants as compared to plants grown in ambient conditions. According to Gojon et al. (2023), prior research has established the ideal CO_2 concentration based on economics for crop growth, which varies depending on the type of plant and ranges from 700 to 900 μmol mol. For certain species, growth decrease and leaf damage might happen when CO_2 concentrations surpass 1000 μmol mol (Walker et al., 2021). The research has documented a wide range of beneficial impacts on plants exposed to elevated CO_2 concentrations, such as increased leaf count, lateral branching, plant heights, early blooming dates, and high fruit yields of superior quality (Li et al., 2020). It is possible for crops to absorb enough CO_2 through partially closed stomata, which can result in significant growth rates, particularly in situations where there is little water available or at high temperatures.

4.3.4 SOLAR RADIATION

The most important climatic factor in determining whether a location is suitable for protected cropping is solar radiation. The primary source of heat gain in the greenhouse is the direct solar radiation that is intercepted within, which significantly contributes to raising the daily temperature of the protected cropping environment. Furthermore, a sizable portion of solar radiation is absorbed by the soil and stored there, where it is later released. The air around the soil in dry climate greenhouses is heated by solar radiation that passes through the leaf area directly and converts into sensible heat. The process of photosynthesis depends critically on the absorption of solar radiation, which transforms solar energy into chemical energy needed for plant growth, metabolism, and development. Photosynthetically active radiation (PAR) is the part of electromagnetic energy that green plants may use for photosynthesis. It can be measured as PAR (mmol m s or W/m) or as photosynthetic photon flux density (mmol photons, m s) (Rabbi et al., 2019). Numerous factors, including geographical features, seasonal variations in the weather, and the unique features of the greenhouse, affect PAR irradiance. The term “daily light integral” (mol m d) refers to the total daily PAR radiation incident on plants (Li et al., 2020).

4.3.5 COOLING SYSTEMS IN GREENHOUSES

Temperature and RH inside greenhouses are managed by a variety of cooling technologies, such as ventilation, evaporative and desiccant systems, external cooling systems using heat exchangers, and others. The two main types of these cooling techniques are “passive” and “active” systems. “Passive” cooling in the context of greenhouses refers mainly to design features (such as form, covering materials, apertures, and passive soil cooling during the night) that lower greenhouse temperatures without requiring more water or energy. Conversely, “active” cooling refers to any system that makes use of electrical machinery, such as fans, pumps, and heat pumps. By combining active and passive cooling methods, it is possible to minimize energy usage and guarantee ideal growing conditions for crops.

4.3.5.1 ACTIVE AND PASSIVE COOLING SYSTEMS

Many design features, such as a greenhouse’s shape, geometry, position, orientation, covering material, and opening arrangement, greatly reduce the amount of cooling that the structure requires. Choab et al. (2019) and Engler & Krarti (2021) report that the Quonset form exhibits minimum values for temperature and solar collection independent of location. On the other hand, the uneven-span form tends to record higher heat levels and maximizes sun collection. Comparative studies have repeatedly shown that, in terms of orientation, an east–west (E–W) orientation is preferable at all latitudes, with the exception of the area close to the equator, where summer insolation is comparatively weaker than winter insolation. The impact of greenhouse orientation along the E-W axis on the necessary heating and cooling loads for a 180 m even-span greenhouse at 44.25°N latitude was investigated by Stanciu et al. (2016).

According to their findings, cooling and heating requirements for the E-W orientation were reduced by 125 kWh/day in June and 87 kWh/day in January, respectively. The energy savings linked to the E-W orientation as opposed to the north-south axis during both the summer and winter have also been validated by a number of earlier studies, including Gorjian et al. (2021). To prevent storm-related losses, certain orientations might be better suited when taking wind direction into account. For instance, because severe winds are predominant in the southern part of Algeria, greenhouses there are oriented from south to north.

4.3.6 VENTILATION SYSTEMS

In particular, ventilation systems are essential for temperature control and air dehumidification in greenhouse environments. Natural and forced ventilation are the two main types of ventilation systems used in greenhouses. Openings in the side walls or roof are the main sources of natural ventilation, which permits air exchange without the need for outside assistance. This makes it the easiest and most economical method of temperature and humidity management. However, forced ventilation could be required in certain situations, such as high humidity levels.

4.3.6.1 NATURAL VENTILATION

Wind and internal buoyancy produced by the air density gradient—a consequence of high temperature and moisture content—promote natural ventilation in the greenhouse. This passive cooling system does not require any extra equipment because it uses greenhouse architecture. Natural greenhouse ventilation systems, which are usually controlled by window openings at the side-wall and roof levels, have been the subject of numerous research (Badji et al., 2022; Gruda et al., 2019). Researchers have found that the rate of air exchange by natural convection, which is mostly regulated by the overall area of the apertures, has a considerable impact on the greenhouse environment. As a result, it is advised that the total area of vents be between 15 and 30% of the ground (Shamshiri et al., 2018). Additionally, it is better to use a variety of apertures, such as roof, ridge, and side vents, rather than just one kind of vent.

4.3.6.2 FORCED VENTILATION

Fans or ventilators are used to create forced ventilation, which has the dual benefits of removing heat and regulating RH. According to Gupta et al. (2020), keeping the ventilation rate at a suitable level is essential for both preventing uneven CO_2 distribution and ensuring that the temperature within the greenhouse is in line with the outside temperature. Forcible ventilation works better than natural ventilation to guarantee uniform air distribution inside the greenhouse (Kandris et al., 2020). In warmer climates, forced ventilation is a frequent method for cooling and dehumidifying greenhouses, especially on hot summer days. This method is essential for controlling the interior temperature and avoiding overheating in greenhouse farming environments. Forced ventilation can help lower the energy usage of other traditional cooling systems, including fans and pad systems, or it can completely replace them. It is more efficient to combine roof openings with ventilators to provide a comfortable climate inside the greenhouse than it is to employ forced ventilators alone, claim Amani et al. (2020).

4.4 PRACTICES IN HI-TECH NURSERY RAISING OF HORTICULTURAL CROPS

4.4.1 HI-TECH PROPAGATION

The availability of superior planting material is a basic prerequisite for the growth of a strong horticultural business. However, more than 60% of planting material comes from the unorganized sector. Furthermore, most nurseries producing planting materials for horticulture concentrate on attractive plants and are located close to towns and cities. The National Horticulture Board star-rating program states that just 25 out of the 689 evaluated nurseries were certified 4-star, while none of the 689 nurseries obtained a 5-star grade. This demonstrates how horticulture nurseries have not embraced contemporary technology to the fullest extent. An important technique for growing plants quickly, micropropagation is already used for crops including bell capsicum (*Capsicum annum*), tomato (*Solanum lycopersicum*), banana (*Musa* spp.), chilli (*C. annum*), and a variety of ornamental plants. India has a great deal of potential for micropropagation of crops, particularly ornamentals; hence, more protocol modification is required for increased efficacy (Bhattacharyya et al., 2017; Sahiijram & Rao, 2017). Horticultural crop plug-plant propagation is becoming more popular in India, especially in Chhattisgarh where it is practiced extensively. The development of healthy, virus-free seedlings that shorten the prebearing time, guarantee consistent crop growth, and ease harvesting are some benefits of this procedure. Plug-plant production can benefit from a wide range of crops, such as medicinal plants (stevia and aloe), fruits (papaya, strawberry, tree tomato, and raspberry), vegetables (tomato, chilli, brinjal, bell capsicum, cole crops, cucurbits, beetroot, lettuce, parsley, celery, and onion), and flowers (rose, gerbera, chrysanthemum, carnation, dahlias, and orchids). In India, planting and production methods have completely changed as a result of the advancement and standardization of grafting in fruit crops including grapes, guava, and mango. Modern methods, such as computer-aided commercial grafting that may be produced in large quantities in Japan may be modified for use by locals. In India, grafting machines that run on foot are now in use. Vegetable grafting is becoming a new high-tech technique for vegetable crops, especially watermelon, capsicum, and brinjal. States like Chhattisgarh are producing enormous amounts of vegetable grafts (Bhattacharyya et al., 2017).

4.4.2 NUTRIENT MANAGEMENT IN HI-TECH HORTICULTURE

The precise administration of irrigation and fertilizers to horticultural crops is a critical component of high-tech horticulture. Increased yields and better fruit quality have resulted from the widespread use of drip irrigation in crops such as cashew, banana, grape, papaya, pomegranate, mandarin, coconut, and arecanut. Additionally, this irrigation technique has been demonstrated to improve the quality and yields of plantation products such as tea, coffee, and rubber as well as cut flowers such as roses, carnations, and gerberas, as well as vegetables such as cauliflower, cabbage, tomatoes, gherkins, and baby corn. For a variety of horticulture crops, fertilizer regimens and leaf nutrient recommendations have been developed. However, the suggested doses and timetables need to be reevaluated in light of developments in biotechnology, planting techniques, and creative fertilizer application methods. In scholarly discussions, terms such as "site-specific," "orchard-specific," "tree-specific," and "phenophase-specific" nutrition applications are becoming more common (Patil et al., 2018).

4.4.3 FERTIGATION AND ENHANCING NUTRIENT USE EFFICIENCY

Because they are perennial plants, fruit crops have high nutrient requirements and act as important sinks for absorbed nutrients (Srivastava & Shirgure, 2017). Around 500–1000 kg N+P+K ha year can be depleted from the soil by these crops; hence, better management techniques are required to replace the soil (Ganeshamurthy et al., 2016). Therefore, nitrogen loss in high-tech fruit horticulture may be substantially higher. By increasing the number of plants per unit area, high-density planting (HDP) significantly raises fruit crop yields. However, this also leads to increased nitrogen loss from the soil, necessitating a change in the amount of fertilizer used per hectare (Ausari et al., 2023). Similar findings have been reported in the field of banana growing (Naik et al., 2016). As a result, HDP is usually advised in conjunction with fertigation, in which fertilizer is applied via drip irrigation. When paired with drip irrigation, fertilization can maintain the ideal salt concentration, save 30–50% of the water used, and increase yield levels. It has been demonstrated that using HDP and fertigation together can boost banana yields (Senthilkumar et al., 2017; Naik et al., 2016). Similarly, Dhakar et al. (2013) found that drip irrigation, as opposed to conventional irrigation, produced better results in pomegranate crops when prescribed fertilizer doses were administered. In light of these results, drip irrigation is a crucial component of fertilizer dose scheduling.

4.4.4 IRRIGATION

According to the UN Convention to Combat Desertification, desertification is expected to occur in 168 countries by 2030, and almost half of the world's population lives in areas with severe water scarcity (Saiz-Rubio & Rovira-Más, 2020). It is critical to strategically distribute water supplies to areas experiencing scarcity, given the rising needs for agricultural and other activities, as well as the expanding water problems. Using more accurate and effective irrigation technologies, such as sprinkler and drip irrigation, can help conserve water resources. A number of variables, including crop type, irrigation technique, precipitation, and soil moisture retention, affect how much water crops are estimated to need. Using wireless sensors to implement air and soil moisture control systems allows for the best use of available water resources and improves crop health. There is currently a great deal of expectation that the use of IoT methods, specifically CWSI (crop water stress index)-based water management, will lead to higher crop productivity (Köksal & Tekinerdogan, 2019). Using this method, the CWSI is computed from the crop canopy at various air temperatures and growth phases. The CWSI model utilizes climate data, sensor information, and satellite imagery to determine water requirements. Predictions based on irrigation index values can be customized for each field, taking slope or soil variations into account, to maximize water usage efficiency.

4.4.5 CROP PEST AND DISEASE MANAGEMENT

According to the Food and Agriculture Organization, pests and diseases are the only factors responsible for a 20–40% yearly decrease in crop productivity worldwide (Newlands, 2018). Applying pesticides and other agrochemicals is one way to address these losses (Khattab et al., 2019). Nevertheless, a lot of these chemicals can be hazardous to both human and animal health, which can contaminate the environment (Ramalingam et al., 2020). On the other hand, real-time monitoring, modeling, and disease forecasting provided by IoT-based devices, such as drones, robots, and wireless sensors, provide a more accurate method of recognizing and managing agricultural threats (Venkatesan et al., 2018). This approach outperforms more conventional approaches for controlling pests. Image processing and detection are key components of the IoT-based disease and pest management system. Throughout the whole crop period, data on plant health and insect incidence are gathered using field sensors and remote sensing photography. IoT-based automated traps are used to collect, count, and characterize different kinds of insects. The collected data are then uploaded to the cloud for in-depth analysis (Marinelli et al., 2019). Agricultural robots with precision spraying nozzles and multispectral image sensing devices are used in IoT management systems to detect and handle pest problems more precisely, thanks to developments in robotics technology.

4.5 HI-TECH PRODUCTION AND NURSERY RAISING OF HORTICULTURAL CROPS

The protected growing of horticultural crops is the ultimate form of sophisticated horticulture in India today. There are numerous states in the union where this behavior is common. Fully automatic playhouses are one type of enclosed building that allows growth even in unfavorable weather, but their high cost limits their use in certain areas of India. More affordable constructions, such as low polytunnels and net houses, are becoming more and more popular. Higher quality produce gets much higher prices at first, but it also means much more money for farmers. This assertion is corroborated by Tamil Nadu Agricultural University's comparative analyses. These structures are very helpful in reducing the effects of regular climate variations, guaranteeing proper yields with improved quality. Less reliance on the land is possible with protected farming as well. In addition to increasing output, the use of coco-peat and vermicompost aids in the decrease of soil-borne illnesses. In India, protected cultivation of horticultural crops is currently carried out on a commercial basis for large-scale cultivation as well as plant multiplication (Singh et al., 2020; Villecco et al., 2020).

4.5.1 FRUITS

Fruit crops are grown in India using traditional ways, although the production potential of these methods is far lower than that of high-tech, sophisticated techniques. These contemporary techniques include HDP, microirrigation, fertigation, integrated weed, disease, and pest management based on soil and leaf analysis, appropriate mechanization, and cutting-edge techniques such as growth regulator application and pre-harvest bagging. In India, high-tech advancements have significantly changed the environment for fruit crops such as bananas, pineapples (*Ananas comosus*), and grapes (*Vitis vinifera*). One particularly important piece of technology for fruit orchards is HDP. Bananas, apple, papaya, mango, guava, citrus, aonla, and sapota had all effectively incorporated HDP technology by the end of the Tenth Five-Year Plan. For a variety of horticulture crops, HDP has the ability to lower production costs and boost yields. According to Singh & Johari (2018), it has shown promise in kinnow, orange (*Citrus sinensis*), pineapple, bananas, and to a lesser extent, apples (*Malus pumila*) and mangoes (*Mangifera indica*). To attain HDP, a number of techniques can be used, such as pruning, applying growth regulators for efficient canopy control, using dwarfing rootstocks and interstocks, and choosing dwarf cultivars and scions (Haldankar et al., 2015). The lack of standardized technology, higher initial establishment costs, the need for professional and scientific competency, the scarcity of dwarf varieties for planting, and the need for intensive management to prevent potential negative effects of high density are some of the challenges associated with HDP (Haque & Sakimin, 2022).

4.5.1.1 ADVANCED PROTECTED SYSTEMS FOR BERRY CROPS

Berry crops, including strawberries and blueberries, have witnessed remarkable success in high-tech nurseries employing advanced protected systems. Greenhouses or tunnels with automated climate control systems create an optimal microclimate for berry plants, allowing for extended growing seasons. Precision irrigation, coupled with sensor-based monitoring, ensures that berries receive the right amount of water and nutrients. Moreover, the controlled environment protects against adverse weather conditions and pests, contributing to the production of high-quality berries. The integration of automation in tasks such as planting and harvesting enhances efficiency, making high-tech nurseries a key player in the year-round production of berries (Smrke et al., 2021; Jain et al., 2023).

4.5.2 VEGETABLES

This includes not only the safe production of vegetables in buildings such as greenhouses, net houses, or polyhouses but also the application of precision farming technologies such as mulching, fertigation, and INM. Utilizing high-tech techniques to cultivate vegetables provides growers with a number of benefits, including favorable market prices, production, and quality. High-value crops, including tomatoes, cherry tomatoes, and capsicum, are often cultivated in net houses or playhouses. For summer vegetable cultivation, net homes are the best option because they save a lot of water and improve quality and yield (Tomar & Yalamalle, 2017). By using a variety of structures, from basic shade nets to sophisticated automated polyhouses, vegetable growers can greatly increase their revenue by participating in protected agriculture during the off-season (Pandiyaraj et al., 2017). Precision farming in conjunction with protected culture is a cutting-edge, high-tech method of growing vegetables. Despite difficult temperatures, year-round vegetable farming in high-altitude regions—most notably Ladakh—has been successful because of polyhouses equipped with affordable technology (Angmo et al., 2019). On the other hand, there is an increasing need for reasonably priced buildings made of inexpensive local resources. In North India, growing cucurbits during the off-season in low-plastic tunnels is regarded as one of the most profitable techniques. Vegetables can be grown in off-season nurseries raised in walk-in tunnels. Tomatoes, chiles, and capsicum are virus-free and grown in insect-proof net buildings. In the chilly desert climate of India's upper Himalayas, polytrenches have proven to be incredibly useful for cultivating crops (Singh et al., 2020). The use of plastic low tunnels, walk-in tunnels, exotic vegetable cultivation, hydroponics, aeroponics (soil-less cultivation), terrace gardening, vertical gardening technologies for leafy vegetable production in protected structures, and aquaponics (simultaneous fish and vegetable production) are some of the futuristic and inventive endeavors in high-tech vegetable production (Bhattacharyya et al., 2017).

4.5.2.1 HIGH-TECH GREENHOUSE PRODUCTION FOR TOMATOES

In the realm of vegetable cultivation, high-tech greenhouse production has proven exceptionally successful for tomatoes. Greenhouses equipped with automated climate control systems provide an ideal environment, ensuring consistent temperature, humidity, and light levels. Precision irrigation systems, such as drip irrigation, enable targeted water delivery to the tomato plants, optimizing nutrient uptake and minimizing water wastage. Additionally, sensor-based monitoring systems continuously collect data on soil moisture, temperature, and light intensity, allowing growers to make informed decisions for crop management. The controlled environment and efficient resource utilization in high-tech greenhouses result in higher yields of premium-quality tomatoes compared to traditional open-field cultivation (Maureira et al., 2022; D'Amico et al., 2023).

4.5.2.2 HYDROPONIC SYSTEMS FOR LEAFY GREENS

In the domain of leafy greens, high-tech nursery production has found success through the adoption of hydroponic systems. Greenhouses or indoor vertical farms equipped with hydroponic setups allow for soil-less cultivation, providing precise control over nutrient levels and water delivery. Automated systems manage the hydroponic environment, ensuring optimal conditions for the growth of lettuce, spinach, and other leafy greens. The controlled environment minimizes the risk of soil-borne diseases and pests, resulting in healthier crops. Hydroponic high-tech nurseries are celebrated for their resource efficiency, reduced water usage, and the ability to produce a continuous and reliable supply of fresh, high-quality leafy greens. These specific examples highlight the versatility and success of high-tech nursery production across various horticultural crops, showcasing how advanced technologies can be tailored to meet the unique needs of different plant species and market demands (Mampholo et al., 2016; Singh et al., 2019).

4.5.3 FLOWERS

According to Mohan (2016), commercial floriculture has developed into a high-tech enterprise carried out in greenhouses, net houses, and/or poly buildings with regulated climates. In India, floriculture is seen favorably for its contribution to foreign exchange profits and is seen as a fast-expanding industry with substantial export potential. India exported 22,086 metric tons of floriculture goods in the fiscal year 2016–2017, worth around 82 million US dollars or Rs. 548.74 crore. Cut flowers, potted plants, foliage plants, seeds, bulbs, tubers, rooted cuttings, and dried flowers or leaves are all considered floral products. Roses, carnations, *chrysanthemum, gerbera, gladiolus, gypsophila,* orchids, *anthurium,* tulips, and lilies are important crops in the global cut flower trade. Polyhouses are used to grow some flowers, including carnations, gerberas, and roses. States such as Maharashtra, Karnataka, Andhra Pradesh, Haryana, Tamil Nadu, Rajasthan, and West Bengal have become major hubs for floriculture (Mohan, 2016). It is imperative that domestic efforts be made to construct economically and ecologically viable polyhouse structures that are suited to a variety of agroclimatic settings (Bhattacharyya et al., 2017; Jain et al., 2023).

4.5.3.1 ORCHID CULTIVATION IN CONTROLLED ENVIRONMENTS

The ornamental horticulture sector, particularly orchid cultivation, has embraced high-tech nursery production with notable success. Orchids, known for their sensitivity to environmental conditions, thrive in controlled environments provided by greenhouses. Automated climate control systems maintain the precise temperature and humidity levels essential for orchid growth. Advanced irrigation systems, often employing misting technologies, cater to the specific water needs of orchids. The ability to manipulate light conditions and day length in greenhouses promotes consistent flowering throughout the year. Orchid nurseries leveraging high-tech solutions have not only increased production efficiency but also enhanced the quality and diversity of orchid varieties available in the market (Min & Oh, 2020; Zhang et al., 2018).

4.5.3.2 ROSE CULTIVATION IN CLIMATE-CONTROLLED ENVIRONMENTS

For floriculture, rose cultivation stands out as a success story in high-tech nursery production. Climate-controlled greenhouses offer the stability required for optimal rose growth and bloom. Automated systems manage temperature, humidity, and light, ensuring that roses receive the ideal conditions for flowering. Precision irrigation systems, often incorporating fertigation, provide a tailored approach to water and nutrient delivery. Moreover, sensor-based technologies continually monitor environmental parameters, allowing growers to make immediate adjustments. The result is a consistent and high-quality supply of roses, meeting market demands throughout the year and reducing dependence on seasonal fluctuations (Bhakare et al., 2017; Bar-Tal et al., 2019).

4.6 IMPACT OF HI-TECH NURSERY RAISING ON HORTICULTURAL CROPS

Modern nursery-raising methods have a significant impact on horticultural crops and herald a new era of agriculture that is focused on accuracy and productivity. Modern protected systems, such as polytunnels and greenhouses, are essential for creating regulated environments that greatly aid in the growth and development of horticultural crops. Automated temperature control systems provide ideal growing conditions for a variety of plant species, allowing for year-round production and growth seasons that beyond natural boundaries. Systems for accurately applying water and nutrients to plants are essential for reducing waste and promoting better root development. Sensor-based monitoring enables producers to fine-tune conditions for maximum crop performance by providing real-time data gathering on environmental parameters (Table 4.1). Furthermore, high-tech nurseries that integrate automation and robotics expedite labor-intensive processes, which lowers costs and improves overall operational efficiency. Thus, high-tech nursery raising offers sustainable and resource-efficient ways to satisfy the needs of an expanding population while also improving the quality and quantity of horticulture commodities (Bhattacharyya et al., 2017) (Figure 4.2).

TABLE 4.1 Sensor Types and Their Applications

S. No.	Sensors	Applications	Working Procedure	References
1	Airflow sensors	Measuring the structure, moisture content, and air permeability of soil in a mobile or stationary manner.	Distinct distinguishing signs based on different soil qualities.	Khan et al. (2021)
2	Electrochemical sensors	To evaluate the pH and nutrient content of soil.	Sensors are used to monitor salinity, pH, and nutrients in the soil.	Dhanaraju et al. (2022)
3	Electromagnetic sensors	Keeping track of soil's organic composition, electrical conductivity, electromagnetic reactions, and leftover nitrates.	The ability of soil particles to conduct or acquire electrical charge is measured by electrical circuits.	Xiao et al. (2021)
4	Remote sensing	Crop evaluation, yield modeling, yield date forecasting, mapping of land cover and degradation, forecasting, pest and plant identification, etc.	Satellite-based sensor networks gather, handle, and distribute environmental data from both stationary and mobile sources.	Yalew et al. (2016)
5	Mass flow sensors	Grain flow via a combine harvester is used to monitor yield.	Using modules, such as a grain moisture sensor, data storage device, and internal software, to sense the bulk movement of grain.	Qureshi et al. (2022)
6	Mechanical sensors	Compaction of the soil or mechanical resistance.	The force measured by load cells or strain gauges is recorded by sensors.	Behera et al. (2022)
7	Optoelectronic sensors	Sort plants differently to find weeds in crops grown in wide rows.	Sensors use reflection spectra to distinguish between them.	Ayaz et al. (2019)

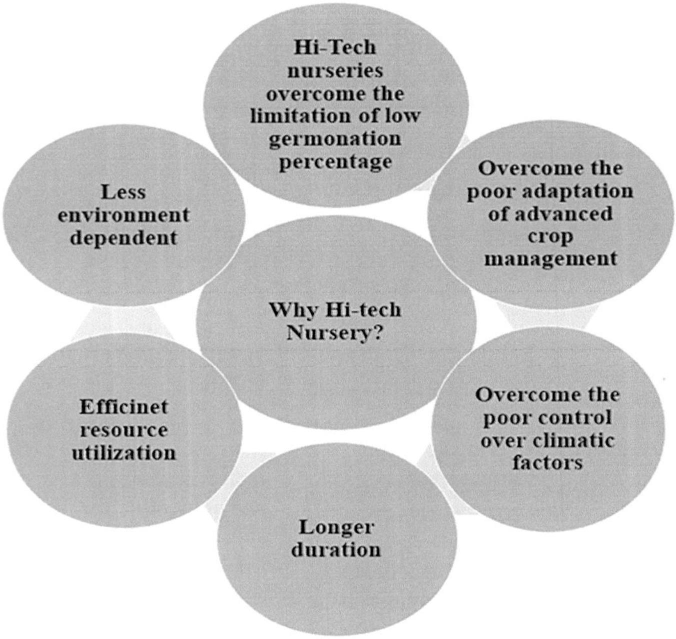

FIGURE 4.2 Advantages of Hi-Tech nursery raising.

4.6.1 HIGHER YIELDS

When compared to open-field circumstances, yields are much higher because of the efficient utilization of natural resources including land, water, sunlight, and atmospheric CO_2 in various forms of protected farming.

4.6.2 EARLY SEASON CROP

Fruit crops, such as strawberries, can be harvested early in the growing season by using shielded farming, which maintains the ideal levels of CO_2, temperature, and sunshine intensity for the crops' growth and development.

4.6.3 ROUND-THE-YEAR PROPAGATION OF PLANTING MATERIAL

Climate management allows for the year-round cultivation of plant material.

4.6.4 PROTECTION FROM FROST AND WIND

A variety of structures, including greenhouses, high tunnels, and low tunnels, are used to cultivate crops in order to shield them from harsh winds and frost.

4.6.5 MINIMIZED CASES OF INSECTS AND DISEASES

When cultivated under protected settings, the presence of insects and illnesses is reduced compared to open field conditions.

4.6.6 HIGHLY USEFUL IN BIOTECHNOLOGY

Structures are widely utilized in protected cultivation to help tissue-cultured plants become acclimated.

4.6.7 STABILIZING MARKET FOR HIGH-QUALITY HORTICULTURAL CROPS

Protected agriculture increases the chances of having high-value horticulture products including fruits, vegetables, and flowers in the early and off seasons by reducing several climatic and weather-related risks. Consequently, this helps to stabilize the market for these crops.

4.7 EMERGING TECHNOLOGIES IN HI-TECH RAISING OF HORTICULTURAL CROPS

4.7.1 IOT-BASED MONITORING

Conventional greenhouse data monitoring techniques frequently encounter difficulties with labor-intensive procedures, low spatiotemporal resolution, lack of data centralization, restricted sharing and accessibility, and organizational management when observing environmental conditions inside a greenhouse. The IoT offers a viable alternative for nondestructive physiological element measurement in cultivated plants as well as environmental monitoring in greenhouses. This allows data sharing among greenhouse producers in a network. The network is a crucial element of IoT applications in greenhouses, covering a wide range of scenarios and enabling innovations in data exchange and communication for greenhouse control and monitoring. According to Kumar et al. (2019), this network consists of hardware, software, and sensors connected via wired or wireless standard communication protocols. IoT applications for greenhouses have piqued the interest of both the academic and industrial sectors (Shafique et al., 2020). In reality, IoT has been used in a variety of research fields, including smart cities, agriculture, and healthcare (Fan et al., 2018). By quickly supplying quantitative data with the useful spatiotemporal resolution, IoT has revolutionized traditional farming observation methods. For example, local greenhouse communities can utilize integrated pest and disease control tactics to stop the spread of related damages by using long-term historical data gathered through IoT apps. Additionally, these kinds of data help improve decision-making and cultural behaviors. The work of Liao et al. (2017) is a noteworthy example, wherein an IoT monitoring system was created to concurrently monitor the environmental aspects of an orchid plant factory and the growth circumstances of Phalaenopsis orchids. Peng et al. (2014) also showed that measurements of the total chlorophyll content of orchids may be made with accuracy using spectroscopic and spectrometer studies. Using a chlorophyll fluorometer, Shafique et al. (2020) evaluated Phalaenopsis's photosynthetic status in a range of illumination scenarios. Even though physiological studies are frequently carried out by hand, these examination methods need highly accurate instruments in order to measure physiological variables in orchids without causing damage. However, farmers might not employ very accurate instruments for these kinds of measurements because of financial constraints. Growers have a big chance to meet these monitoring needs because to the IoT's scalability and adaptability.

4.7.2 ARTIFICIAL INTELLIGENCE (AI)

AI applications in hi-tech nurseries involve the analysis of vast datasets to make informed decisions in real time. AI algorithms can predict crop diseases, optimize resource allocation, and recommend personalized care plans for each plant. Smart monitoring systems, powered by AI, can detect subtle changes in plant health and identify potential issues before they become visible to the naked eye. AI-driven automation enhances the decision-making process, allowing for more precise and timely interventions (Koukounaras, 2020).

4.7.3 NANOTECHNOLOGY

Nanotechnology involves the manipulation of materials at the nanoscale and has promising applications in agriculture. Nanosensors can be utilized to monitor nutrient levels in plants and soil, providing valuable insights for precise nutrient management. Nanomaterials can enhance the efficiency of delivery systems for fertilizers, pesticides, and growth regulators, ensuring targeted and controlled release. Nanotechnology also holds the potential to improve water-use efficiency by developing smart materials that regulate water absorption and release in plant roots (Chaupoo et al., 2023). The integration of precision agriculture, robotics, AI, and nanotechnology in hi-tech nurseries holds tremendous potential for optimizing resource use, improving crop yields, and fostering sustainable practices in horticultural crop cultivation under protected structures. The ongoing advancements in these technologies are expected to shape the future of nursery management, contributing to a more resilient and productive agricultural sector (Lastochkina et al., 2022).

4.7.4 PRECISION AGRICULTURE

Precision agriculture involves the use of technology to optimize crop yields, minimize waste, and enhance resource efficiency. In hi-tech nurseries, precision agriculture can be employed through the integration of sensors, GPS, and data analytics. Sensors can monitor soil conditions, nutrient levels, and environmental variables. Precision irrigation systems ensure that crops receive the right amount of water, reducing water wastage and promoting water-use efficiency. By precisely managing inputs such as fertilizers and pesticides, precision agriculture helps minimize environmental impact while maximizing productivity (Shafi et al., 2019).

4.8 AUTOMATION/ROBOTICS IN HIGH-TECH NURSERY PRODUCTION

A robot is a programmable machine that can carry out a number of intricate tasks on its own. Robots are artificial agents that usually take the shape of electromechanical systems. Robotics play a crucial role in automating various tasks in horticultural nurseries, leading to increased efficiency and reduced labor costs. Automated planting, pruning, and harvesting systems can be implemented to streamline nursery operations. Robots equipped with sensors can navigate through the nursery, collecting data on plant health, growth patterns, and environmental conditions. The use of robotics also addresses labor shortages in agriculture, providing a reliable and consistent workforce (Hemming, 2019). Robot control can be integrated into the system or directed externally. While horticulture or industrial crops are the main focus of current crop robotics research, grains and oilseeds may not be as popular among robotics entrepreneurs because of how mechanized their production is currently (Pekkeriet et al., 2015; Lytridis et al., 2021). Robots can be used for a wide range of jobs, including managing explosives, postlasting underground to stabilize mine roofs, or mining in hostile environments where it is difficult for humans to survive. Pneumatic and robotic mechanisms are necessary for precision farming in India, even if traditional farm equipment has been developed by researchers and farm equipment firms. For example, using robots to eliminate weeds will help reduce the amount of herbicide used, which will result in more organic products. Likewise, robots can be used to minimize intensive labor when transplanting seedlings. Reports indicate that the number of farmers in India is declining daily, despite significant developments in the field. Many factors, including declining profitability, perceived hazards, and the indifference of younger generations in farming, are blamed for this loss. Instead of staying in agriculture, many farmers are choosing to work in construction enterprises as alternative careers. While manually handled tools such as hoes and weeders are frequently utilized for intercultural operations, particularly in horticultural crops, tractors continue to be the principal equipment for many farm operations. Robots are widely used in the horticulture industry for harvesting, drone spraying, field monitoring, sorting, grading, and packing finished products. In nurseries and greenhouses, they are also used to some extent (Marshall et al., 2016).

4.8.1 TYPES OF ROBOTS USED IN HORTICULTURE

4.8.1.1 DEMETER

Demeter, a robotic harvester, possesses the ability to autonomously navigate and cut crops without requiring human supervision. Resembling a conventional harvester, it incorporates cameras that discern the distinction between cut and uncut crops. This visual information guides its navigation, determines the placement of its cutter head, and signals when it reaches the end of a crop row for a turnaround. Additionally, the robot can be remotely controlled. It is capable of learning a predefined path and subsequently following that path with the aid of its onboard sensors and computer control systems (Roldán et al., 2018; Reddy et al., 2016) (Figure 4.3).

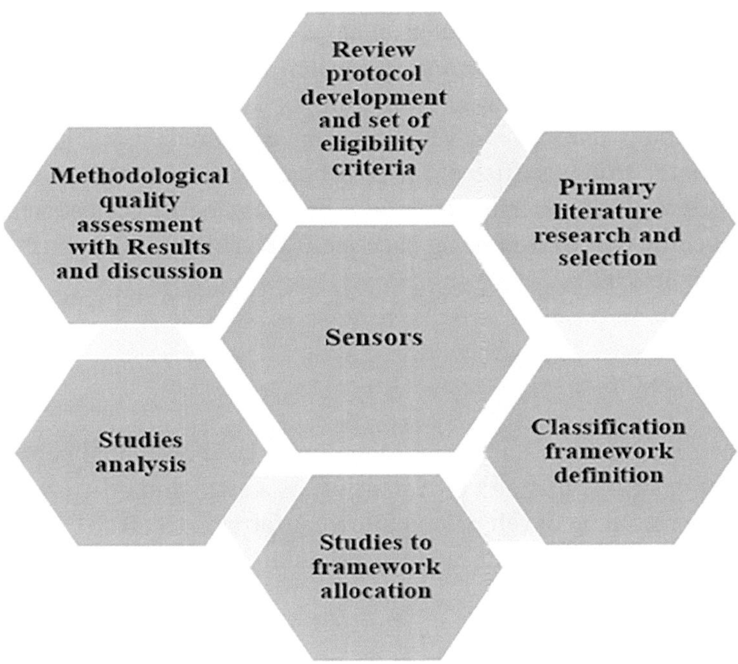

FIGURE 4.3 Robotics in agriculture for performing various field works.

4.8.1.2 WEED CONTROLLER

A weed-seeking robot with four-wheel drive capability was created, designed to eliminate and eradicate weeds. For crops arranged in rows, a hoe is employed to weed between them. An intelligent hoe utilizes a vision system to detect crop rows, autonomously steering itself with precision between the rows, thereby minimizing the reliance on herbicides. Weed identification relies on color photography. These equipped robots contribute to the generation of weed maps that identify specific plants (Fountas et al., 2020; Steward et al., 2019; Zhang et al., 2019).

4.8.1.3 FRUIT-PICKING ROBOT

These robots are required to harvest fruits without causing harm to the tree's branches or leaves, and they need to have access to all parts of the tree during the harvesting process. The robot employs video image capturing to differentiate between fruits and leaves. A camera, mounted on the robot arm, captures colors that are then compared to predefined properties stored in the memory. If a match is found, the robot proceeds to pick the fruit. In cases where the fruit is obscured by leaves, an air jet is used to blow the leaves away, providing a clearer view. The pressure applied to the fruit is sufficient for detachment from the tree but carefully controlled to avoid damaging the fruit. The design of the gripper is tailored to the specific type of fruit being harvested (Tang et al., 2020; Onishi et al., 2019; Bogue, 2020).

4.8.1.4 DRONES

A drone, functioning as an aerial robot, provides an elevated perspective of the land, akin to that of a bird. This offers a swift and efficient method for monitoring crop development, allowing farmers to identify areas that might require replanting or targeted pesticide application (Mogili et al., 2018; Kurkute et al., 2018).

4.9 OPPORTUNITIES, CURRENT APPLICATIONS, AND FUTURE DIRECTIONS IN HI-TECH NURSERY PRODUCTION

Integration of these technologies offers a holistic approach to sustainable and efficient nursery raising. The combination of precision agriculture, robotics, AI, and nanotechnology is expected to lead to "smart nurseries" capable of adapting dynamically to changing environmental conditions. Increased connectivity and data sharing among devices and systems will facilitate a more comprehensive understanding of plant growth dynamics. Continuous innovation in these fields is likely to result in cost-effective solutions, making hi-tech nursery practices accessible to a broader range of horticulturists. Robotic arms equipped with vision systems can precisely sow seeds, place seedlings in trays, and optimize plant spacing (Cobb et al., 2020). This improves uniformity, reduces waste, and minimizes human error. Automated robots can handle delicate grafting procedures and transplant fragile seedlings with minimal damage (Adigüzel et al., 2018). This increases efficiency, reduces labor costs, and ensures consistent quality. Automated irrigation systems utilize sensors to monitor soil moisture, temperature, and nutrient levels, delivering precise water and fertilizer doses based on plant needs (Król et al., 2019). This optimizes resource utilization and promotes sustainable production. Specialized robots equipped with vision and tactile sensors can autonomously prune branches, harvest ripe fruits, and sort produce based on size and quality (Zhou et al., 2020). This improves yield, reduces harvest losses, and enhances labor safety. The upfront cost of automation and robotics can be a significant barrier for smaller nurseries (Gómez-Gil et al., 2020). Government subsidies and adaptable system designs could address this challenge. Integrating robotic systems with existing infrastructure and data management systems requires expertise and ongoing maintenance (Chacko et al., 2019). User-friendly interfaces and modular designs are crucial for wider adoption. Adapting robots to handle diverse plant shapes, sizes, and delicate structures remains a challenge (Nadal-Serrano et al., 2018). Advancements in computer vision and AI are necessary for flexible and adaptable robots. Balancing the benefits of year-round production with the environmental impact of energy-intensive systems poses a dilemma. Efforts toward developing energy-efficient technologies and sustainable practices are essential to address this concern (Kaushik, 2021).

4.10 CONCLUSION

Modern horticulture has the potential to double crop production and increase farmers' income (DFI) through the application of high-tech techniques. (1) Climate-resilient production technology through high-tech interventions; (2) conservation of existing germplasm and utilization of underutilized plant and land use with development of new varieties/ hybrids suited for high-tech horticulture; (3) application of biotechnology and nanotechnology; and (4) precision farming focused on targeted yield, crop and region-specific nutrient management, and irrigation resource conservation are some of the broad themes to carry forward the goal of high-tech horticulture. Real-time data from the crops must be recorded and instantly presented to the decision-makers as a successful high-tech horticulture business in order to stay up with the current information-based decisions. In the industrialized world, significant progress has been made in the areas of automated irrigation control, decision-support systems, and novel sensor technologies. These have reduced the cost of acquisition and operation while also greatly enhancing the technologies' efficiency and dependability.

KEYWORDS

- **high-tech nursery raising**
- **protected cultivation**
- **nursery management**
- **fruits**
- **vegetables**

REFERENCES

Adigüzel, N., Adigüzel, N. F., & Güvenç, A. (2018). A robotic system for automatic tomato grafting. *Biosystems Engineering,* 169, 311–321.

Ahmad, S., Ashraf, I., & Anjum, M.A. (2017). Fruit and vegetable nurseries: Establishment and management. In: *Horticulture: Science and Technology*. Faisalabad: University of Agriculture, pp. 133–159.

Amani, M., Foroushani, S., Sultan, M., & Bahrami, M. (2020). Comprehensive review on dehumidification strategies for agricultural greenhouse applications. *Applied Thermal Engineering,* 181, 115979.

Angmo, P., Dolma, T., Namgail, D., Tamchos, T., Norbu, T., Chaurasia, O.P., & Stobdan, T. (2019). Passive solar greenhouse for round the year vegetable cultivation in trans-Himalayan Ladakh region, India. *Defence Life Science Journal,* 4(2), 103–116.

Ausari, P.K., Gharate, P.S., Saikanth, D.R.K., Devi, O.B., Bahadur, R., & Singh, Y.S. (2023). High-tech farming techniques in fruit crops to secure food demand: A review. *International Journal of Environment and Climate Change,* 13(11), 2716–2730.

Ayaz, M., Ammad-Uddin, M., Sharif, Z., Mansour, A., & Aggoune, E.H.M. (2019). Internet-of-Things (IoT)-based smart agriculture: Toward making the fields talk. *IEEE Access*, 7, 129551–129583.

Badji, A., Benseddik, A., Bensaha, H., Boukhelifa, A., & Hasrane, I. (2022). Design, technology, and management of greenhouse: A review. *Journal of Cleaner Production,* 373, 133753.

Bar-Tal, A., Saha, U.K., Raviv, M., & Tuller, M. (2019). Inorganic and synthetic organic components of soilless culture and potting mixtures. In: *Soilless Culture*. Elsevier, pp. 259–301.

Behera, S.K., Adamchuk, V.I., Shukla, A.K., Pandey, P.S., Kumar, P., Shukla, V., Thiyagarajan, C., Rai, H.K., Hadole, S., Sachan, A.K, Singh, P., Trivedi, V., Mishra, A., Butail, N.P., Kumar, P., Prajapati, R., Tiwari, K., Suri, D., & Sharma, M. (2022). The scope for using proximal soil sensing by the farmers of India. *Sustainability,* 14(14), 8561.

Benke, K., & Tomkins, B. (2017). Future food-production systems: Vertical farming and controlled-environment agriculture. *Sustainability: Science, Practice and Policy*, 13(1), 13–26.

Bhakare, B.D., Shinde, J.B., Tuwar, S.S., Tambe, A.D., Dhonde, M.B., & Bodake, P.S. (2017). Development of deficit irrigation practices under drip for marigold-rabi Sorghum crop sequence for varied planting techniques in changing climate scenario. *Journal of Agriculture Research and Technology*, 42(3), 142.

Bhattacharyya, T., Haldankar, P.M., Patil, V.K., Salvi, B.R., Haldavanekar, P.C., Pujari, K.H., & Dosani, A.A. (2017). Hi-tech horticulture: Pros and cons. *Indian Journal of Fertilisers*, 13(12), 46–58.

Bogue, R. (2020). Fruit picking robots: Has their time come? *Industrial Robot: The International Journal of Robotics Research and Application*, 47(2), 141–145.

Chacko, J., Singh, A.K., & Singh, N.P. (2019). Precision agriculture through sensor based IoT driven smart nursery. *Computers and Electronics in Agriculture,* 165, 104999.

Chaupoo, A.S., Dhakad, A., Beese, S., Kumar, A., Kumar, P., Pandey, S.K., & Kurdekar, A.K. (2023). Transforming horticulture: The influence of nanotechnology on crop enhancement. *International Journal of Environment and Climate Change,* 13(11), 3381–3400.

Choab, N., Allouhi, A., El Maakoul, A., Kousksou, T., Saadeddine, S., & Jamil, A. (2019). Review on greenhouse microclimate and application: Design parameters, thermal modeling and simulation, climate controlling technologies. *Solar Energy*, 191, 109–137.

Cobb, C.J., Aurich, D., Hermsmeier, M., & Hemming, J.D. (2020). A robotic system for transplanting seedlings grown in individual cell trays. *Computers and Electronics in Agriculture,* 174, 105404.

D'Amico, A., De Boni, A., Palmisano, G.O., Acciani, C., & Roma, R. (2023). Environmental analysis of soilless tomato production in a high-tech greenhouse. *Cleaner Environmental Systems*, 11, 100137.

Dhakar, M.K., Kaushik, R., Sarolia, D., & Bana, M. (2013). Effect of fertigation through low cost drip irrigation system on vegetative growth characteristics in pomegranate cv. Bhagwa. *Green Farming*, 4(1), 33–36.

Dhanaraju, M., Chenniappan, P., Ramalingam, K., Pazhanivelan, S., & Kaliaperumal, R. (2022). Smart farming: Internet of Things (IoT)-based sustainable agriculture. *Agriculture*, 12(10), 1745.

Engler, N., & Krarti, M. (2021). Review of energy efficiency in controlled environment agriculture. *Renewable and Sustainable Energy Reviews,* 141, 110786.

Fan, K., Jiang, W., Li, H., & Yang, Y. (2018). Lightweight RFID protocol for medical privacy protection in IoT. *IEEE Transactions on Industrial Informatics*, 14(4), 1656–1665.

Farooq, F., Kubrevi, S.S., Peer, Q.J.A., Ganaie, N.A., Ali, A., & Wani, F.J. (2023). Constraints faced by fruit nursery growers in district Kulgam of Jammu and Kashmir, India. *Journal of Experimental Agriculture International*, 45(10), 18–21.

Fountas, S., Mylonas, N., Malounas, I., Rodias, E., Hellmann Santos, C., & Pekkeriet, E. (2020). Agricultural robotics for field operations. *Sensors*, 20(9), 2672.

Ganeshamurthy, A.N., & Rupa T.R. (2016) *Nutrient Management in Nurseries. Compendium of Lectures*. Indian Institute of Horticultural Research.

Ghani, S., Bakochristou, F., ElBialy, E.M.A.A., Gamaledin, S.M.A., Rashwan, M.M., Abdelhalim, A.M., & Ismail, S.M. (2019). Design challenges of agricultural greenhouses in hot and arid environments a review. *Engineering in Agriculture, Environment and Food*, 12(1), 48–70.

Gojon, A., Cassan, O., Bach, L., Lejay, L., & Martin, A. (2023). The decline of plant mineral nutrition under rising CO2: Physiological and molecular aspects of a bad deal. *Trends in Plant Science*, 28(2), 185–198.

Gómez-Gil, J.M., Peña-Roríguez, A.Á., & Reca, I.P. (2020). Economic feasibility of robotic automation in greenhouses: A case study in southern Spain. *Computers and Electronics in Agriculture,* 179, 105650.

Gorjian, S., Calise, F., Kant, K., Ahamed, M.S., Copertaro, B., Najafi, G., Zhang, X., Aghaei, M., & Shamshiri, R.R. (2021). A review on opportunities for implementation of solar energy technologies in agricultural greenhouses. *Journal of Cleaner Production,* 285, 124807.

Gruda, N., Bisbis, M., & Tanny, J. (2019). Influence of climate change on protected cultivation: Impacts and sustainable adaptation strategies—A review. *Journal of Cleaner Production,* 225, 481–495.

Gupta, D., Santosh, D.T., & Debnath, S. (2020). Modeling and simulation application for greenhouse microclimatic studies and structural analysis. In: S. Maitra, D.J. Gaikwad, T. Shankar, Eds., *Protected Cultivation and Smart Agriculture*. pp. 300–312.

Haldankar, P.M., Parulekar, Y.R., Kireeti, A., Kad, M.S., Shinde, S.M., & Lawande, K.E. (2015). Studies on influence of bagging of fruits at marble stage on quality of mango cv. Alphonso. *Journal of Plant Studies,* 4(2), 12.

Haque, M.A., & Sakimin, S.Z. (2022). Planting arrangement and effects of planting density on tropical fruit crops—A review. *Horticulturae,* 8(6), 485.

Hassan, S., Bhat, K.M., Jan, A., Mehraj, S., Wani, S.A., Khanday, M.U.D., & Bisati, I.A. (2018). Managing genetic resources in temperate fruit crops. *Economic Affairs,* 63(4), 987–996.

Hemming, J. (2019). Current developments in greenhouse robotics and challenges for the future. In: *International Symposium on Advanced Technologies and Management for Innovative Greenhouses: GreenSys* 2019, 1296, pp. 975–986.

Iddio, E., Wang, L., Thomas, Y., McMorrow, G., & Denzer, A. (2020). Energy efficient operation and modeling for greenhouses: A literature review. *Renewable and Sustainable Energy Reviews,* 117, 109480.

Jain, S., Kore, D.S., GK, K., Mohapatra, A., Baksh, H., Kumar, V., Mohanty, S., & Haokip, S.W. (2023). A comprehensive review on protected cultivation of horticultural crops: Present status and future prospects. *International Journal of Environment and Climate Change,* 13(11), 3521–3531.

Kandris, D., Nakas, C., Vomvas, D., & Koulouras, G. (2020). Applications of wireless sensor networks: An up-to-date survey. *Applied System Innovation,* 3(1), 14.

Kaushik, P. (2021). Precision vegetable farming technologies: An update. In: *Technology in Agriculture*, p. 155.

Khan, N., Ray, R.L., Sargani, G.R., Ihtisham, M., Khayyam, M., & Ismail, S. (2021). Current progress and future prospects of agriculture technology: Gateway to sustainable agriculture. *Sustainability,* 13(9), 4883.

Khattab, A., Habib, S.E., Ismail, H., Zayan, S., Fahmy, Y., & Khairy, M.M. (2019). An IoT- based cognitive monitoring system for early plant disease forecast. *Computers and Electronics in Agriculture,* 166, 105028.

Köksal, Ö., & Tekinerdogan, B. (2019). Architecture design approach for IoT-based farm management information systems. *Precision Agriculture,* 20, 926–958.

Koukounaras, A. (2020). Advanced greenhouse horticulture: New technologies and cultivation practices. *Horticulturae,* 7(1), 1.

Król, D., Kostrubiec, M., & Wójcik, A. (2019). Sensor-based intelligent irrigation system with a fuzzy logic controller. *Sensors,* 19(2), 328.

Kumar, S., Tiwari, P., & Zymbler, M. (2019). Internet of Things is a revolutionary approach for future technology enhancement: A review. *Journal of Big Data,* 6(1), 1–21.

Kurkute, S.R., Deore, B.D., Kasar, P., Bhamare, M., & Sahane, M. (2018). Drones for smart agriculture: A technical report. *International Journal for Research in Applied Science and Engineering Technology,* 6(4), 341–346.

Kwok, A.G., & Grondzik, W. (2018). *The Green Studio Handbook: Environmental Strategies for Schematic Design*. Routledge.

Lakshmanan, R., Djama, M., Selvaperumal, S.K., & Abdulla, R. (2020). Automated smart hydroponics system using internet of things. *International Journal of Electrical and Computer Engineering (IJECE),* 10(6), 6389–6398.

Lalk, G.T., Bi, G., Zhang, Q., Harkess, R.L., & Li, T. (2020). High-tunnel production of strawberries using black and red plastic mulches. *Horticulturae,* 6(4), 73.

Lastochkina, O., Aliniaeifard, S., SeifiKalhor, M., Bosacchi, M., Maslennikova, D., & Lubyanova, A. (2022). Novel approaches for sustainable horticultural crop production: Advances and prospects. *Horticulturae,* 8(10), 910.

Li, K., Sha, Z., Xue, W., Chen, X., Mao, H., & Tan, G. (2020). A fast modeling and optimization scheme for greenhouse environmental system using proper orthogonal

decomposition and multi-objective genetic algorithm. *Computers and Electronics in Agriculture,* 168, 105096.
Li, T., & Bi, G. (2019). Container production of southern highbush blueberries using high tunnels. *HortScience,* 54(2), 267–274.
Liao, M.S., Chen, S.F., Chou, C.Y., Chen, H.Y., Yeh, S.H., Chang, Y.C., & Jiang, J.A. (2017). On precisely relating the growth of Phalaenopsis leaves to greenhouse environmental factors by using an IoT-based monitoring system. *Computers and Electronics in Agriculture,* 136, 125–139.
Lowry, G.V., Avellan, A., & Gilbertson, L.M. (2019). Opportunities and challenges for nanotechnology in the agri-tech revolution. *Nature Nanotechnology*, 14(6), 517–522.
Lytridis, C., Kaburlasos, V.G., Pachidis, T., Manios, M., Vrochidou, E., Kalampokas, T., & Mampholo, B.M., Maboko, M.M., Soundy, P., & Sivakumar, D. (2016). Phytochemicals and overall quality of leafy lettuce (Lactuca sativa L.) varieties grown in closed hydroponic system. *Journal of Food Quality,* 39(6), 805–815.
Mândru, M.C., Hoza, D., & Cristea, S. (2018). Pathogens with economic importance for tomato crops growing in the field and their control. *Scientific Papers. Series B, Horticulture,* 62, 499–505.
Marinelli, M.V., Scavuzzo, M., Giobellina, B., & Scavuzzo, M. (2019). Geoscience and remote sensing on horticulture as support for management and planning. *Agronomy Research*, 2(2), 43.
Marshall, J.A., Bonchis, A., Nebot, E., & Scheding, S. (2016). Robotics in mining. In: *Handbook of Robotics*, Springer, pp. 1549–1576.
Maureira, F., Rajagopalan, K., & Stöckle, C.O. (2022). Evaluating tomato production in open- field and high-tech greenhouse systems. *Journal of Cleaner Production,* 337, 130459.
Min, S.Y., & Oh, W. (2020). Effects of nutrient solution application methods and rhizospheric ventilation on vegetative growth of young moth orchids without a potting medium in a closed-type plant factory. *Journal of People, Plants, and Environment*, 23(5), 545–554.
Mogili, U.R., & Deepak, B.B.V.L. (2018). Review on application of drone systems in precision agriculture. *Procedia Computer Science*, 133, 502–509.
Mohan, V. 2016. Flower Fascination: India set to be floriculture trade leader. *The Times of India* (Delhi Edition), 14.02.2016.
Nadal-Serrano, N., Suárez-Vico, R., Álvarez-Fontalvo, A., & Gil-Ribes, J. M. (2018). Robot technology applied to tree trimming in citrus orchards. *Agronomy*, 8(5), 110.
Naik, M.H., Vanajalatha, K., Ramesh, T., & Prasanth, P. 2016. Influence of high density plantingand fertigation on physiological parameters, flowering and yield of banana (Musa acuminata L.) cv. Grand Naine of Main and ratoon crop. *International Journal of Advanced Research in Science, Engineering and Technology*, 3, 2800–2809.
Nainabasti, A., Thapa, R., Subedi, B., Lamsal, A., & Malla, S. (2022). Protected cultivation of horticultural crops and its status in Nepal, *Journal of Agricultural Research Pesticides and Biofertilizers*, 4 (1);. http://doi.org/org/06.2022/1.1077.
Newlands, N.K. (2018). Model-based forecasting of agricultural crop disease risk at the regional scale, integrating airborne inoculum, environmental, and satellite-based monitoring data. *Frontiers in Environmental Science*, 6, 63.

Onishi, Y., Yoshida, T., Kurita, H., Fukao, T., Arihara, H., & Iwai, A. (2019). An automated fruit harvesting robot by using deep learning. *Robomech Journal*, 6(1), 1–8.

Pandiyaraj, P., Vijayakumar, S., & Nimbolkar, P.K. (2017). Modern nursery raising systems in vegetables. *International Journal of Agriculture Sciences*, 0975–0371.

Patil, K.D., More, S.S., Wahane, M.R., Puranik, U.Y., & Khobragade, N.H. (2018). Micronutrients importance in hi-tech horticulture. *Journal of Pharmacognosy and Phytochemistry,* 7(3), 628–635.

Pekkeriet, E.J., Van Henten, E.J., & Campen, J.B. (2015). Contribution of innovative technologies to new developments in horticulture. *Acta Hortic,* 1099, 45–54.

Peng, G., Lahlali, R., Hwang, S.F., Pageau, D., Hynes, R.K., McDonald, M.R., Gossen, B.D., & Strelkov, S.E. (2014). Crop rotation, cultivar resistance, and fungicides/biofungicides for managing clubroot (*Plasmodiophora brassicae*) on canola. *Canadian Journal of Plant Pathology*, 36(sup1), 99–112.

Planning Commission. 2007. Report of the Working Group on Horticulture, Plantation Crops and Organic Farming for the XI Five Year Plan (2007–2012). Planning Commission, Government of India.

Qureshi, T., Saeed, M., Ahsan, K., Malik, A.A., Muhammad, E.S., & Touheed, N. (2022). Smart agriculture for sustainable food security using internet of things (IoT). *Wireless Communications and Mobile Computing*, 2022, 1–10.

Shamshiri, R., Kalantari, F., Ting, K.C., Thorp, K.R., Hameed, I.A., Weltzien, C., Ahmed, D., & Shad, Z.M. (2018). Advances in greenhouse automation and controlled environment agriculture: A transition to plant factories and urban agriculture. *International Journal of Agricultural and Biological Engineering*, 2018; 11(1): 1–22.

Rabbi, B., Chen, Z.H., & Sethuvenkatraman, S. (2019). Protected cropping in warm climates: A review of humidity control and cooling methods. *Energies*, 12(14), 2737.

Ramalingam, B., Mohan, R.E., Pookkuttath, S., Gómez, B.F., Sairam Borusu, C.S.C., Wee Teng, T., & Tamilselvam, Y.K. (2020). Remote insects trap monitoring system using deep learning framework and IoT. *Sensors*, 20(18), 5280.

Ramanatha, R.V., Rajasekharan, P.E., & Anushma, P.L. (2023). Conservation and management of fruit and nut genetic resources. In: *Fruit and Nut Crops*. Singapore: Springer Nature Singapore, pp. 1–22.

Reddy, N.V., Reddy, A.V.V.V., Pranavadithya, S., & Kumar, J.J. (2016). A critical review on agricultural robots. *International Journal of Mechanical Engineering and Technology*, 7(4), 183–188.

Roldán, J.J., del Cerro, J., Garzón-Ramos, D., Garcia-Aunon, P., Garzón, M., De León, J., & Barrientos, A. (2018). Robots in agriculture: State of art and practical experiences. In: *Service Robots*, pp. 67–90.

Rouphael, Y., Spíchal, L., Panzarová, K., Casa, R., & Colla, G. (2018). High-throughput plant phenotyping for developing novel biostimulants: From lab to field or from field to lab? *Frontiers in Plant Science*, 9, 1197.

Sahiijram, L. and Rao, M. 2017. Micro-propagation for mass production of horticultural crops. In: K.L. Chadha, S.K. Singh, P. Kalia, W.S. Dhillon, T.K. Behera, & J. Prakash, Eds., *Doubling Farmers Income through Horticulture.* New Delhi: The Horticulture Society of India and Daya Publishing House, pp. 29–33.

Saiz-Rubio, V., & Rovira-Más, F. (2020). From smart farming towards agriculture 5.0: A review on crop data management. *Agronomy*, 10(2), 207.

Sambo, P., Nicoletto, C., Giro, A., Pii, Y., Valentinuzzi, F., Mimmo, T., Lugli, P., Orzes, G., Mazzetto, F., Astolfi, S., Terzano, R, & Cesco, S. (2019). Hydroponic solutions for soilless production systems: Issues and opportunities in a smart agriculture perspective. *Frontiers in Plant Science*, 10, 923.

Senthilkumar, M., Ganesh, S., Srinivas, K., Panneerselvam, P., Nagaraja, A., & Kasinath, B.L. (2017). Fertigation for effective nutrition and higher productivity in banana—A review. *International Journal of Current Microbiology and Applied Sciences*, 6(7), 2104–2122.

Shafi, U., Mumtaz, R., García-Nieto, J., Hassan, S.A., Zaidi, S.A.R., & Iqbal, N. (2019). Precision agriculture techniques and practices: From considerations to applications. *Sensors*, 19(17), 3796.

Shafique, K., Khawaja, B.A., Sabir, F., Qazi, S., & Mustaqim, M. (2020). Internet of things (IoT) for next-generation smart systems: A review of current challenges, future trends and prospects for emerging 5G-IoT scenarios. *IEEE Access*, 8, 23022–23040.

Singh, A.P., & Johari, D. (2018). Scope of ferns in horticulture and economic development. In: *Current Advances in Fern Research*, pp. 153–175.

Singh, B. 2014. Protected cultivation of horticultural crops in India—Challenges and opportunities. *Agrotechnology*, 2(4), 1.

Singh, H., Dunn, B., & Payton, M. (2019). Hydroponic pH modifiers affect plant growth and nutrient content in leafy greens. *Journal of Horticultural Research*, 27(1), 31–36.

Singh, S., Singh, F., & Bhatnagar, P. (2020). Hi-tech vegetable nursery: Key gradient for economic development. *IOSR Journal of Agriculture and Veterinary Science*, 13, 38–41.

Smrke, T., Veberic, R., Hudina, M., Zitko, V., Ferlan, M., & Jakopic, J. (2021). Fruit quality and yield of three highbush blueberry (*Vaccinium corymbosum* L.) cultivars grown in two planting systems under different protected environments. *Horticulturae,* 7(12), 591.

Soussi, M., Chaibi, M.T., Buchholz, M., & Saghrouni, Z. (2022). Comprehensive review on climate control and cooling systems in greenhouses under hot and arid conditions. *Agronomy*, 12(3), 626.

Srivastava, A.K., & Shirgure, P.S. (2017). Soil fertility management for better productivity of fruit crops. In: K.L. Chadha, S.K. Singh, P. Kalia, W.S. Dhillon, T.K. Behera, & Jai Prakash, Eds., *Doubling Farmers Income through Horticulture*. New Delhi: The Horticulture Society of India and Daya Publishing House, pp. 307–338.

Stanciu, C., Stanciu, D., & Dobrovicescu, A. (2016). Effect of greenhouse orientation with respect to EW axis on its required heating and cooling loads. *Energy Procedia*, 85, 498–504.

Steward, B., Gai, J., & Tang, L. (2019). The use of agricultural robots in weed management and control. In: *Robotics and Automation for Improving Agriculture.* Burleigh Dodds Science Publishing, pp. 161–186.

Sultan, M., Miyazaki, T., Saha, B.B., & Koyama, S. (2016). Steady-state investigation of water vapor adsorption for thermally driven adsorption based greenhouse air-conditioning system. *Renewable Energy*, 86, 785–795.

Tang, Y., Chen, M., Wang, C., Luo, L., Li, J., Lian, G., & Zou, X. (2020). Recognition and localization methods for vision-based fruit picking robots: A review. *Frontiers in Plant Science*, 11, 510.

Tegen, H., Dessalegn, Y., & Mohammed, W. (2016). Influence of mulching and varieties on growth and yield of tomato under polyhouse. *Journal of Horticulture and Forestry*, 8(1), 1–11.

Tomar, B.S., & Yalamalle, V.R. (2017). Improving the availability and quality of vegetable seeds for higher profitability. In: *Advances in Production Technology of Commercial Vegetable Crops.* Department of Vegetable Science, Dr YS Parmar University of Horticulture and Forestry. Solan, Himachal Pradesh, pp. 11–25.

Ummyiah, H.M., Wani, K.P., Khan, S.H., & Magray, M.M. (2017). Protected cultivation of vegetable crops under temperate conditions. *Journal of Pharmacognosy and Phytochemistry,* 6(5), 1629–1634.

Venkatesan, R., Kathrine, G.J.W., & Ramalakshmi, K. (2018). Internet of Things based pest management using natural pesticides for small scale organic gardens. *Journal of Computational and Theoretical Nanoscience*, 15(9–10), 2742–2747.

Villecco, D., Pane, C., Ronga, D., & Zaccardelli, M. (2020). Enhancing sustainability of tomato, pepper and melon nursery production systems by using compost tea spray applications. *Agronomy*, 10(9), 1336.

Walker, A.P., De Kauwe, M.G., Bastos, A., Belmecheri, S., Georgiou, K., Keeling, R.F., G., Bastos, A., Belmecheri, S., Georgiou, K., Keeling, R.F., McMahon, S.M., Medlyn, B.E., Moore, D.J.P., Norby, R.J., Zaehle, S., Anderson-Teixeira, K.J., Battipaglia, G., Brienen, R.J.W., Cabugao, K.G., Cailleret, M., Campbell, E., Canadell, J.G., Ciais, P., Craig, M.E., Ellsworth, D.S., Farquhar, G.D., Fatichi, S., Fisher, J.B., Frank, D.C., Graven, H., Gu, L., Haverd, V., Heilman, K., Heimann, M., Hungate, B.A., Iversen, C.M., Joos, F., Jiang, M., Keenan, T.F., Knauer, J., Körner, C., Leshyk, V.O., Leuzinger, S., Liu, Y., MacBean, N., Malhi, Y., McVicar, T.R., Penuelas, J., Pongratz, J., Powell, A.S., Riutta, T., Sabot, M.E.B., Schleucher, J., Sitch, S., Smith, W.K., Sulman, B., Taylor, B., Terrer, C., Torn, M.S., Treseder, K.K., Trugman, A.T., Trumbore, S.E., van Mantgem, P.J., Voelker, S.L., Whelan, M.E, & Zuidema, P.A. (2021). Integrating the evidence for a terrestrial carbon sink caused by increasing atmospheric CO2. *New Phytologist*, 229(5), 2413–2445.

Xiao, X., Fang, Y., Xiao, X., Xu, J., & Chen, J. (2021). Machine-learning-aided self-powered assistive physical therapy devices. *ACS Nano*, 15(12), 18633–18646.

Yalew, S.G., Van Griensven, A., Mul, M.L., & van der Zaag, P. (2016). Land suitability analysis for agriculture in the Abbay basin using remote sensing, GIS and AHP techniques. *Modeling Earth Systems and Environment*, 2, 1–14.

Yang, W., Feng, H., Zhang, X., Zhang, J., Doonan, J.H., Batchelor, W.D., Xiong, L., & Yan, J. (2020). Crop phenomics and high-throughput phenotyping: Past decades, current challenges, and future perspectives. *Molecular Plant*, 13(2), 187–214.

Zhang, Q., Karkee, M., & Tabb, A. (2019). The use of agricultural robots in orchard management. arXiv preprint arXiv:1907.13114.

Zhang, S., Yang, Y., Li, J., Qin, J., Zhang, W., Huang, W., & Hu, H. (2018). Physiological diversity of orchids. *Plant Diversity*, 40(4), 196–208.

Zhou, R., Tang, Y., Wu, F., & Zhu, X. (2020). A novel strawberry harvesting robot with high efficiency and fruit protection. *Computers and Electronics in Agriculture,* 175, 105496.

CHAPTER 5

GIS and DGPS System for Enhanced Productivity in Horticultural Crops

S. K. PANDEY, P. K. MISHRA, T. PANDIARAJ,* and SHASHANK VERMA

Acharya Narendra Deva University of Agriculture and Technology, Ayodhya, Uttar Pradesh, India

*Corresponding author

ABSTRACT

The integration of geographic information systems (GIS) and differential global positioning system (DGPS) technologies has revolutionized the agricultural sector, particularly in the domain of horticultural crop management. This chapter explores the applications and benefits of GIS and DGPS systems in enhancing productivity, sustainability, and efficiency in horticultural crop production. GIS technology provides a powerful platform for spatial data analysis, allowing farmers and researchers to map and monitor various aspects of horticultural crop cultivation. Through GIS, spatial variability in soil properties, climate conditions, and crop health can be accurately assessed, leading to targeted management practices such as precision irrigation, nutrient management, and pest control. DGPS complements GIS by providing high-precision location information, enabling farmers to precisely position equipment and apply inputs with accuracy.

5.1 INTRODUCTION

Horticultural crops demand meticulous attention and monitoring on a per-plant basis to restrict the emergence of pests and diseases to an acceptable

Smart Horticulture Production: Integrating Advanced Technologies.
Rajni Rajan and Duniya Ram Singh (Eds.)

DOI: 10.1201/9781998511679-5

level. This approach has given way to technological advancements that have automated and optimized farm operations, bringing in more precision and reliability. Precision agriculture has ushered in better weed control, soil health, and higher yields. Precision agriculture has arisen as a solution to the sustainable food production challenge. This approach leverages technology to deliver crops with precise amounts of nutrients and treatments for optimal yield. By automating the application of agrochemicals, overuse is eliminated, resulting in cost savings and reduced environmental spillover, such as nutrient leaching. Additionally, predictive technologies have the potential to enhance crop performance by enabling the customization of crop varieties.

5.2 AUTOMATION FOR PLANT OBSERVING

Crop scouting is a vital stage in pest and disease management. It permits the farmer to make suitable decisions and take precautionary actions. It reduces the waste of time and money, allows movements to be recognized, and forecasts upcoming threats and potential problems. Typical equipment for plant observation in a field or greenhouse contains a magnifying glass, plastic sack for plant samples, clipboard, marker, and camera. Nowadays, adhesive traps and mobile apps can give a distant more precise picture of the pests and diseases in a crop; several apps even make available suggestions for possible solutions.

A recent advancement involves the creation of digital scouting applications tailored for mobile devices, offering a monitoring platform to detect diseases and propose biological remedies for horticulture. For instance, employing precise image recognition, a scanner can tally the insects captured on sticky cards. This enables scanning of sticky traps within a crop, while a dashboard on a desktop/laptop provides a direct schematic overview of the findings. The collected data can then be integrated with other metrics, facilitating a comprehensive analysis to suggest biological solutions. Additionally, the application may incorporate machine or deep learning capabilities to enhance its predictive accuracy over time.

A geographic information system (GIS) is similar to a conventional database system, but with a key distinction: it requires that all data be associated with a geographic (spatial) reference, such as latitude/longitude or other spatial coordinates.

There are several meanings of geographical indications, as researchers stress various aspects of their significance (Figure 5.1). However, the word GIS is tough to define because it signifies integrating numerous subject areas. There is not any universally agreed-upon definition for a GIS, as acknowledged by Demers's in 1997. The National Centre of Geographic Information and Analysis provided established a widely embraced definition of GIS: "A GIS is a system including hardware, software, and procedures designed to aid in the management, manipulation, analysis, modeling, representation, and display of geo-referenced data for addressing intricate issues related to resource planning and management."

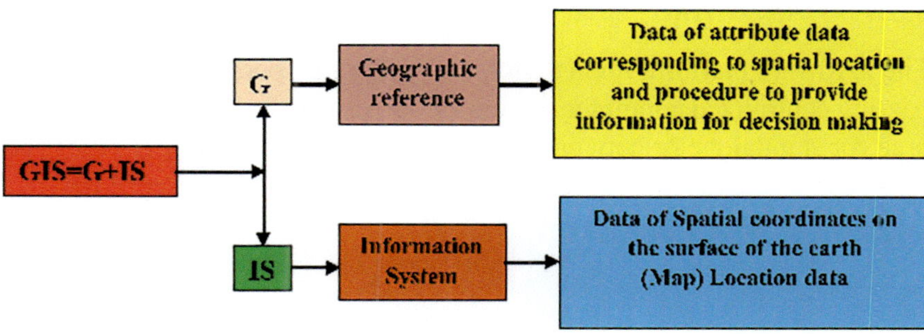

FIGURE 5.1 Comprehension of GIS meaning.

Environmental Systems Research Institute outlined GIS as a structured assemblage of computer hardware, software, geographic data, and personnel tailored to efficiently capture, store, update, manipulate, analyze, and exhibit geographically referenced information. GIS represents a specific instance of information systems, wherein the database comprises observations on spatially distributed features, activities, or events, classifiable in space as points, lines, or areas. Through a GIS, data concerning these points, lines, or areas can be manipulated to retrieve information for ad hoc inquiries and analyses.

The United States Geological Survey (USGS) provided a definition of GIS as a computer hardware and software system devised for the collection, management, analysis, and visualization of geographically referenced data. This definition is sufficiently comprehensive and applicable for agricultural uses of GIS. According to Jachimski (2001), "geoinformatics" is "the science of the gathering, processing and dissemination of information which is spatially defined within the Earth's system."

Geoinformatics has been defined as "the science and the technology dealing with the structure and character of spatial information, its capture, its classification, its storage, processing, portrayal and dissemination, including the infrastructure necessary to secure optimal use of this information" or "the art, science and technology dealing with the acquisition, storage, processing, production, presentation and dissemination of geoinformation" (Oledzki, 2004).

5.3 GIS COMPONENTS

5.3.1 HARDWARE

The hardware platform upon which GIS software operates varies, encompassing centralized computer servers as well as desktop computers configured for standalone or networked environments (Figure 5.2).

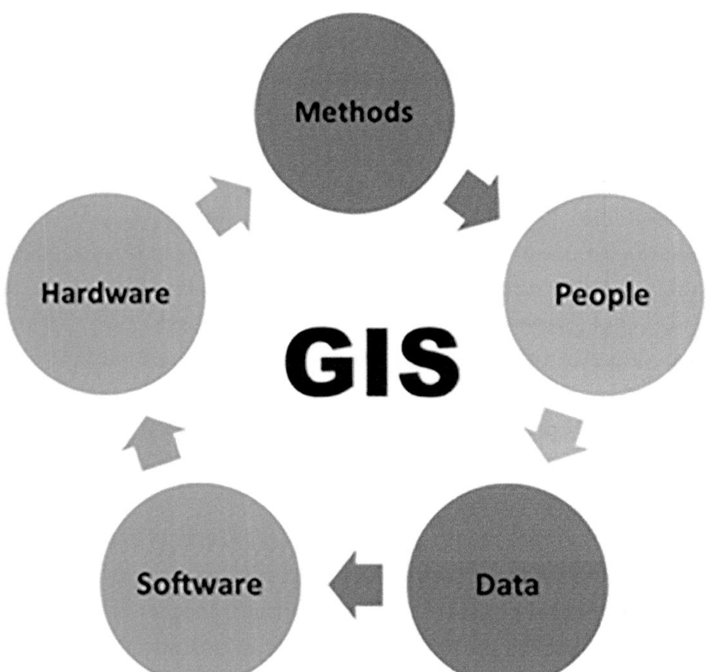

FIGURE 5.2 Components of GIS.

5.3.2 SOFTWARE

GIS software makes available the functions and tools to store, examine, and show geographic information. The main software components are as following:

- Input and manipulation tools for geographic information
- Contains a database management system
- Contains tools that support geographic analysis, query, and visualization (Figure 5.3)
- Access tools for the graphical user interface

FIGURE 5.3 Software components of GIS.

5.3.3 DATA

Data serves as the vital foundation of a GIS. It includes geographic information and associated tabular data, which can be obtained either internally or through commercial data vendors. GIS systems seamlessly integrate spatial data with other datasets, often leveraging database management systems, a common organizational tool for spatial data organization and management.

5.3.4 PEOPLE

GIS technology's effectiveness is constrained without the involvement of individuals who oversee its management and formulate strategies for its application in addressing real-world challenges. Users of GIS encompass a spectrum, including technical experts responsible for system design and upkeep, as well as those who utilize it to facilitate their operational tasks.

5.3.5 METHODS

A GIS achieves operational success based on a meticulously crafted plan and specific business regulations, encompassing models and operational protocols tailored to the distinct needs of each entity. Essentially, general-purpose GIS systems execute five fundamental processes or tasks.

5.3.6 GIS SOFTWARE: COMMERCIAL VERSUS OPEN ACCESS/ FREE

GIS software represents a bottleneck in the GIS industry, with a significant portion of resources, approximately 50% annually, allocated toward its procurement and maintenance. This substantial investment often deters users from transitioning from conventional methods to GIS. However, there has been a recent shift in the landscape of GIS software usage. Numerous new open-source and free software options have emerged in the market. Free software is readily accessible via the internet but lacks the capability for modification or updating without access to the program's code. In contrast, open-source software not only comes at no cost but also grants users full access to the program's code, enabling them to modify and update it according to their specific requirements.

5.4 COMPONENTS OF PRECISION FARMING

5.4.1 DATA MANAGEMENT SYSTEMS

Data management systems play a crucial role in collecting, storing, and analyzing data obtained from various technologies such as Global Positioning System (GPS), GIS, remote sensing, among others. This data is then utilized to inform decision-making processes related to agricultural operations, ultimately enhancing the efficiency and productivity of farms.

5.4.2 AUTOMATED MACHINERY

Automation in machinery, such as auto-driving tractors and drones, is used to complete tasks such as planting, spraying, and harvesting. This technology allows farmers to save time and reduce labor costs while also refining the accuracy and efficiency of farming processes. Sensors like Lidar, Radar, Cameras (color space, Stereo, NDVI, IR), ultrasound, thermal imaging, multi and hyper-spectral readings, and yield monitors are used in tractors for computing crop properties in orchards.

5.4.3 VARIABLE RATE TECHNOLOGY

Variable rate technology (VRT) enables farmers to apply inputs like fertilizers and pesticides at different rates throughout their fields. This precision application is informed by data collected through GPS, GIS, and remote sensing technologies, allowing farmers to apply inputs precisely where they are most needed, optimizing resource utilization and minimizing waste.

5.4.4 REMOTE SENSING

Remote sensing includes the use of satellites and additional aerial platforms to gather data about crops and the environment. This data could be used to monitor crop health, detect pest and disease outbreaks, and make decisions about irrigation and fertilization. A remote sensing system provides four basic components for measuring and recording data about a remote area. These components contain a power source, transmission line, target, and satellite sensor. Remote sensing has significant utility in estimating area cover mapping and classifying land use and land cover features, such as vegetation, soil, water, forests, and human activities. It can create information about an image's colour, texture, tone, pattern, alignment, and more. The near-infrared band (>700 nm) of satellite images is not sensitive to the human eye, but it is the band that is reflected at a higher rate by incoming sunlight from healthy sources of plants because this band is sensitive to leaf cell structure and leaf water content.

It is truly remarkable what innovative approaches researchers can employ with satellite imagery and remote sensing technology. Some of these include:

- Crop insurance
- Soil moisture
- Crop stands
- Yield estimation
- Detect the presence of pests and diseases

5.4.5 GEOGRAPHIC INFORMATION SYSTEM

GIS is utilized to generate maps depicting the diversity in soil type, topography, and other attributes of the field. This data aids in devising a tailored plan for each field, considering its distinct conditions.

5.4.6 GLOBAL POSITIONING SYSTEM

GPS is used in horticulture crops to accurately locate the position of the farming equipment and the crops in the field (Figure 5.4). This assists farmers in efficiently navigating the field and applying inputs, such as fertilizers and pesticides, precisely where they are required.

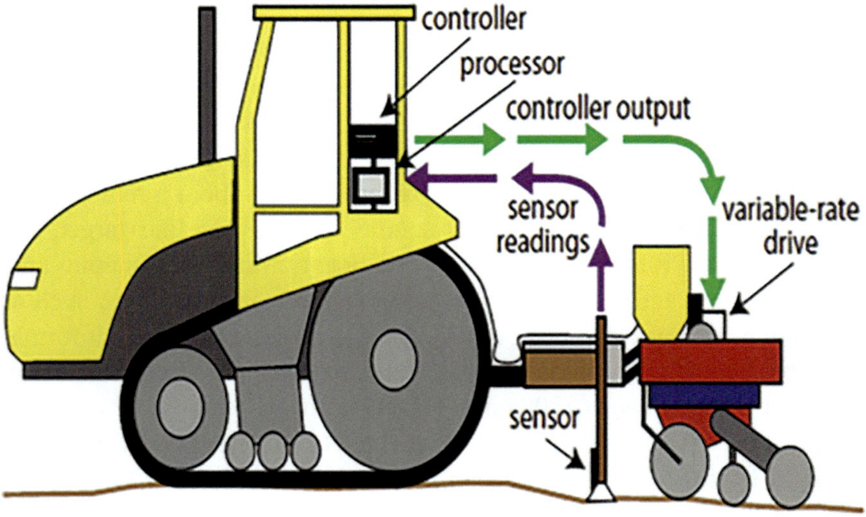

FIGURE 5.4 Device of global positioning system (GPS).

Source: Adapted from Mohan et al. (2021). https://creativecommons.org/licenses/by/4.0/

5.4.6.1 POSITION CONCEPTS OF GPS

- Latitude
- Longitude
- Altitude

5.4.6.2 LATITUDE AND LONGITUDE

Latitude and longitude are spherical coordinates on the surface of the Earth. Latitude denotes the angular distance north or south of the equator, while longitude represents the angular distance east or west of the prime meridian, which passes through Greenwich, England. GPS systems utilize latitudes and longitudes to pinpoint locations accurately.

5.4.6.3 COMPONENTS OF THE GPS

1. Space Segment
2. Control Segment
3. User Segment

5.4.6.3.1 Space Segment

- 24 GPS space vehicles.
- Satellites orbit the globe in 12 h.
- 6 orbital planes inclined at 55° with the equator.
- This constellation makes available 5–8 SVs from any point on the Earth.

5.4.6.3.2 Control Segment

- The control segment involves five stations.
- They calculate the distances of the overhead satellites every 1.5 sec and send the revised data to Master control.
- Here, the satellite orbit, clock routine, and health of the satellite are estimated, and whether repositioning is essential is determined.
- This gathered information is sent to the three uplink stations.

The tasks of the Control Segment are as follows:

- To monitor and control the satellite system regularly.
- To forecast the satellite ephemerides and the behavior of satellite clocks.
- To update intermittently the navigation message for the satellite.

5.4.6.3.3 User Segment

It consists of receivers that decode the signals from the satellites. The receiver accomplishes the following tasks:

- Selecting one or more than one satellites
- Acquiring GPS signals
- Computing and tracking
- Recovering navigation data
- Precision oscillator
- Power supply
- User interface, command, and display panel
- Memory data storage

5.5 ROLE OF GPS IN HORTICULTURE

GPS and GIS technologies are extensively employed in precision agriculture for various tasks such as farm planning, field mapping, soil sampling, crop scouting, and yield mapping (Figure 5.5). Additionally, GPS technology enables tractor guidance, enabling farmers to maneuver tractors and equipment even in conditions of low visibility.

FIGURE 5.5 Role of GPS in horticulture.

Precision farming, also well-known as precision agriculture or site-specific farming, involves using advanced technology to manage crops more precisely and efficiently. It encompasses gathering and analyzing data to make informed decisions regarding planting, irrigation, fertilization, pest control, and other agricultural activities. The primary components of precision farming include:

5.6 GEOGRAPHIC INFORMATION SYSTEM (GIS)

GIS is an influential tool used in the modern era of horticulture for precision farming. It is a computer-based technology that uses geospatial data to analyze and visualize various land-use and management phenomena. GIS comprises various components, including hardware, software, data, and people. The hardware component of GIS includes computers, scanners, GPS receivers, and printers, among other devices. The software component includes various GIS software, such as ArcGIS, QGIS, and GRASS GIS (Table 5.1). Data is another essential component of GIS. It includes various

TABLE 5.1 List of GIS Software Available Commercially/Freely to the User Community

S. No.	Software	Functionality/Remarks
		Commercial software/proprietary software
1.	ArcGIS	Core modules, market leader but high cost, many more to be bought for other applications
2.	Geomedia	Core modules of GIS support education and research institutions
3.	MapInfo	Moderate cost
4.	AutoCAD	Map better input and database creation facility
5.	J T Maps (India)	Quite economical and works in a vector model
		Open source
6.	GRASS GIS	Satellite Data Analysis & GIS (http://grass.itc.it/)
7.	Quantum GIS	Desktop GIS supports all OS (http://qgis.org/)
8.	ILWIS	Satellite Data Analysis & GIS (www.itc.nl)
9.	JUMP	Read shp and gml format, display facility and support for wms and wfs, limitations of working with large data files (http://jump-project.org/)
10.	Post GIS	With Spatial extensions for the open source. PostgreSQL database, letting geospatial probes. http://postgis.refractions.net/
11.	Map server	Web server GIS software (http://mapserver.gis.umn.edu/)

data types, including geographical, topographical, soil, climate, and vegetation. The accuracy and quality of the statistics used in GIS directly impact the effectiveness of the analysis and decision-making processes. Individuals constitute the ultimate component of GIS. This encompasses GIS professionals tasked with designing, implementing, and overseeing GIS systems. They also include farmers and other stakeholders who use GIS to make informed assessments about crop management and land use.

In hi-tech horticulture, GIS is used to gather data about the land, such as soil types, topography, and microclimate, to identify the best areas for crop production. The data is analyzed using a variety of tools to generate crop suitability maps, aiding farmers in making informed decisions regarding crop selection, fertilization, irrigation, and pest management. GIS also monitors crop growth and yield, allowing farmers to make timely decisions regarding crop management practices. Additionally, GIS helps monitor environmental factors, such as water quality and soil erosion, helping farmers manage the land sustainably. To summarize, GIS serves as a vital element in advanced horticulture, furnishing farmers with a robust instrument to analyze, visualize, and administer geospatial data for precision agriculture.

5.7 GIS PREDICTION OF RISK MAPS

- Metrological data
- Digital elevation model
- Longitude
- Latitude
- Temperature
- Relative humidity
- Precipitation
- Simulation model
- Risk assessment

5.8 NEED FOR GIS IN HORTICULTURE

- The user or operator can access and select data by area or theme to combine one data set with another, examine spatial characteristics of statistics, search for particular features, update rapidly and economically, and measure alternatives.

- Allows the user to know geographic information easily without having to go through large volumes of puzzling data in tabular format.
- Envisioning the geography of a particular location is no doubt easier than examining raw data.
- The potential and substantial profits of using GIS make it a necessary tool, making the work relaxed and more productive.
- Decreasing sets of manual maps and associated storage costs present an opportunity.
- Enhanced speed and breadth of access to geographic information.
- Enhanced analytical capabilities, such as assessing areas, distances, and patterns of horticultural crops.
- Streamlined communication of information to public officials and stakeholders.
- Enhancement of horticultural crop quality.
- Increased productivity in horticultural crops and dissemination of public information.
- Improved efficiency in map updates.
- Capability to monitor growth and development trends over time.
- Enhanced ability to aggregate data for specific subregions.

5.9 FUNCTIONS OF GIS IN AGRICULTURE

The functions of GIS define the phases that have to be taken to implement a GIS. These steps should be followed to get a systematic and effective system. The steps involved are discussed in the following subsections:

5.9.1 DATA CAPTURE

- GIS data often originates from diverse origins. Data sources commonly involve manual digitization and scanning of aerial photographs, paper maps, and existing digital datasets.
- Remote-sensing satellite imagery and GPS technology offer promising avenues for data input into GIS systems.
- Digitization, a conversion process that transforms paper maps into numerical data suitable for computer storage, occurs in this phase.
- Digitization simplifies map data into points, lines, or cells that can be stored in GIS computer systems.
- There are two primary methods of digitization: manual digitizing and scanning.

5.9.2 DATA COMPILATION

- Following the digitization of map features, the user concludes the compilation phase by associating all spatial features with their corresponding attributes and rectifying any errors introduced during the data conversion process.
- The outcome of the compilation process is a collection of digital files, each faithfully depicting the spatial and attribute data from the original map documents.
- These digital files include geographic coordinates for spatial entities (such as points, lines, polygons, and cells) representing the mapped features.

5.9.3 DATA STORAGE

- Following digital compilation, GIS stores digital map files on magnetic or other digital storage media.
- Data storage relies on a standard data model that transforms map data into a digital format.
- The two predominant types of data models are Raster and Vector. These models streamline map data into simplified forms suitable for efficient computer storage.

5.9.4 APPLICATION OF GIS IN HORTICULTURAL CROP

- GIS serves as a tool to create visual representations of data and conduct spatial analyses, aiding in informed decision-making.
- GIS finds practical application in traditional cartography, such as mapping fire hydrant locations along roads or delineating boundaries, like different fields of horticultural crops on a farm.
- However, the true strength of GIS lies in its capability to analyze multiple data layers or variables.
- Examples in agriculture might include maps displaying farm injury rates by county or crop acreage lost to floods by tax map parcel.
- In horticulture, more intricate spatial analyses might examine variables like soil type, wind direction, rainfall, slope, aspect, topography, or elevation to support crop management, site suitability

assessment, and drainage planning, while also mitigating risks such as flood, drought, erosion, and disease.

- GIS aids farmers in adapting to diverse variables, monitoring individual crop health, estimating yields, and optimizing production.
- Remote sensing, employing satellites, drones, and manned aircraft, gathers surface information from high altitudes.
- Landsat 8, a collaboration between USGS and NASA, orbits the Earth every 16 days, capturing visible light spectrum data useful for assessing factors like plant health, nutrient status, pest infestation, and moisture levels in horticultural crops.
- Additionally, it detects thermal infrared radiation (TIR) beyond human vision, with varying intensity wavelengths emitted by different vegetation types and landscapes.
- Recorded data is converted to digital imagery, useful for tasks such as managing irrigation water or detecting plant diseases.
- Specific applications include evaluating fruit or crop maturity.
- Aerial and TIR imagery, collected through remote sensing, are commonly used GIS layers, with Landsat 8 data freely accessible to the public.
- Higher resolution imagery from low-altitude aircraft, collected over longer cycles, supplements this.
- Leveraging data from remote sensors and farm machinery-mounted sensors, farmers enhance decision-making, considering factors like past yields, terrain, soil properties, and real-time crop quality metrics (Figure 5.6).
- GPS-equipped combine harvesters precisely measure crop yields, water content, and chlorophyll levels in real-time at harvest locations.

5.9.5 ADVANTAGES OF GIS

- Examining both the spatial and thematic aspects of data comprehensively.
- Emphasizing the geographic dimensions of a research inquiry.
- Dealing with substantial amounts of data.
- Incorporating data from diverse and disparate origins.
- Enabling numerous visualization techniques.

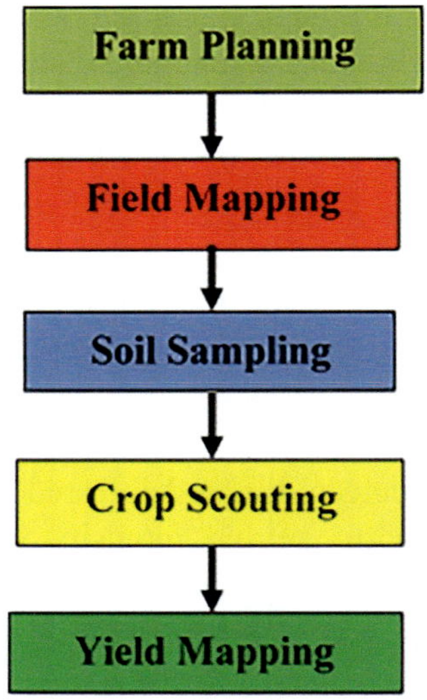

FIGURE 5.6 Application of GPS in horticultural crops.

5.9.6 DISADVANTAGES OF GIS

- Data are costly.
- The learning curve on GIS software could be lengthy.
- Shows spatial connections but does not make available utter solutions.
- Origins in the Earth sciences and computer science. Solutions may not be suitable for humanities research.

5.10 DIFFERENTIAL GEO-POSITIONING SYSTEM (DGPS)

Differential global positioning system (DGPS) represents a satellite-based navigation framework renowned for providing exceptionally precise location data tailored for precision agriculture applications (Figure 5.7). DGPS mitigates inaccuracies inherent in GPS signals due to atmospheric and other environmental factors, thereby enhancing the reliability of GPS data. This system relies on a network of base stations to transmit corrective signals to receivers mounted on farming equipment or vehicles. These receivers juxtapose correction signals with incoming GPS signals to compute more accurate positions. Leveraging this heightened precision, DGPS guides various precision agriculture tasks such as seeding, spraying, and harvesting. The accuracy of DGPS hinges on factors like the density of base stations in the vicinity and the distance between the receiver and the nearest base station. Typically, DGPS can achieve positioning accuracies within the range of 1–2 cm, far surpassing those achievable with standalone GPS. The adoption of DGPS in precision agriculture yields numerous benefits, including enhanced mapping accuracy, streamlined field operations, minimized input wastage, and improved crop yield and quality. Nonetheless, the cost of DGPS equipment remains a notable deterrent, and its efficacy might be compromised in heavily forested areas or regions with significant signal obstructions.

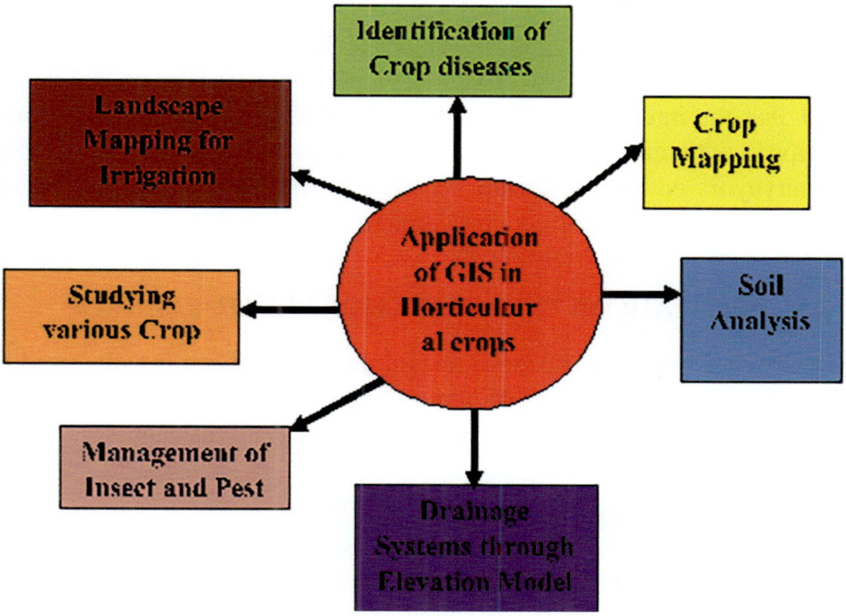

FIGURE 5.7 Use of DGPS in fertilizer application.

The pursuit of precision and accuracy remains a perpetual aspiration within the dynamic domain of GISs. DGPS emerges as a pivotal tool that has revolutionized mapping and surveying methodologies. This chapter delves into the intricacies of DGPS surveys, shedding light on its applications, underlying principles, and transformative impact on the GIS sector. DGPS, an enhancement to the GPS, aims to bolster the accuracy of location data. Despite the commendable accuracy of GPS, its susceptibility to various errors—ranging from atmospheric disturbances to satellite clock discrepancies and signal reflections—necessitates refinement. DGPS addresses these issues by incorporating a reference station at a known location to compute differentials and refine GPS-derived positions.

The technical framework of DGPS encompasses a control hub, an array of ground-based reference stations, and mobile GPS receivers. Each reference station comprises a GPS receiver and a reliable power source, strategically distributed across designated regions. These stations continuously monitor signals from overhead GPS satellites, discerning disparities between their known precise positions and their observed locations. A pivotal element of DGPS is the control hub, responsible for computing correction factors, collating correction data from reference stations, and disseminating results to mobile GPS receivers deployed in the field. Corrected information is relayed to receivers via various mediums, including the internet, satellite communication, and radio transmissions. Equipped with this corrected data, mobile GPS receivers, whether wielded by surveyors or integrated into vehicles, substantially enhance location accuracy by applying correction factors to raw GPS data.

5.11 APPLICATIONS OF DGPS IN HORTICULTURAL CROP

The DGPS can provide accurate spatial data; it has been broadly used in many different areas of the GIS sector. The following are some important areas in which DGPS surveying has had a significant influence:

5.11.1 LAND SURVEYING

DGPS has emerged as a vital instrument for land surveyors, facilitating the acquisition of accurate data pertaining to topography, land characteristics, and boundaries. This technology plays a pivotal role in enabling effective mapping for various purposes such as urban planning, property assessment, and cadastral surveys.

5.11.2 PRECISION AGRICULTURE

DGPS plays a crucial role in precision agriculture within the farming sector. Utilizing equipment like tractors equipped with DGPS capabilities, farmers can precisely carry out tasks such as seed planting, fertilizer application, and irrigation control. This targeted approach enhances overall agricultural productivity, reduces environmental footprint, and optimizes resource utilization.

5.11.3 ENVIRONMENTAL SURVEILLANCE

Environmental monitoring and research greatly benefit from DGPS surveys. Scientists leverage DGPS technology to map vegetation patterns, monitor ecosystem shifts, and observe wildlife habitats. The precise spatial data obtained through DGPS aids in the protection and management of natural resources.

5.11.4 INFRASTRUCTURE DEVELOPMENT

DGPS surveying proves valuable for infrastructure endeavors such as utility installations and road construction projects. Accurate location data ensures precise execution of construction tasks, minimizing errors and the need for rework. This aspect holds particular significance in urban planning, where new infrastructure must seamlessly integrate with existing regulations and frameworks.

5.11.5 EMERGENCY RESPONSE AND DISASTER MANAGEMENT

Disaster management and emergency response heavily rely on DGPS technology. First responders strategize evacuation routes, coordinate rescue efforts, and navigate disaster-stricken areas with precision using DGPS-equipped devices. The invaluable accuracy provided by DGPS is indispensable in urgent and critical situations.

5.11.6 ADVANTAGES OF DGPS SURVEY TECHNIQUES

5.11.6.1 SUBMETER ACCURACY

DGPS surpasses standard GPS with its submeter accuracy, achieved by rectifying inaccuracies in satellite signals caused by orbital variations, clock disparities, and atmospheric conditions. Real-time correction signals from ground-based reference stations enable DGPS receivers to attain submeter accuracy, revolutionizing GIS applications reliant on precise location data.

5.11.6.2 REMOVAL OF SYSTEMATIC MISTAKES

The capability of DGPS to mitigate the systematic inaccuracies present in traditional GPS measurements stands out as one of its primary advantages. Such errors, encompassing issues like ephemeris and selective availability, can introduce inaccuracies in location data. DGPS employs reference stations to consistently monitor and rectify these errors, ensuring the elimination of systematic issues in the resultant output.

5.11.6.3 ENHANCED PRODUCTIVITY AND EFFICIENCY

DGPS surveying enhances efficiency by reducing the time required for data collection and analysis. With increased precision, GIS specialists can gather more reliable spatial data in shorter durations, accelerating project timelines. This boost in productivity holds particular importance for time-critical applications such as precision agriculture and emergency response planning.

5.11.6.4 BETTER ASSET ADMINISTRATION

Within the GIS framework, DGPS technology is indispensable for asset management. Accurate mapping and monitoring of utilities, infrastructure, and land parcels facilitate efficient resource allocation and maintenance scheduling. This precision is especially critical for industries prioritizing asset longevity and infrastructural integrity.

5.11.6.5 ENHANCED GEOREFERENCING

DGPS excels in providing reliable spatial reference points, which are crucial for GIS applications. Accurate georeferencing is essential, especially in tasks such as remote sensing, where precise alignment of satellite imagery and aerial surveys with ground features is necessary. DGPS significantly reduces georeferencing errors, resulting in more precise and informative spatial analytics.

5.11.6.6 PRECISION FARMING

DGPS technology has revolutionized precision farming, fundamentally transforming the agricultural sector. Through precise field boundary mapping, crop health surveillance, and optimized fertilizer and pesticide application in horticultural crops, DGPS enhances crop yields while mitigating environmental impacts. DGPS's outstanding accuracy simplifies the creation of precise prescription maps for variable rate applications.

5.11.6.7 ENHANCED PRODUCTIVITY AND EFFICIENCY

DGPS surveying enhances efficiency by reducing the time required for data gathering and analysis. With enhanced precision, GIS specialists can collect more reliable spatial data in shorter durations, expediting project completion, particularly in agriculture.

5.11.7 AUTOMATIC VARIABLE RATE FERTILIZER APPLICATOR

Fertilizers account for 50% of the total agricultural expenditure. The technology thus variably and robotically applies fertilizer as per requirement and highly cuts down the fertilizer wastage by up to 35%, as in conventional methods (Mohan et al., 2021).

- Traditional application methods often lead to significant wastage, such as through broadcasting.
- Accurately estimating soil nutrient levels in real-time has posed a significant challenge for precise application.
- VRT utilizing a map-based strategy, sensors, and GPS has been developed for agricultural purposes (Figure 5.8).
- Machinery assesses fertilizer needs at specific locations, applying precise amounts through mechatronic actuation controlled by GPS and a controller.

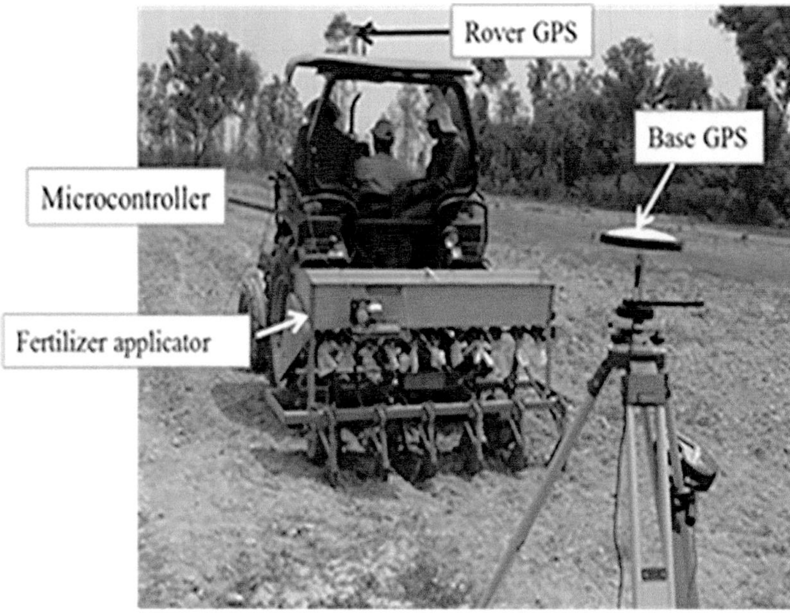

FIGURE 5.8 An illustration of a sensor based system for varying crop inputs.

5.11.8 BLOCK DIAGRAM OF UNIFORM SEED RATE APPLICATION SYSTEM

- Utilizing radar sensors to determine operational velocity.
- The operator inputs the desired seed spacing and the number of seed cells per metering plate according to the planting requirements.
- Using the operational speed and operator-provided data, the microcontroller, coupled with a motor controller, adjusts the DC motor to ensure consistent seed spacing.
- Infrared sensors identify any missing seeds.
- Adjustments to the DC motor are made by the microcontroller in response to seed discrepancies, ensuring uniformity (Figure 5.9).

5.12 NEW DEVELOPMENTS IN ANTENNA AND RECEIVER TECHNOLOGY

As advancements continue in these fields, DGPS devices are expected to become more compact and efficient in their operations. This evolution

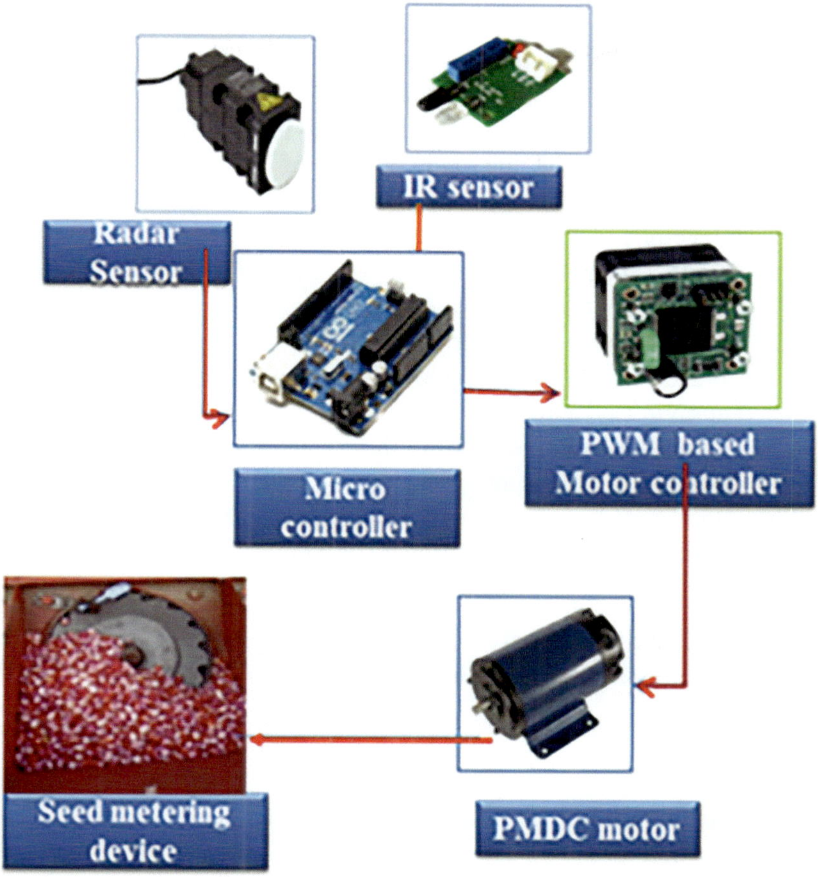

FIGURE 5.9 Block diagram of uniform seed rate application system.

may lead to the integration of DGPS capabilities into a broader array of consumer products, fostering greater acceptance. Spearheading precise geospatial data collection, DGPS surveying offers sectors reliant on accurate location data a transformative solution. The adoption of DGPS technology is anticipated to proliferate as the GIS industry evolves, reshaping approaches to environmental monitoring, mapping, and surveying. With ongoing technological progress and problem-solving efforts, DGPS is poised to remain a cornerstone in the pursuit of utmost accuracy within GISs.

5.12.1 AUTOMATIC SPRAYING SYSTEMS USING ULTRASONIC SENSORS FOR ORCHARDS

- Traditional methods of eradicating infections in orchard crops have posed risks due to the overuse of chemicals.
- An automatic sprayer based on an ultrasonic sensing system was developed for an exact use.
- Robust ultrasonic sensors identify the existence of a tree canopy, and the controller activates the application system.
- The automated sprayer operates solely when the tree canopy is present, remaining inactive otherwise, a departure from continuous spraying systems.
- Extensive involvement of the mechatronic system makes it a durable and efficient applicator.
- This automatic sprayer reduces over 30% of chemical spray compared with the traditional sprayers.

Ultrasonic sensors identify individual trees, causing the spraying process to pause temporarily in the spaces between trees. These outcomes in cost and time savings, as the holding tank requires less frequent refilling.

5.12.2 ADVANTAGES

- Reduction of cost and time by less usage of pesticides in horticulture.
- Durable and robust sensor solution.
- Extensive capabilities for precise identification of objects.
- Easy installation and parameter adjustment within orchards.

KEYWORDS

- **geographic information systems**
- **differential global positioning system**
- **horticultural crop cultivation**
- **data management systems**
- **variable rate technology**

REFERENCES

Demers, M.N. (1997). *Fundamentals of Geographic Information Systems.* John Wiley & Sons.

Jachimski, J. (2001). *Dyskusjato Matters of Terminology in Geoinformatics* [in Polish]. Kraków: Geoinformatyka Polonica, No. 3, pp. 57–79.

Mohan, S.S., Ajay, A., & Jayan, P.R. (2021). GPS and sensor based technologies in variable rate fertilizer application. *International Journal of Agriculture Environment and Biotechnology,* 14(1), 21–27

Oledzki, J.R. (2004). An integrated spatial research tool geoinformatics—An integrated spatial research tool. In: *Miscellania Geographica.* J.M. Kwiatowska (English translation), vol. 11, pp. 323–331. www.wgsr.uw.edu. pl/pub/uploads/mcg04/35oledzki. pdf.

CHAPTER 6

Interrelation of Nanotechnology With Smart Horticulture Production

PRATAP KADAM[1], BHARAT TAYNATH[1], KHALANGRE ATUL[2,*], AHAMED SHABEER T. P.[1], and HRUTIK SHEVKAR[3]

[1]*National Referral Laboratory, ICAR-National Research Centre for Grapes, Pune, India*

[2]*Department of Horticulture, Lovely Professional University, Phagwara, Punjab, India*

[3]*Department of Agriculture-Food Science and Technology (DISTAL), University of Bologna, Bologna, Italy*

*Corresponding author

ABSTRACT

Nanotechnology's integration into smart horticultural practices marks a pivotal advancement in agriculture, transforming traditional methods into efficient, precise, and sustainable processes. This comprehensive exploration delves into the intricate interplay between nanotechnology and smart horticulture, elucidating its applications across diverse domains. In the realm of pest and disease management, nanotechnology has redefined crop protection. Nanoencapsulated pesticides and fertilizers have emerged as potent tools, delivering targeted solutions while minimizing environmental impact. Concurrently, nanoscale sensors and biosensors offer real-time insights, empowering farmers with data-driven strategies for crop and soil management. Smart irrigation systems, driven by nanosensors, optimize water usage, addressing a critical agricultural challenge. Nanomaterial coatings have elevated the durability of agricultural equipment, ensuring

Smart Horticulture Production: Integrating Advanced Technologies.
Rajni Rajan and Duniya Ram Singh (Eds.)

DOI: 10.1201/9781998511679-6

sustainability and reducing waste. Ethical considerations, including equitable access and responsible innovation, underscore the ethical dimensions of nanotechnology integration. This chapter showcases practical applications through compelling two case studies. From nanoencapsulated pesticides enhancing apple orchard management to nanocomposite coatings extending fruit shelf life, these examples illustrate the tangible benefits of nanotechnology. Looking forward, this chapter explores future trajectories, including innovations in nanomaterials, precision agriculture, environmental sustainability, and data-driven farming.

6.1 INTRODUCTION

Nanotechnology, the discipline focused on manipulating matter at the molecular or atomic scale, has instigated a transformative wave across diverse industries. This has fundamentally reshaped the boundaries of scientific understanding and spurred technological innovation (Roco, 2003). At the heart of nanotechnology lies the manipulation of materials and their properties at the nanoscale, typically at dimensions less than 100 nm (Feynman, 1960). In order to establish context, it is important to note that a nanometer is roughly 100,000 times tinier than the diameter of a human hair. At this scale, we delve into the domain of individual molecules and atoms, where materials exhibit unique properties that differ significantly from their characteristics in larger quantities.

The inception of nanotechnology can be traced back to the visionary physicist Richard Feynman, who, in his famous 1959 lecture, said, "There's Plenty of Room at the Bottom," speculated about ability to manipulate individual atoms and molecules (Dresselhaus et al., 2001). However, it was not until the advent of powerful microscopes and advanced experimental techniques in the latter part of the 20th century that scientists could delve into this minuscule world with precision. Nanotechnology's allure lies in the extraordinary phenomena exhibited by materials at this scale. Quantum effects and surface area-to-volume ratios become paramount, leading to materials with properties vastly different from their macroscale equivalents. As an illustration, gold nanoparticles (NPs) may exhibit colors such as red or purple, a phenomenon referred to as surface plasmon resonance. This characteristic holds significant importance in a variety of biological and chemical applications, as highlighted by Haiss et al. (2007).

Meanwhile, carbon nanotubes (CNTs), which are cylindrical formations composed of carbon atoms, showcase exceptional strength and electrical conductivity owing to their distinct atomic configuration. These attributes make CNTs a focal point in research within the fields of materials science and electronics, as emphasized by Baughman et al. (2002).

In the field of horticulture, the integration of nanotechnology has introduced a revolutionary era characterized by heightened precision and efficiency. Nanoscale materials, such as NPs and nanocomposites, have been engineered to enhance soil quality, improve nutrient delivery, and even deliver targeted treatments to plants(Khodakovskaya et al., 2009). These applications harness the inherent characteristics of nanomaterials, like their large surface area for interaction and their ability to cross biological barriers, allowing for precise and controlled interventions at the cellular and molecular levels. The multifaceted nature of nanotechnology in horticulture extends beyond traditional practices. It encompasses nanoscale sensors that can detect minute changes in environmental conditions, nanocarriers that deliver nutrients precisely when and where they are needed, and nanodevices that enable real-time evaluation of plant health (Cui et al., 2017). With such capabilities, nanotechnology stands as a beacon of hope in addressing the problems posed by an increasing global population, climate change, diminishing natural resources.

As we delve deeper into this chapter, we will unravel the intricacies of nanotechnology's applications in horticulture. From enhancing fertilizers to developing nanosensors for smart agriculture, this chapter will explore the synergistic relationship between nanotechnology and smart horticultural production, paving the way for a sustainable and technologically advanced agricultural future.

6.2 SMART HORTICULTURAL PRODUCTION: AN OVERVIEW

The evolution of horticulture into the realm of smart horticultural production marks a monumental shift in the way we cultivate crops. This transformation is not just about adopting cutting-edge technologies but represents a holistic reimagining of traditional farming practices through the integration of intelligent systems. At its core, smart horticultural production is driven by the amalgamation of various technologies, including sensors, data analytics, automation, and artificial intelligence (AI).

These components synergize to create an intricate web of interconnected processes, revolutionizing the way we approach agriculture (Recha & Kimathi, 2019). Sensors serve as the sensory organs of smart horticultural systems, capturing data from the environment and the crops themselves. Soil moisture sensors, for instance, measure the hydration levels in the soil, providing crucial insights into irrigation needs. These sensors are often combined with weather stations that collect meteorological data, enabling farmers to correlate weather patterns with plant growth and make informed decisions about watering and protection against adverse weather conditions(Lu & Wang, 2019).

Data analytics plays a pivotal role in translating raw data into actionable insights. Advanced algorithms process data from different sources, such as satellites, sensors, and drones, to identify patterns and trends. By analyzing these data, farmers gain valuable information about crop health, pest infestations, and resource utilization. Predictive models, powered by data analytics, enable proactive measures against potential threats, significantly enhancing the resilience of horticultural systems (Khot et al., 2019). Automation is the backbone of smart horticultural production, optimizing labor and resource use. Automated systems control irrigation pumps, fertilization processes, and even harvesting equipment. Robotics, equipped with computer vision systems, can precisely plant seeds, apply fertilizers, and harvest crops. These technologies not only reduce the physical strain on farmers but also ensure precision in every aspect of cultivation, leading to higher yields and resource efficiency (Khaliq et al., 2018).

AI adds a cognitive layer to smart horticultural systems. Machine learning algorithms, a subset of AI, process vast datasets, learning and adapting to new information over time. In agriculture, AI-driven applications range from predicting disease outbreaks based on environmental conditions to optimizing the composition of fertilizers for specific soil types. AI also enables the integration of diverse data sources, providing a comprehensive view of the agricultural ecosystem and guiding farmers in decision-making processes (Duro & Khot, 2016). This convergence of technologies in smart horticultural production transcends the traditional limitations of farming. It empowers farmers with real-time, data-driven insights, enabling precise interventions that enhance productivity while minimizing environmental impact. Moreover, it fosters a more sustainable agricultural future by promoting optimal resource utilization, reducing waste, and mitigating the risks associated with climate change.

6.3 NANOTECHNOLOGY IN HORTICULTURE

6.3.1 NANOMATERIAL-ENHANCED FERTILIZERS

Nanomaterial-enhanced fertilizers represent a groundbreaking advancement in agriculture, offering very specific and targeted delivery of nutrients in system. These fertilizers utilize nanotechnology to encapsulate nutrients, protecting them from leaching and environmental factors, ensuring their release precisely where and when the plant needs them most. Traditionally, fertilizers often suffer from inefficiencies. Nutrients can leach into the soil, becoming inaccessible to plants, or they might be applied in excess, leading to environmental pollution. Nanomaterial-enhanced fertilizers address these issues by encapsulating nutrients within nano-sized carriers (He et al., 2017). These carriers, often made from biodegradable materials or polymers, protect the nutrients during transport and gradually release them based on environmental cues, such as soil moisture or pH levels (Wang et al., 2019a). Moreover, the nanoscale size of these carriers facilitates their interaction with plant roots at the cellular level. Nutrients can be absorbed more efficiently, leading to improved crop yields while reducing the overall fertilizer usage. This not only benefits farmers economically but also mitigates the environmental impact associated with excessive fertilizer application.

Several studies have demonstrated the effectiveness of nanomaterial-enhanced fertilizers. For instance, nanoencapsulated nitrogen and phosphorus fertilizers have shown increased nutrient use efficiency and reduced environmental contamination (Hussain et al., 2016). Additionally, the slow and controlled release of nutrients by these nanomaterials promotes healthier plant growth and root development, ultimately enhancing overall crop productivity (Rico et al., 2011). The integration of nanomaterial-enhanced fertilizers into agricultural practices signifies a significant step toward sustainable and precision agriculture (PA). By optimizing nutrient delivery and minimizing waste, these nanotechnology-driven solutions contribute to the global efforts for food security and environmental conservation (Figure 6.1).

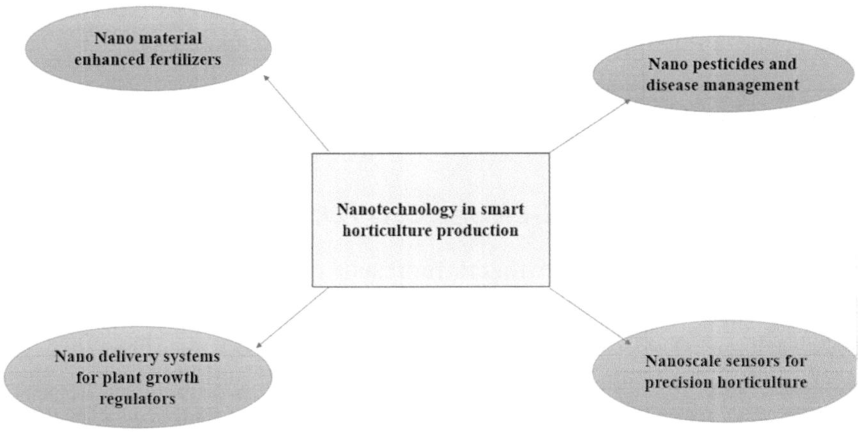

FIGURE 6.1 Nanotechnology in smart horticulture production.

6.3.2 NANOPESTICIDES AND DISEASE MANAGEMENT

Nanotechnology has revolutionized pest and disease management in agriculture through the development of nanopesticides. These innovative formulations leverage the unique properties of nanomaterials to enhance the efficacy of pesticides while minimizing environmental impact and health risks. Traditional pesticides often face challenges such as rapid degradation, uneven distribution, and harm to nontarget organisms. Nanopesticides address these issues by encapsulating active pesticide ingredients within nanocarriers, protecting them from degradation and ensuring controlled release (Kah et al., 2018). The nanoscale size of these carriers allows for better adhesion to plant surfaces, enhancing the coverage and retention of the pesticide on leaves and stems.

Moreover, nanopesticides can be engineered for targeted delivery, where they are designed to respond to specific cues such as pest presence or environmental conditions (Kumar et al., 2017). This targeted release ensures that pesticides are deployed only when and where they are needed, minimizing their dispersion into the environment. By reducing the quantity of pesticides needed and limiting their exposure to nontarget organisms, nanopesticides contribute significantly to sustainable pest management practices. Nanomaterials also offer a platform for the development of biopesticides, which are derived from natural sources. Biopesticides, when combined with nanotechnology, exhibit enhanced stability and prolonged efficacy. Nanocarriers protect biopesticides from degradation by ultraviolet (UV) light and enzymes, ensuring their effectiveness over an extended period (Zhang et al., 2019). This combination of biological control methods with nanotechnology represents a promising approach for sustainable and eco-friendly pest management in agriculture.

Several studies have demonstrated the effectiveness of nanopesticides in various crops. For example, nanoencapsulation of neem oil, a popular biopesticide, has shown increased persistence and efficiency against a range of agricultural pests (Pavela & Benelli, 2016). Additionally, nanopesticides have been developed for specific crop diseases, effectively controlling pathogens while minimizing the environmental footprint (Tripathi & Yadav, 2019). The adoption of nanopesticides in agriculture marks a significant step forward, offering a potent tool for integrated pest management. By harnessing the power of nanotechnology, farmers can enhance the effectiveness of pest control measures while promoting environmental sustainability and food safety.

6.3.3 NANOSCALE DELIVERY SYSTEMS FOR PLANT GROWTH REGULATORS

Nanoscale delivery systems have revolutionized the way plant growth regulators (PGRs) are utilized in agriculture. PGRs are essential for modulating plant development and ensuring optimal growth and yield. However, their efficient delivery to plants has been a challenge. Nanotechnology offers innovative solutions by encapsulating PGRs within nanocarriers, enhancing their stability, targeted delivery, and overall effectiveness. Nanocarriers, typically made of biodegradable polymers or lipids, protect PGRs from degradation due to environmental factors like light, heat, and enzymes. Moreover, their nanoscale size allows them to penetrate plant tissues effectively, facilitating the controlled release of PGRs at the cellular level (Raliya & Tarafdar, 2013). This precise delivery ensures that the PGRs exert their effects exactly where needed, promoting specific plant responses without wastage or overuse.

Additionally, nanoscale delivery systems can be engineered to respond to specific triggers such as pH, temperature, or enzymes. This responsiveness enables on-demand release of PGRs in response to particular environmental conditions or plant developmental stages (Hussain et al., 2016). Such controlled release mechanisms enhance the efficiency of PGRs by synchronizing their actions with the plant's growth cycle. Furthermore, nanotechnology enables the combination of multiple PGRs or other active compounds within a single nanocarrier. This synergistic approach allows for the targeted regulation of multiple plant processes simultaneously, enhancing crop quality and yield (Rajput & Minkina, 2019). Studies have shown the efficacy of nanoscale delivery systems for PGRs in various crops. For example, nanoencapsulation of auxins, a class of PGRs, has been demonstrated to improve root development and enhance stress tolerance in plants. Similarly, nanocarriers delivering cytokinins, another group of PGRs, have been effective in promoting flowering and fruiting in several fruit crops (Suh et al., 2014; Raskar & Laware, 2019).

6.3.4 NANOSCALE SENSORS FOR PRECISION HORTICULTURE

Nanoscale sensors have emerged as transformative tools in the realm of precision horticulture, offering unprecedented capabilities for real-time monitoring and data acquisition. These sensors operate at the nanometer scale, enabling the detection of minuscule changes in environmental parameters and plant conditions. Their integration into horticultural practices revolutionizes decision-making processes, allowing farmers to respond promptly to the dynamic needs of crops and environmental fluctuations.

Nanoscale sensors are designed to detect specific signals, such as variations in soil moisture, nutrient levels, pH, or the presence of pathogens. Utilizing principles from fields like nanoelectronics and nanophotonic, these sensors can capture data at an incredibly fine resolution (Giraldo & Landry, 2019). For instance, nanosensors can precisely measure the moisture content of soil, helping farmers determine optimal irrigation schedules and prevent water wastage (Shrivastava et al., 2018). Similarly, they can detect subtle changes in nutrient concentrations, allowing for precise adjustments in fertilization practices to ensure optimal plant nutrition(Wang et al., 2018). Moreover, nanoscale sensors are often equipped with wireless communication capabilities, enabling them to form networks within horticultural fields. These sensor networks create a comprehensive monitoring system, often referred to as the Internet of Things (IoT) in horticulture (Hu et al., 2019). The data collected from these sensors are transmitted in real time to central databases or mobile devices, providing farmers with instant access to critical information about their crops.

Additionally, nanosensors can be engineered to detect specific plant stresses, such as the presence of diseases or pests. Through early identification of these stress factors, farmers can employ specific interventions, thereby lessening the reliance on wide-ranging treatments and mitigating environmental impact (Qiao & Zhang, 2019). Nanoscale sensors thus enhance the efficiency of pest and disease management strategies, promoting sustainable horticultural practices.

6.4 SYNERGY BETWEEN NANOTECHNOLOGY AND SMART HORTICULTURE

Smart horticulture signifies an agricultural approach that uses advanced techniques and latest tools for decision-making derived from data. Its main goals are to increase crop yields, optimize resource management, and prioritize waste reduction, improved efficiency, and enhanced profitability. In contrast, nanotechnology is a scientific and technological field focused on materials and structures at the nanoscale. Nanotechnology holds the potential to transform smart horticulture by providing innovative techniques and tools for amplifying production of crops. Despite their distinct focuses, nanotechnology and smart horticulture share interconnected elements. Nanotechnology has the potential to bring forth novel tools and materials that enhance smart horticulture practices. Nanosensors, for instance, can provide continuous, specific, and real observations of soil and plant health, facilitating correct and specific crop management. Additionally, NPs can enhance the delivery of BOTH nutrients and pesticides to plants, lowering the waste and improving the impact.

In the realm of smart horticulture, the integration of nanotechnology represents a transformative force that yields numerous advantages, including increased crop yields, waste reduction, and a decreased environmental footprint. Through the nanoscale modification of materials, unique characteristics are harnessed, offering substantial benefits over traditional farming methods. Nanosensors, for instance, facilitate real-time monitoring of soil and plant health, providing farmers with precise data on optimal water and fertilizer application, thereby enhancing crop management (Lu & Wang, 2016). Furthermore, NPs have proven to be effective carriers for pesticides and fertilizers, thereby reducing their environmental impact. Additionally, nanotechnology plays a role in enhancing agricultural materials like plant fibers and seeds. Enhancing these materials on the nanoscale renders them more resilient against pests and environmental stressors (Lu & Wang, 2016).

Nevertheless, it is essential to recognize that unlocking the complete potential of nanotechnology in smart horticulture requires a cautious approach, considering potential risks and environmental impacts on soil, water, and human health. As research progresses, the integration of nanotechnology into smart horticulture holds the promise of revolutionizing crop cultivation, fostering efficiency, sustainability, and waste reduction, all while diminishing the environmental footprint.

6.4.1 IMPROVED NUTRIENT UTILIZATION

The increasing global demand for enhanced food production requires innovative agricultural approaches to increase crop productivity while minimizing adverse environmental impacts. This escalating global demand for increased food production underscores the necessity for inventive agricultural strategies that can improve crop productivity while mitigating environmental consequences. The use of nanoscale materials, specifically nano-fertilizers (NFs), has demonstrated efficacy in enhancing the efficiency of crops to utilize the supplied nutrients and alleviating the impacts of extra application of fertilizers (Khoot et al., 2019). Nanotechnology holds the potential to significantly enhance the efficiency of delivery of nutrients to crops by encapsulating in NPs, allowing precise and controlled release (Khaliq et al., 2018). NFs are typically manufactured by encapsulating nutrients within nanoscale carriers like metal oxide NPs or polymeric nanocapsules, facilitating controlled nutrient release and reducing losses from leaching or volatilization (Duro & Khot, 2016).

6.4.2 ENHANCED PEST CONTROL

Intelligent horticulture, facilitated by nanotechnology, presents a promising alternative to conventional methods of pest control by addressing challenges related to effectiveness, environmental impact, and safety (Liu et al., 2015). This innovative technique involves the creation of materials and technologies at the nanoscale to precisely and effectively administer pesticides and other chemicals used in pest management. For example, by enhancing the solubility and stability of active compounds, nanopesticides enable targeted administration and regulated release. Because of this, less pesticide is used, reducing pollution to the environment and effects on nontarget areas (Rai et al., 2016).

The use of nanopesticides in intelligent horticulture improves pest management by altering the way pesticides are administered by real-time observation of insect populations and environmental factors (Kah et al., 2018). NPs can be made to target specific pests or plant structures in order to assure effective pesticide delivery and the development of new, potent insecticides at lower dosages (Kah et al., 2018). Furthermore, intelligent horticulture based on nanotechnology minimizes its environmental impact by enabling more precise pesticide administration and lowering the total amount released into the environment. Examples include the focused delivery of insecticides using nanocapsules, which maximizes performance while limiting environmental release, and the delivery of ribonucleic acid (RNA) interference molecules for highly targeted pest management using NPs (Qaim, 2016).

Furthermore, by developing less hazardous pesticides, intelligent horticulture based on nanotechnology promotes the safety of farmers and consumers by minimizing direct pesticide exposure. By reducing the need for substantial pesticide treatments, the use of nanobiosensors for early pest and disease detection improves consumer and farmer safety (Klaine et al., 2008). Besides these benefits, intelligent horticulture based on nanotechnology encourages sustainable farming methods and raises crop yields. Using NPs to transport pesticides and other control agents precisely lowers the amount of chemicals applied, limiting the emergence of pests resistant to pesticides and guaranteeing the long-term efficacy of pest management strategies for the sustainability of agriculture as a whole. This tactic also encourages the use of cutting-edge methods for managing pests, such as combining multiple pesticides into nanoformulations for a synergistic effect that enhances pest control and reduces the requirement for chemical pesticides (Siegrist et al., 2007). Nanotechnology also aids in the monitoring and control of pest populations through the use of advanced sensing and diagnostic techniques. The combination of nanobiosensors and remote sensing technology provides farmers with real-time data on insect populations, crop health, and environmental conditions, allowing them to make informed decisions about when and where to apply pesticides (Klaine et al., 2008).

6.4.3 AUTOMATED MACHINERY

Automated technology plays a major role in the precision and effectiveness of smart horticulture production. Its integration with nanosensors increases the precision and effectiveness of agricultural activities. Nanosensors are minuscule instruments utilized in automated machinery to identify certain materials or environmental circumstances. They improve accuracy and productivity in a range of automated machinery uses. One common application is soil moisture monitoring, where automated irrigation systems can quickly adjust the water delivery rate thanks to nanosensors. By providing accurate and timely information on soil moisture, nanosensors assist to prevent overwatering or underwatering, hence reducing the likelihood of adverse effects on crop growth. Nanosensors can measure temperature, humidity, and nutrient concentrations in addition to soil moisture (Wang et al., 2019b). These data enable automated machinery to make real-time adjustments, optimizing crop growth and minimizing resource waste.

The incorporation of nanosensors into automated machinery has numerous advantages. Its main advantage is improved operational precision and effectiveness, which reduces waste and encourages optimal crop growth. Moreover, it reduces the need for human intervention, freeing up labor resources for alternative uses. Furthermore, it provides farmers with real-time information on environmental conditions, empowering them to make informed decisions about crop management.

6.4.4 DATA ANALYTICS

In order to produce smart horticulture, data analytics is essential since it gives farmers the ability to make well-informed decisions about crop planting, fertilization, and harvesting. NPs have the potential to improve data-collecting accuracy and precision, which would boost analytics reliability. In a hypothetical application, farmers might employ NPs to test soil or water samples for particular contaminants or diseases in order to have a thorough understanding of their crops. Because NPs are sensors, they are useful for detecting poisons in soil and water.

For example, heavy metals in soil samples can be detected using gold NPs. In a study, researchers constructed an electrochemical sensor based on rGO/AuNPs/tetraphenyl porphyrin nanoconjugates, and the sensor demonstrated good sensitivity and selectivity for identifying cadmium ions in soil and food samples (Kumar et al., 2019). Similarly, viruses and bacteria are found and removed from water samples using magnetic NPs. In a different study, researchers developed a biosensor based on magnetic NPs that effectively identified *Escherichia coli* in water samples with high sensitivity and specificity (Pavela & Banelli, 2016). With the innovative use of NPs in clever horticulture, precise data analytics is advanced and farmers are able to make better informed decisions that will enhance crop management.

6.4.5 NANOMATERIALS USE IN PLANT GROWTH

Because nanomaterials have the potential to improve plant growth and production, their use in horticulture has drawn more interest. Of the different types of nanomaterials, nanocellulose, nanolignocellulose, and nanocarbons are particularly notable because of their established benefits for plant growth. CNTs, graphene, and fullerenes are a few nanocarbon forms that show great promise in promoting plant growth. Applying CNTs to tomato plants, for example, significantly accelerated their development, according to one study (Tripathi & Yadav, 2019). The use of graphene oxide (GO) has also been shown to promote plant growth, as evidenced by the increased height, root length, and dry weight of treated wheat seedlings (Raliya & Tarafdar, 2013).

Plant development has been shown to be aided by nanocellulose, which includes cellulose nanofibrils (CNFs) and cellulose nanocrystals (CNCs) (Hussain et al., 2016; Rajput & Minkina, 2019). Specifically, it has been shown that CNFs enhance root growth and augment plant uptake of water and nutrients. Additionally, it has been shown that cellulose anionic hydrogel-based nanofibers have benefits for sesame seed germination (Suh et al., 2014). Nanolignocellulose, a combination of lignin and cellulose NPs, has been demonstrated to enhance plant root hair growth and enhance nutrient and water absorption. The hydrophobicity of soybean leaf surfaces can be altered by cellulose nanofibers, and they can also confer resistance against the fungus *Phakopsora pachyrhizi* (Raskar & Laware, 2019).

PA and nanotechnology have the potential to fully revolutionize sustainable agricultural growth through increased sustainability and efficiency.

The combination of modern technology and data-driven decision-making tools with nanoscale materials and devices has the potential to benefit several significant sectors. These sectors include data analytics, automated machinery, pest control, and nutrition supply. This integration may result in improved nutrient utilization, precise and effective pesticide distribution, real-time monitoring of soil and plant properties, and precise input delivery by maximizing agricultural yields, reducing waste, and minimizing environmental effects (Table 6.1).

TABLE 6.1 Application of Nanomaterials in Smart Horticulture Production

Sr. No.	Applications	Nanomaterials
01	Water and soil management	Nano-clay zeolites, hydrogels, etc.
02	Plant protection (pest and disease)	Nanosensors, smart delivery system, nano capsules, and nanoparticles
03	Fertilizer application	Nano capsules and nanoparticles
04	Control soil erosion	Nano-zeolite, nano binder, and nano clay
05	Waste management	Nanofibers, nano encapsulation, and nano sensors

6.5 FUTURE DIRECTIONS IN NANOTECHNOLOGY-ENABLED SMART HORTICULTURAL PRODUCTION

The future of smart horticultural production lies in the continued integration and innovation of nanotechnology, offering promising avenues for sustainable agriculture and food security. Several key areas will shape the future landscape of nanotechnology in horticulture.

6.5.1 NANOMATERIAL INNOVATIONS

Nanopesticides and fertilizers: Ongoing research aims to develop nanoformulations of pesticides and fertilizers that ensure targeted delivery, minimizing environmental impact and maximizing efficiency(Kah et al., 2018).

Nanomaterial coatings: Advanced nanocoatings for seeds and agricultural equipment will enhance durability, pest resistance, and overall performance, contributing to sustainable farming practices(Li et al., 2021).

6.5.2 PRECISION AGRICULTURE ADVANCEMENTS

Nanobiosensors: The development of highly sensitive nanobiosensors will enable real-time monitoring of plant health, soil conditions, and environmental parameters, allowing for precise, data-driven decision-making in agriculture(Jain & Roychoudhary, 2019).

Nanorobotics: Progress in nanorobotics will lead to autonomous, precise interventions within plant cells, enabling targeted delivery of nutrients or removal of infected plant parts (Balasubramanian et al., 2011).

6.5.3 DATA-DRIVEN AGRICULTURE

Big data analytics: Utilizing big data analytics in conjunction with nanoscale sensors will enable farmers to analyze vast amounts of information, leading to optimized resource utilization, increased yields, and reduced environmental impact (Gebbers & Adamchuk, 2010).

Machine learning: Machine learning algorithms will process data from nanosensors, weather stations, and crop models, providing predictive insights and enhancing agricultural productivity (Friedl et al., 2010).

6.5.4 ETHICAL AND SOCIETAL IMPLICATIONS

Ethics and governance: Continued research into the ethical implications of nanotechnology in agriculture will inform governance frameworks, ensuring responsible innovation and equitable access to nanotechnological advancements (Macnaghten et al., 2005).

Public engagement: Engaging the public and stakeholders in discussions about the risks and benefits of nanotechnology in agriculture is essential for building trust and addressing concerns, fostering socially responsible agricultural practices (Siegrist & Cvetkovich, 2000). As nanotechnology continues to advance, its integration into smart horticultural production holds the promise of revolutionizing agriculture. By addressing environmental concerns, ensuring ethical practices, and harnessing the power of data, the future of smart horticulture will be characterized by efficiency, sustainability, and increased food security.

6.6 CONCLUSION

Nanotechnology-enabled smart horticultural techniques optimize nutrient delivery, pest management, and water usage, leading to increased crop yields and improved produce quality. PA, facilitated by nanoscale sensors and data analytics, ensures optimal use of resources such as water, fertilizers, and pesticides. This sustainable approach mitigates environmental impact and conserves valuable resources. Smart horticultural practices, driven by nanotechnology, enable the cultivation of crops in controlled environments. This resilience is crucial in the face of climate change, ensuring consistent food production even in challenging environmental conditions. Ethical considerations, including equitable access, environmental justice, and public engagement, are integral parts of nanotechnology-driven horticulture. Responsible innovation ensures that the benefits of advanced technologies are accessible to all while safeguarding societal and environmental well-being. The future of smart horticultural production lies in the continuous exploration of nanotechnological innovations, big data integration, and ethical governance. These elements will drive sustainable agricultural practices, ensuring food security for growing global populations. In conclusion, the integration of nanotechnology into smart horticultural production offers unprecedented opportunities for sustainable agriculture. By embracing these innovations responsibly, society can navigate the challenges of the future, ensuring a consistent supply of high-quality food while preserving our natural resources.

KEYWORDS

- **nanoparticles**
- **nanotechnology**
- **nanosensors**
- **nutrient delivery systems**
- **smart horticulture production.**

REFERENCES

Balasubramanian, S., Kagan, D., Hu, C.M., Campuzano, S., Lobo-Castañon, M.J., Lim, N., Kang, D.Y., Zimmerman, M., Zhang, L., Wang, J., & Zhang, L. (2011). Micromachine-enabled capture and isolation of cancer cells in complex media. *Angewandte Chemie International Edition*, 50(18), 4161–4164.

Baughman, R.H., Zakhidov, A.A., & de Heer, W.A. (2002). Carbon nanotubes—The route toward applications. *Science*, 297(5582), 787–792.

Cui, H., Wang, Y., Yang, M., Huang, J., Xing, J., Zou, D., & Wei, Z. (2017). Development of a portable nanosensor platform for the rapid detection of pesticides in water. *Nanoscale*, 9(38), 14419–14426.

Dresselhaus, M.S., Dresselhaus, G., & Eklund, P.C. (2001). *Science of Fullerenes and Carbon Nanotubes: Their Properties and Applications*. Academic Press.

Duro, D.C., & Khot, L.R. (2016). Precision agriculture and food security in the 21st century: A review. *Agricultural Systems*, 146, 202–220.

Feynman, R.P. (1960). There's plenty of room at the bottom. *Engineering and Science*, 23(5), 22–36.

Friedl, M.A., Sulla-Menashe, D., Tan, B., Schneider, A., Ramankutty, N., Sibley, A., & Huang, X. (2010). MODIS collection 5 global land cover: Algorithm refinements and characterization of new datasets. *Remote Sensing of Environment*, 114(1), 168–182.

Gebbers, R., & Adamchuk, V.I. (2010). Precision agriculture and food security. *Science*, 327(5967), 828–831.

Giraldo, J.P., & Landry, M.P. (2019). Plant nanobionics approaches to augment plant photosynthesis and biochemical sensing. *Nature Materials*, 18(12), 1154–1163.

Haiss, W., Thanh, N.T., Aveyard, J., & Fernig, D.G. (2007). Determination of size and concentration of gold nanoparticles from UV-vis spectra. *Analytical Chemistry*, 79(11), 4215–4221.

He, L., Liu, Y., He, H., Chen, Y., Shi, J. (2017). Nanocomposites for plant growth promotion and disease management. *Chemical Reviews*, 117(23), 13566–13601.

Hu, J., Li, Z., & Qu, J. (2019). Application of Internet of Things technology in agricultural environmental monitoring. In: *Internet of Things (IoT) for Automated and Smart Applications*, Academic Press, pp. 155–176.

Hussain, I., Singh, N.B., Singh, A., Singh, H., & Singh, S.C. (2016). Green synthesis of nanoparticles and its potential application. *Biotechnology Letters*, 38(4), 545–560.

Jain, R., & Roychoudhury, S. (2019). Agriculture 4.0: A review. *Computers and Electronics in Agriculture*, 157, 205–227.

Kah, M., Kookana, R.S., & Gogos, A. (2018). A critical evaluation of nanopesticides and nanofertilizers against their conventional analogues. *Nature Nanotechnology*, 13(8), 677–684.

Khaliq, A., Irshad, F., Ahmed, E., Kim, D.S., & Rho, S. (2018). A Review on Internet of Things (IoT), Internet of Everything (IoE) and Internet of Nano Things (IoNT) sensors role in precision agriculture. *Agricultural Research & Technology: Open Access Journal*, 16(4), 555926.

Khodakovskaya, M., Dervishi, E., Mahmood, M., Xu, Y., Li, Z., Watanabe, F., & Biris, A.S. (2009). Carbon nanotubes are able to penetrate plant seed coat and dramatically affect seed germination and plant growth. *ACS Nano*, 3(10), 3221–3227.

Khot, L.R., Sankaran, S., Carter, A.H., & Kerševan, R. (2019). Machine learning applications in agriculture. *Computers and Electronics in Agriculture*, 161, 288–301.

Kumar, P., Mahato, A.K., Kamle, M., Mohanta, T.K., & Kang, S.G. (2017). Aflatoxins: A global concern for food safety, human health and their management. *Frontiers in Microbiology*, 7, 2170.

Kumar, V., Park, D., & Gurung, M. (2019). Green and environment friendly approach for the efficient delivery of Ag-based nanopesticides. *Journal of Environmental Management*, 231, 281–289.

Li, Y., Zhu, H., Shen, Y., & Lv, F. (2021). Self-healing polymer nanocomposites with enhanced mechanical and conductive properties based on graphene nanoplatelets. *Composites Science and Technology*, 207, 108737.

Liu, R., Lal, R., & Potdar, N. (2015). Effect of nanoparticle dispersions on soil-water characteristic curve. *Geoderma*, 259, 134–142.

Lu, Y., & Wang, L. (2016). Internet of Things (IoT) for smart precision agriculture. In: *Internet of Things and Data Analytics Handbook*. John Wiley & Sons, pp. 475–506.

Macnaghten, P., Kearnes, M., & Wynne, B. (2005). Nanotechnology, governance, and public deliberation: What role for the social sciences? *Science Communication*, 27(2), 268–291.

Pavela, R., & Benelli, G. (2016). Essential oils as ecofriendly biopesticides? Challenges and constraints. *Trends in Plant Science*, 21(12), 1000–1007.

Qaim, M. (2016). Role of new plant breeding technologies for food security and sustainable agricultural development. *Applied Economic Perspectives and Policy*, 38(1), 211–236.

Qiao, Y., & Zhang, P. (2019). Nanotechnology-based intelligent fertilizers and smart delivery systems: A review. *Environmental Chemistry Letters*, 17(4), 1473–1486.

Rai, M., Ingle, A., & Birla, S. (2016). Nano-pesticides: Emerging prospects for agriculture. In: *Plant Nanotechnology*. Cham: Springer, pp. 221–233.

Rajput, V., & Minkina, T. (2019). Nanotechnology: A new approach in agriculture. In: *Nanotechnology: An Agricultural Paradigm*. Singapore: Springer, pp. 1–34.

Raliya, R., & Tarafdar, J.C. (2013). Biosynthesis and characterization of zinc, magnesium and titanium nanoparticles: An eco-friendly approach. *International Nano Letters*, 3(1), 30.

Raskar, S., & Laware, S.L. (2019). Chapter 16—nanoparticles in plant growth and development. In: *Nanotechnology in Agriculture*. Academic Press, pp. 381–400.

Recha, J.W., & Kimathi, A.N. (2019). Harnessing the Fourth Industrial Revolution (4IR) Technologies for Agri-Food Systems Transformation in Africa. CTA Working Paper.

Rico, C.M., Majumdar, S., Duarte-Gardea, M., & Peralta-Videa, J.R. (2011). Interaction of nanoparticles with edible plants and their possible implications in the food chain. *Journal of Agricultural and Food Chemistry*, 59(8), 3485–3498.

Roco, M.C. (2003). Nanotechnology: Convergence with modern biology and medicine. *Current Opinion in Biotechnology*, 14(3), 337–346.

Shrivastava, P., Saxena, S., & Singh, R. (2018). Nanotechnology: A sustainable tool for enhancing soil fertility and plant growth. In: *Nanotechnology: Soil and Water Engineering*. Singapore: Springer, pp. 93–121.

Siegrist, M., & Cvetkovich, G. (2000). Perception of hazards: The role of social trust and knowledge. *Risk Analysis*, 20(5), 713–720.

Siegrist, M., Keller, C., & Kastenholz, H. (2007). Frequent consumption of organic vegetables. *Food Quality and Preference*, 18(6), 772–778.

Suh, S.C., Choi, S., Lee, J.S., & Yang, J.W. (2014). Nano-formulation improves the penetration and deposition of cytokinins into plant cells. *Journal of Nanoscience and Nanotechnology*, 14(9), 6846–6849.

Tripathi, D., & Yadav, V. (2019). Nanotechnology for plant disease management. In: *Nanotechnology: An Agricultural Paradigm*. Singapore: Springer, pp. 93–114.

Wang, Z., Gerstl, Z., Malakouti, M.J., & Ganjegunte, G.K. (2018). Smart nutrient management using nanotechnology. In: *Nanotechnology: Soil and Water Engineering*. Singapore: Springer, pp. 193–216.

Wang, H., Zhao, P., Kong, X., Sun, Y., Li, H., & Yan, X. (2019a). Hydrogel-based smart materials for application in agriculture: A review. *Colloids and Surfaces B: Biointerfaces*, 173, 581–590.

Wang, Z., Xie, L., Xu, P., Liu, W., Liu, Z., Tang, W., & Jin, Y. (2019b). Nanotechnology promotes the sustainable agriculture: Current achievements and future perspectives. *Journal of Controlled Release*, 296, 140–161.

Zhang, Y., Wang, Y., Wang, S., Fang, X., & Wang, X. (2019). Stability enhancement of Azadirachtin A in Azadirachtin formulation using nanoparticles. *Pesticide Biochemistry and Physiology*, 157, 125–131.

CHAPTER 7

Smart Nutrient Delivery: System, Mechanism, and Utilization

PRASANN KUMAR* and JAISURIYAN K.

Department of Agronomy, School of Agriculture, Lovely Professional University, Phagwara, Punjab, India

*Corresponding author

ABSTRACT

Smart nutrient delivery (SND) represents a paradigm shift in precision-controlled nutrient distribution systems. It makes use of advanced technologies to maximize both efficiency and responsiveness. SND is the subject of this research, which investigates its complex system design, the mechanisms that lie beneath it, and its multiple applications. Many different kinds of SND systems are examined in this study. Some systems are based on nanotechnology, others on the Internet of Things, and others on bioresponsive approaches. Mechanisms such as sensing, monitoring, and controlled release are some of the mechanisms that are investigated to shed light on the inner workings of SND. In addition, the study examines the application of SND in various fields, including agriculture, medicine, healthcare, and environmental management, among others. SND has far-reaching implications because it is pivotal in improving crop yields, targeted drug delivery, personalized nutrition, and ecological sustainability. There are also many other applications for SND. This chapter aims to provide a concise overview of the significance of SND, which will provide a foundation for understanding its evolving role in various fields and stimulate further exploration and innovation.

Smart Horticulture Production: Integrating Advanced Technologies.
Rajni Rajan and Duniya Ram Singh (Eds.)

DOI: 10.1201/9781998511679-7

7.1 INTRODUCTION

7.1.1 IMPORTANCE OF EFFICIENT NUTRIENT DELIVERY

The efficient delivery of nutrients is an essential component in a wide variety of industries, including agriculture, medicine, and environmental management, to name a few (Chakraborty et al., 2023; Paul et al., 2022). One of the most critical aspects of this process is its direct impact on productivity, sustainability, and the well-being of individuals. In agriculture, ensuring that crop yields are maximized and food security is maintained, precise nutrient delivery is pivotal. When it comes to medicine, effective nutrient delivery has the potential to boost the efficacy of treatments, ultimately resulting in better patient outcomes. In addition, environmental applications depend on efficient nutrient delivery to achieve sustainable resource management. These applications include soil remediation and water systems. As the demand for improved performance and sustainability continues to rise, there is a growing recognition of the necessity of developing novel approaches to the delivery of nutrients. Traditional delivery methods frequently face difficulties associated with imprecision, inefficient resource utilization, and environmental impact. Intelligent nutrient delivery (smart nutrient delivery [SND]) systems came into existence due to the search for more efficient methods of delivering nutrients (Polaka et al., 2024; Prajapati & Roy, 2023; Zhu et al., 2023).

7.1.2 CHALLENGES IN TRADITIONAL DELIVERY SYSTEMS

Although fundamental, traditional nutrient delivery systems face several obstacles that prevent them from functioning at their full potential. Conventional farming practices frequently result in either excessive or insufficient application of fertilizers, which in turn causes pollution of the environment and a reduction in soil fertility. In medicine, nontargeted drug delivery has the potential to result in adverse effects and reduce the therapeutic efficacy of treatments. Furthermore, in environmental applications, ineffective nutrient management may be a contributing factor in the deterioration of water quality and the health of the soil. The necessity of transitioning toward nutrient delivery systems that are more intelligent and responsive is brought into focus by these challenges. The limitations of traditional approaches underscore the urgency of exploring and adopting innovative technologies to address these issues and usher in a new era of efficiency and sustainability (El-Husseiny et al., 2024; Gebresenbet et al., 2023; Wanyama et al., 2023).

7.1.3 EXPLORE SMART NUTRIENT DELIVERY SYSTEMS

The primary objective of this research is to delve into the realm of smart nutrient delivery (SND) systems. This investigation aims to investigate the numerous kinds of existing SND systems thoroughly. Each of these systems makes use of cutting-edge technologies to bring about improvements in terms of precision, control, and responsiveness in the delivery of nutrients. To unlock the potential benefits these systems could provide across various industries, it is essential to have a thorough understanding of their complexities. The purpose of this research is to acquire an understanding of the mechanisms that are responsible for brilliant nutrient delivery. This research is being conducted concurrently with the investigation of SND systems. When it comes to optimizing the performance of these systems, having a more in-depth understanding of how they function is very important. This includes everything from sensing and monitoring technologies to controlled release mechanisms. By dissecting these mechanisms, we can acquire additional knowledge regarding the technical aspects that contribute to the efficiency of SND systems (Avşar & Mowla, 2022; Chen et al., 2022; Dong et al., 2023).

7.1.4 EXPLORE SMART NUTRIENT DELIVERY SYSTEMS

Aside from exploring and comprehending the subject matter, this research aims to investigate the application of SND systems in various industries. One of the most critical areas where SND has the potential to revolutionize nutrient delivery practices is in the fields of agriculture, medicine, and environmental management. To gain valuable insights into the versatility and potential of SND to address sector-specific challenges, it is essential to evaluate the impact and effectiveness of SND in these various sectors. This introduction lays the groundwork for a more in-depth exploration of SND systems. It highlights the critical importance of efficient nutrient delivery, the challenges faced by traditional systems, and the many objectives this research aims to accomplish. In the following sections, we will delve into the various types of SND systems, the mechanisms that lie beneath them, and the diverse applications that have the potential to revolutionize industries and advance sustainability (Aminu et al., 2023; Logesh et al., 2021; Thorn et al., 2021).

7.2 SMART NUTRIENT DELIVERY SYSTEMS: PRECISION, CONTROL, AND RESPONSIVENESS

SND systems represent a paradigm shift in the manner in which nutrients are distributed across a variety of industries, including agriculture, medicine, and environmental management, among others. These systems address the limitations of conventional methods, which use cutting-edge technologies to improve nutrient delivery's precision, control, and responsiveness. The definition of SND systems and their characteristics will be investigated in depth during this all-encompassing investigation, emphasizing accuracy, control, and responsiveness (Kaur et al., 2023; Liu et al., 2023a; Türk et al., 2021) (Figure 7.1).

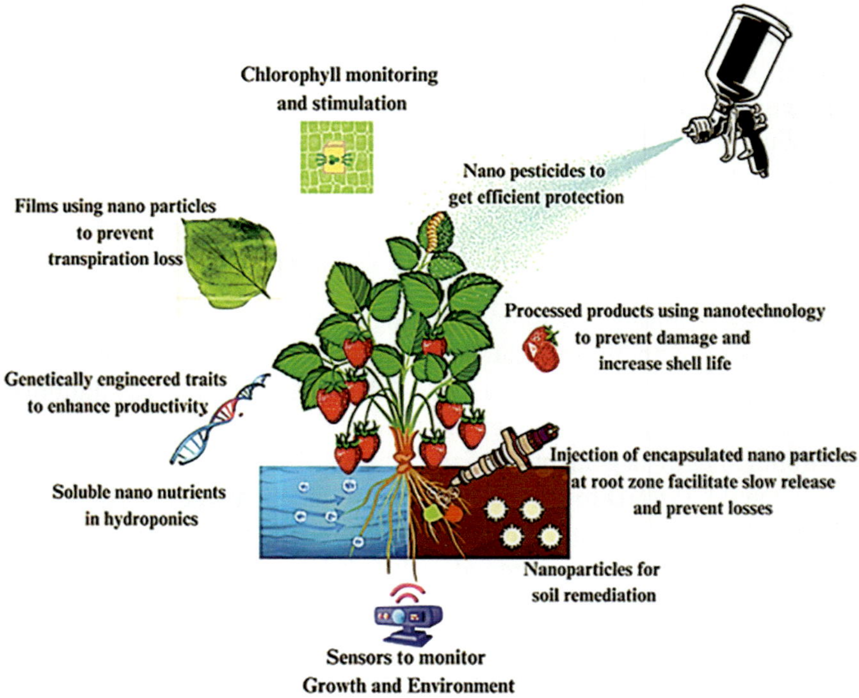

FIGURE 7.1 Smart nutrient delivery.

Source: Created by authors © permission is not required from other sources.

7.2.1 PRECISION

Standardized nutrient delivery (SND) systems are distinguished from more conventional nutrient delivery methods by their emphasis on precision. In agriculture, precision is paramount to maximize crop yields while minimizing environmental impact. It is possible to measure the levels of soil nutrients, crop requirements, and ecological conditions precisely with the help of SND systems, which use advanced sensors and technologies. This approach, driven by data, makes it possible to deliver nutrients in a targeted manner, ensuring fertilizers are applied in the appropriate quantities, at the correct times, and in the proper locations. Precision in SND extends beyond agriculture and into the medical field. Precision is essential in the delivery of drugs to maximize therapeutic efficacy while simultaneously minimizing adverse effects. Incorporating sensors and feedback mechanisms into intelligent drug-delivery systems enables the monitoring of physiological parameters, allowing real-time adjustments to dosage and delivery rates. The effectiveness of treatments is improved due to this precision, leading to better patient outcomes. Furthermore, precision in environmental applications, such as soil remediation, involves the delivery of nutrients in a targeted manner to restore the health of the soil without negatively impacting the ecosystems in the surrounding area. In the field of environmental management, SND systems make use of advanced sensing technologies to evaluate the particular requirements of the environment. This enables precise interventions for sustainably managing resources (Nandini & Geetha, 2021; Saxena et al., 2022; Srivastava et al., 2024; Thananukul et al., 2021).

7.2.2 CONTROL

One of the most essential features of SND systems is control, which gives users the ability to manage and regulate the delivery of nutrients with high precision. When it comes to agriculture, traditional methods frequently lack the necessary control mechanisms, which can result in either an excessive or insufficient application of fertilizers. On the other hand, SND systems offer granular control over the application of nutrients by utilizing automated and adaptive technologies. Integration of Internet of Things (IoT) devices and data analytics to make decisions about nutrient delivery in real-time is what is meant by the term "automated control" in the agricultural industry. These systems can adjust nutrient levels according to the moisture content of the soil, the temperature, and the stage of crop growth. The utilization of resources is optimized, waste is reduced, and the impact on the environment is mitigated thanks to this dynamic control mechanism. Control is essential in drug delivery for tailoring treatments to the specific requirements of individual patients. Intelligent drug-delivery systems utilize programmable mechanisms to regulate the release of nutrients or medications. This ensures that the delivery is both consistent and under control. This level of control improves the safety and effectiveness of the therapeutic interventions offered. Regarding environmental applications, control mechanisms that enable precise adjustments to the delivery of nutrients in response to changing conditions are highly beneficial. Intelligent sensors and actuators allow real-time control over the release of nutrients, encouraging environmentally sustainable ways of doing things (Daund et al., 2021; Lu et al., 2024; Wang et al., 2024b; Wu et al., 2022).

7.2.3 RESPONSIVENESS

One of the defining characteristics of SND systems is their responsiveness, which enables them to adjust to shifting environmental conditions and deliver nutrients promptly. Responding quickly and effectively is essential in agriculture because it allows for managing dynamic factors such as weather patterns, crop stress, and shifting soil conditions. Real-time monitoring and predictive analytics are incorporated into SND systems to implement proactive responses to these variables. This helps to ensure that nutrient delivery is optimized for the conditions that are currently present. Regarding medical care, the responsiveness of intelligent drug-delivery systems makes it possible to provide personalized and adaptable healthcare. Continuous feedback can be provided by sensors embedded in the body or worn on wearable devices. This enables the system to react to changes in the body's physiological state. Because of this responsiveness, patients can receive the appropriate nutrients or medications when required. Applications that deal with the environment can reap the benefits of the responsiveness of SND systems in meeting the ever-changing requirements of ecosystems. These systems can adapt to changes in nutrient levels, pollution, or other environmental factors, which contributes to implementing sustainable resource management practices. This is true whether the systems are used for water or soil management. The landscape of nutrient distribution is being reimagined by innovative nutrient delivery systems, which embody precision, control, and responsiveness. SND is positioned as a transformative force across agriculture, medicine, and environmental management due to its distinguishing characteristics from traditional methods. In the following sections, we will delve deeper into the various types of SND systems, the mechanisms responsible for their operation, and the extensive applications of these systems across multiple industries (Dastghaib et al., 2020; Feng et al., 2023; Huang et al., 2024; Takeshita et al., 2023; Wang et al., 2024).

7.3 TYPES OF SMART NUTRIENT DELIVERY SYSTEMS

SND systems represent a significant advancement in various industries, offering precision, control, and responsiveness. This section will delve into three distinct types of SND systems: IoT-based systems, nanotechnology applications, and bioresponsive delivery. Each type harnesses unique technologies to optimize nutrient delivery, addressing specific agriculture, medicine, and environmental management challenges.

7.3.1 IOT-BASED SYSTEMS

IoT-based SND systems leverage the power of the IoT to create intelligent and interconnected networks for efficient nutrient management. These systems incorporate sensors, actuators, and communication devices to collect and transmit real-time data, enabling precision agriculture and other applications.

7.3.2 SENSORS AND DATA ANALYTICS

IoT systems utilize sensors to monitor environmental conditions, soil health, and crop growth. The sensors can quantify moisture levels, nutrient composition, and temperature. Subsequently, the gathered data are examined using sophisticated analytics to produce valuable observations regarding the most advantageous nutrient needs (Aliakbar Ahovan et al., 2022; Chojnacka & Chojnacki, 2024; Mazidi et al., 2022; Song et al., 2023; Zhang et al., 2023).

7.3.3 AUTOMATED DECISION-MAKING

The instantaneous data produced by IoT sensors enables automated decision-making in the delivery of nutrients. Algorithms process the information and activate actions such as modifying fertilizer quantities, initiating irrigation, or deploying drones for precise nutrient application. Automated decision-making improves accuracy and minimizes the requirement for human involvement.

7.3.4 REMOTE MONITORING

An essential benefit of IoT-based systems is the ability to monitor from a distance. Web interfaces or mobile applications enable remote access and control of nutrient delivery systems for farmers, healthcare professionals, and environmental managers, providing real-time data. This functionality allows for prompt interventions and modifications, enhancing resource management's effectiveness.

7.3.5 INTEGRATION WITH PRECISION AGRICULTURE

IoT-based soil nutrient detection (SND) systems effortlessly combine with precision agriculture methods. IoT technologies provide precise control advantageous for precision farming practices, such as variable rate applications. This integration enhances the delivery of nutrients by customizing it to the particular soil and crop conditions, resulting in increased yields and decreased environmental impact (Pandya et al., 2023; Raj et al., 2024; Singh et al., 2024; Tian & Liu, 2023; Wang et al., 2022).

7.3.6 NANOTECHNOLOGY APPLICATIONS

Nanotechnology has revolutionized SND by providing innovative methods for accurate and focused delivery of nutrients. Nanoparticles, nanocarriers, and other materials at the nanoscale are utilized to transform nutrient delivery systems in agriculture, medicine, and environmental applications.

7.3.7 NANOENCAPSULATION FOR AGRICULTURE

Nanoencapsulation techniques encompass the process of encapsulating nutrients within carriers that are on the nanoscale. These carriers safeguard nutrients from degradation, increase solubility, and facilitate controlled release. This guarantees that plants obtain nutrients in an optimal form for absorption, thereby enhancing nutrient utilization and minimizing environmental runoff.

7.3.8 NANOMEDICINE FOR TARGETED DRUG DELIVERY

Nanotechnology is essential in medicine to develop SND systems that enable precise drug delivery to specific targets. Nanocarriers, such as liposomes or nanoparticles, can encapsulate nutrients or medications, enabling targeted delivery to particular cells or tissues. This focused strategy reduces adverse effects and improves the effectiveness of treatment (Abadi et al., 2021; Dahiya & Dahiya, 2023; Hossain et al., 2024; Shahbazi et al., 2023; Shi et al., 2020).

7.3.9 NANOSENSORS FOR ENVIRONMENTAL MONITORING

Environmental applications benefit from nanosensors that enable precise monitoring of nutrient levels in water, soil, or air. These nanosensors can detect variations in nutrient concentrations and transmit data in real-time. Such monitoring facilitates responsive and adaptive nutrient management strategies, contributing to environmental sustainability.

7.3.10 SMART NANOFERTILIZERS

Smart nanofertilizers are a novel implementation of nanotechnology in agriculture. Nanofertilizers exhibit controlled nutrient release in response to particular environmental cues, such as soil pH or moisture-content variations. Implementing controlled release mechanisms guarantees plants receive nutrients appropriately, reducing both wastage and ecological consequences.

7.3.11 BIORESPONSIVE DELIVERY

Innovative nutrient delivery systems utilize biological cues and responses to regulate the release of nutrients. These systems are designed to interact with living organisms, adapting their delivery methods based on specific physiological conditions. Bioresponsive delivery is employed in targeted drug delivery, agriculture, and environmental management.

7.3.12 RESPONSIVE POLYMERS IN DRUG DELIVERY

Bioresponsive polymers are utilized in pharmaceutical drug-delivery systems. These polymers experience conformational alterations due to physiological factors such as pH, temperature, or enzyme activity. This ability to respond promptly allows for the deliberate discharge of nutrients or medications at the specific location, thereby enhancing the effectiveness of the treatment (Karimi et al., 2023; Liu et al., 2024; Moheb Afzali et al., 2024; Qiao et al., 2022; Singh et al., 2023).

7.3.13 PLANT–MICROBE INTERACTIONS IN AGRICULTURE

Bioresponsive delivery exploits the mutually beneficial interactions between plants and microbes. Microorganisms can increase the availability of nutrients for plants by dissolving minerals or converting atmospheric nitrogen into a usable form. Bioresponsive systems utilize these interactions to enhance the absorption of plant nutrients, thereby supporting the implementation of sustainable agricultural methods.

7.3.14 BIODEGRADABLE NANOCARRIERS IN ENVIRONMENTAL APPLICATIONS

Bioresponsive nanocarriers designed to degrade or release nutrients in response to specific environmental conditions have advantageous implications for ecological applications. These environmentally friendly nanocarriers reduce the adverse environmental effects and offer a sustainable method for managing nutrients in water systems and soil (Figure 7.2).

7.3.15 CELLULAR-LEVEL RESPONSIVENESS IN MEDICINE

Within advanced medical applications, bioresponsive delivery systems function at the cellular level. Nanoparticles equipped with biological triggers can react to specific cellular signals, resulting in the precise release of nutrients or medications at the required time. Such a high level of responsiveness improves the effectiveness and security of therapeutic interventions. The study of IoT-based systems, nanotechnology applications, and bioresponsive delivery in SND systems highlights the various creative methods influencing the future of nutrient management. Each category offers distinct benefits to agriculture, medicine, and environmental management,

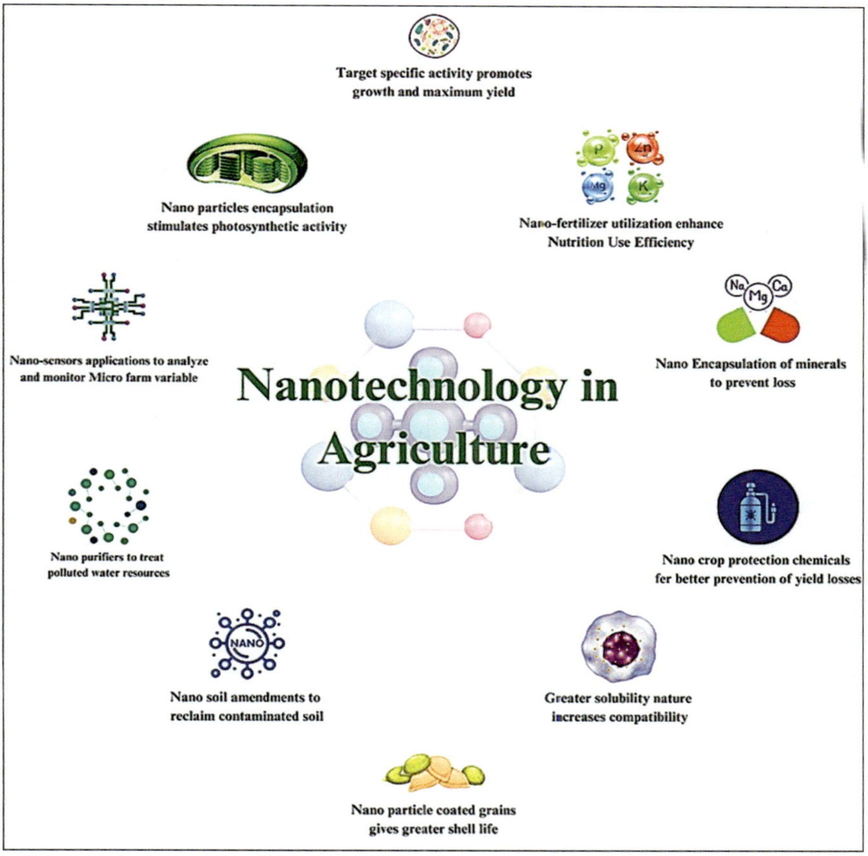

FIGURE 7.2 Nanotechnology in agriculture for smart nutrient delivery.

Source: Created by authors © permission is not required from other sources.

thereby enhancing the sustainability and efficiency of resource utilization. The following sections will explore the fundamental mechanisms that propel these SND systems, offering a more profound comprehension of their functioning and potential (Chen et al., 2024; Fan et al., 2023; Khanna et al., 2023; Shreya et al., 2023; Zehra et al., 2023).

7.4 MECHANISMS OF SMART NUTRIENT DELIVERY

SND mechanisms represent a sophisticated integration of cutting-edge technologies designed to optimize nutrient distribution's precision, control, and responsiveness. This multifaceted approach involves intricate processes that can be broadly categorized into two main components: sensing and monitoring and controlled release.

7.4.1 SENSING AND MONITORING

7.4.1.1 SENSOR TECHNOLOGIES

Sensor technologies form the foundation of SND mechanisms, providing real-time data acquisition and environmental monitoring. Various sensors measure critical parameters such as soil nutrient levels, crop conditions, or physiological factors in medical applications. For agriculture, soil sensors may include moisture sensors, nutrient sensors, and temperature sensors, allowing farmers to assess the specific needs of their crops. In medical contexts, biosensors and physiological sensors are crucial in monitoring patient conditions. The advancement of sensor technologies has led to the development of highly specialized and accurate devices. In agriculture, for instance, optical sensors can assess crop health by analyzing the reflection of light, while in medicine, wearable sensors can continuously monitor vital signs. These sensors contribute to creating a data-driven foundation, facilitating the precision of nutrient delivery by providing real-time insights into the dynamic conditions of the targeted environment (El-Husseiny et al., 2022; Raikwar & Jain, 2022; Sharma et al., 2022, 2024; Tan et al., 2023).

7.4.1.2 REAL-TIME MONITORING

Real-time monitoring is pivotal to SND mechanisms, enabling immediate responsiveness to changing conditions. In agriculture, real-time monitoring allows farmers to adapt nutrient delivery strategies based on weather patterns, soil moisture levels, and crop growth stages. In healthcare, real-time tracking of a patient's physiological parameters ensures that nutrient or medication delivery can be adjusted promptly in response to deviations from the desired state. The integration of real-time monitoring in SND systems relies on the seamless communication between sensors, actuators, and control systems. The ability to receive and process data in real-time empowers SND systems to make dynamic decisions, optimizing nutrient delivery for the specific needs of the system in focus (He et al., 2023; Liu et al., 2021; Parekh et al., 2023; Prange et al., 2023; Riseh et al., 2024).

7.4.2 CONTROLLED RELEASE

7.4.2.1 TRIGGER MECHANISMS

Controlled release mechanisms are instrumental in regulating the timing and dosage of nutrient delivery. These mechanisms ensure nutrients are released precisely when and where needed, preventing wastage and minimizing environmental impact. In agriculture, controlled release mechanisms may involve using innovative fertilizers encapsulated in materials that respond to specific triggers. For instance, the release of nutrients could be triggered by soil moisture levels, ensuring that fertilizers are only released during optimal conditions for plant uptake. In medicine, drug-delivery systems may utilize trigger mechanisms such as pH-responsive materials, allowing for the controlled release of medications in specific physiological environments. The choice of trigger mechanisms depends on the targeted application and the desired responsiveness of the SND system. Integrating innovative materials and responsive technologies ensures that nutrient release aligns with the specific conditions identified through sensing and monitoring (Chen, 2020; Norouzi & Abdouss, 2023; Pachuau, 2020; Raja et al., 2023; Saha & Bharadwaj, 2023).

7.4.3 FEEDBACK LOOPS

Feedback loops play a vital role in maintaining the dynamic equilibrium of SND systems. These loops enable continuous adjustments based on sensor information and the system's real-time performance. Feedback mechanisms create a closed-loop system where the output influences the input, allowing adaptive and self-regulating nutrient delivery. In agriculture, feedback loops may involve adjusting nutrient levels based on the growth and health of crops. For instance, if sensors detect a deficiency or excess of certain nutrients, the feedback loop triggers appropriate adjustments in the nutrient delivery system. In healthcare, feedback loops can be implemented to respond to changes in patient conditions, ensuring that nutrient or medication delivery is optimized. The effectiveness of feedback loops relies on the seamless integration of sensing technologies, actuators, and control systems. The iterative nature of feedback mechanisms ensures that SND systems can adapt to evolving conditions, providing a level of responsiveness essential for efficient nutrient delivery. The mechanisms of

SND systems, encompassing sensing and monitoring as well as controlled release, represent a convergence of advanced technologies aimed at revolutionizing nutrient distribution. The intricate interplay of sensor technologies, real-time monitoring, trigger mechanisms, and feedback loops creates a dynamic framework that optimizes nutrient delivery across diverse applications, from precision agriculture to targeted drug delivery in medicine. These mechanisms form the backbone of SND systems, contributing to their efficacy, sustainability, and potential for transformative impact in various fields (Alipio et al., 2019; Beheshtizadeh et al., 2023; Khan et al., 2024; Ni et al., 2023; Teng et al., 2023).

7.5 UTILIZATION IN DIFFERENT SECTORS

In the 21st century, technological progress has deeply influenced all aspects of our lives, fundamentally changing our approach to different domains. Agriculture, medicine, and environmental management are also undergoing a significant revolution due to advanced technologies. This essay examines the influence of technology in various fields, specifically precision farming and crop improvement in agriculture, targeted drug administration and personalized nutrition in medicine and healthcare, and nutrient management in water systems and soil remediation in environmental applications. Agricultural practice that utilizes advanced technology and data analysis to optimize crop production and resource management (Ismail et al., 2022; Kamsani et al., 2021; Ramasamy et al., 2017; Saif et al., 2022; Xu et al., 2024). Precision farming, or precision agriculture, is an innovative method that uses technology to optimize different aspects of farming, ranging from planting to harvesting (Figure 7.3).

This approach entails the utilization of cutting-edge technologies such as GPS guidance systems, sensors, robotics, and data analytics to make well-informed decisions regarding agricultural practices. Precision farming offers farmers immediate and accurate information regarding the well-being of their crops, the state of the soil, and the patterns of weather. These data enable correct resource management, encompassing water, fertilizers, and pesticides. Precision farming techniques enable farmers to optimize crop yields, minimize resource inefficiencies, and mitigate environmental consequences. Using drones with sensors and cameras is crucial in precision farming. These uncrewed aerial vehicles (UAVs) can hover above agricultural areas, capturing detailed images and gathering

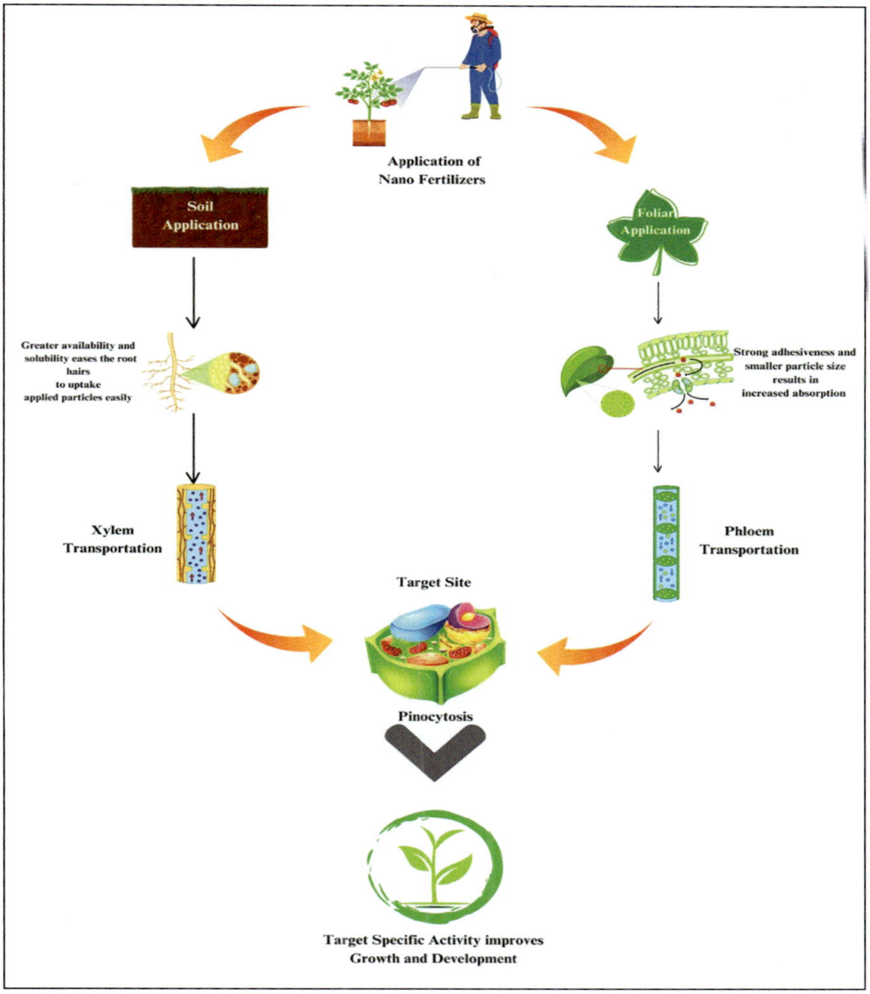

FIGURE 7.3 Mode of action of nanofertilizers and mechanism.

Source: Created by authors © permission is not required from other sources.

information regarding crops' condition, pests' presence, and nutrient levels. Subsequently, farmers can analyze these data to implement focused interventions, such as applying pesticides exclusively in areas where necessary or modifying irrigation practices according to the specific needs of each crop. Integrating artificial intelligence (AI) and machine learning algorithms significantly enhances the capabilities of precision farming. These technologies can analyze extensive data to detect patterns and forecast crop performance. AI algorithms can forecast disease outbreaks, allowing farmers to proactively implement preventive measures before the occurrence of a widespread infestation. Crop enhancement refers to the application of genetic modifications, innovative breeding techniques, and advanced biotechnological tools to improve the quality and quantity of agricultural produce. These advancements tackle the difficulties of climate change, population expansion, and sustainable agriculture requirements. Genetically modified organisms have been a central focus in improving crops. Scientists can incorporate particular genes into crops to give them characteristics such as resistance to pests, the ability to withstand harsh environmental conditions or improved nutritional value. Genetically modified crops can be manipulated to generate pesticides, decreasing reliance on external chemical inputs. Advanced breeding techniques, such as CRISPR-Cas9, have significantly transformed the creation of new crop varieties alongside genetic modifications (Ahmed et al., 2023; Balu et al., 2023; Bharathi et al., 2022; Bhavsar et al., 2023; El-Husseiny, Mady, Hamabe, et al., 2022; Zou et al., 2022). The CRISPR technology facilitates accurate gene editing, empowering scientists to make specific modifications to the plant's DNA without introducing exogenous genes. This can expedite breeding, producing more efficient crops with the desired characteristics. Biotechnology has a broader scope in improving crops than just genetic modifications. Scientists are currently investigating the application of nanotechnology for targeted delivery of nutrients to plant cells, enhancing the efficiency of nutrient absorption and utilization. Nanoparticles can be engineered to achieve a controlled and gradual release of nutrients, guaranteeing a continuous provision to the crops and reducing nutrient runoff, a prevalent environmental issue in conventional farming methods. Targeted drug delivery has revolutionized medicine by providing more accurate and effective methods to deliver therapeutic agents to specific cells or tissues. Conventional drug-delivery methods typically entail the systemic administration of drugs, resulting in their distribution throughout the entire body, impacting both healthy and diseased tissues. Targeted drug delivery aims to reduce adverse effects and improve the therapeutic effectiveness of medications. Nanotechnology is essential for the development of precise drug-delivery systems. Nanoparticles can be designed to transport medications precisely to the targeted location, avoiding unaffected tissues and reducing overall exposure. This focused strategy is invaluable in cancer therapy, aiming to administer chemotherapy drugs exclusively to cancerous cells while preserving healthy cells. Liposomes, micelles, and polymeric nanoparticles are nanocarriers employed for the precise administration of drugs. These structures can enclose many therapeutic substances, such as small-molecule drugs, proteins, and nucleic acids. By altering the surface characteristics of these nanoparticles, scientists can regulate their interactions with biological tissues, guaranteeing adequate drug transportation to the desired destination. Medical imaging advancements and nanotechnology contribute to developing targeted drug-delivery strategies. Advanced imaging modalities, such as magnetic resonance imaging and positron emission tomography, enable accurate visualization of pathological tissues. This information is vital for developing drug-delivery systems capable of navigating intricate anatomical structures and targeting specific regions within the body. Personalized nutrition signifies a fundamental change in healthcare, acknowledging that individuals possess distinct dietary requirements determined by their genetic composition, lifestyle, and state of health. Progress in genetic testing, nutritional science, and digital health technologies has facilitated the development of customized nutrition interventions to meet an individual's specific needs. Genomic data are crucial in the field of personalized nutrition (Arif et al., 2022; de León et al., 2023; Henn et al., 2024; Liu et al., 2023b; Shanmugavel et al., 2023; Wang et al., 2023b). Genetic testing can unveil an individual's inclination toward nutritional deficiencies, intolerances, or metabolic traits (Figure 7.4).

These data enable healthcare professionals to create customized dietary regimens that target particular requirements and maximize nutrient consumption. Mobile applications and wearable devices in digital health allow the immediate gathering of data on individuals' dietary habits, physical activity, and general well-being. These data are crucial for developing adaptable, individualized dietary schemes that can be modified in response to shifts in lifestyle, health circumstances, or nutritional objectives. The incorporation of AI into personalized nutrition amplifies its efficacy. To produce customized suggestions, AI algorithms can examine extensive datasets encompassing genetic data, dietary inclinations, and medical records. The recommendations may encompass nutritional plans, recipes, and lifestyle modifications tailored to an individual's dietary needs. Optimizing the management of nutrients in water systems: managing nutrients in water systems is a crucial part of environmental conservation, as it tackles the issues related to nutrient contamination in rivers, lakes, and oceans.

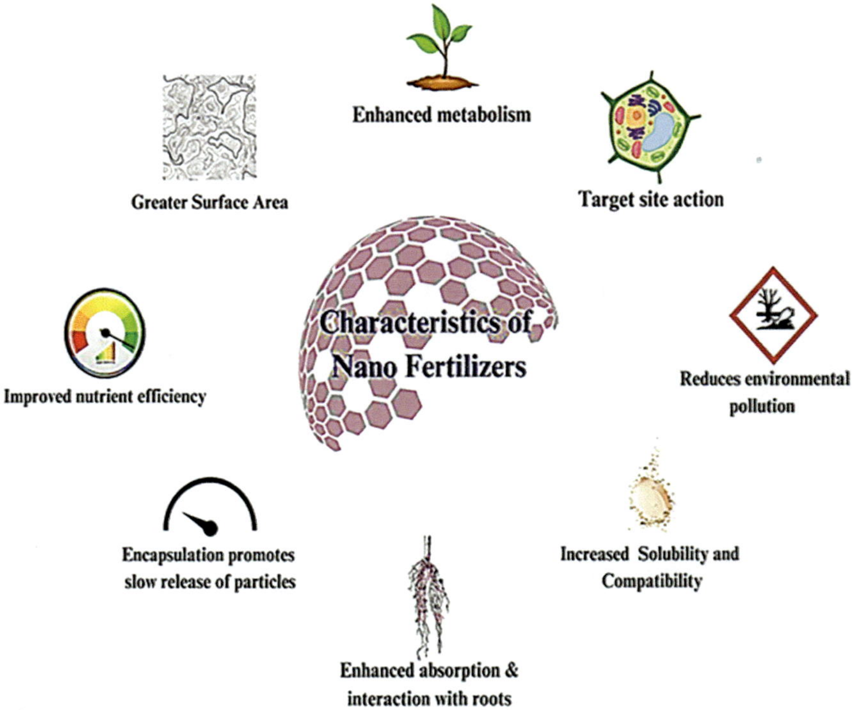

FIGURE 7.4 Characteristics of smart nutrients for horticultural crops.

Source: Created by authors © permission is not required from other sources.

Water quality degradation, harmful algal blooms, and ecosystem imbalances can occur due to excessive nutrient runoff from agricultural activities, wastewater discharge, and industrial processes. Technology is crucial in surveilling and controlling nutrient levels in water systems. Remote sensing technologies, such as satellite imagery and UAVs, allow for evaluating water quality on a broad scope. These technologies can detect variations in nutrient concentrations, enabling authorities to pinpoint areas with heightened nutrient levels and carry out focused interventions. Sensor-based monitoring systems offer instantaneous data on the concentration of nutrients in aquatic environments. These sensors can be strategically positioned in rivers, lakes, and coastal regions to measure parameters such as nitrogen and phosphorus concentrations continuously (Biondo et al., 2022; Eissa et al., 2021; Hani Shukri et al., 2023; Jia et al., 2022; Liu et al., 2023c, 2023d; Machchhar et al., 2022; Wang et al.,

2023b). The data produced by these sensors facilitate prompt decision-making and the execution of measures to alleviate nutrient pollution. Innovative technologies also enable the advancement of intelligent water management systems. These systems combine data from different sources, such as weather forecasts, agricultural practices, and water-quality monitoring, to maximize the efficiency of applying nutrients in agriculture. Intelligent water management systems mitigate the risk of nutrient runoff into water bodies by synchronizing nutrient inputs with crop requirements and environmental conditions. Soil remediation technologies are designed to restore soils that have been contaminated or degraded to a state that is healthy and sustainable. Soil contaminants can originate from industrial activities, improper waste disposal, or agricultural practices. The remediation process entails extracting, degrading, or immobilizing pollutants to guarantee the soil's capacity to sustain plant growth and ecological functions. Phytoremediation is an environmentally friendly and economical method for treating polluted soil, which involves using plants to extract, accumulate, or break down harmful substances. Hyperaccumulators are specific types of plants that can absorb significant amounts of metals or organic pollutants from the soil. These plants can be strategically placed in polluted areas to expedite the extraction of contaminants. Bioremediation is an eco-friendly method for treating soil pollution by utilizing the metabolic processes of microorganisms to break down or convert harmful substances (Chen, 2020; El-Husseiny et al., 2022; Liu et al., 2021; Moheb Afzali et al., 2024; Saha & Bharadwaj, 2023; Sharma et al., 2024; Tan et al., 2023; Xu et al., 2024; Zehra et al., 2023). Microorganisms like bacteria and fungi can decompose organic pollutants into nontoxic substances or transform toxic metals into less harmful compounds. *In situ* bioremediation can facilitate the natural remediation of soil by microorganisms, eliminating the need for extensive excavation. Advanced soil sensing technologies enhance the efficacy of soil remediation strategies. Soil sensors can evaluate the degree of pollution, observe variations in soil characteristics, and monitor the advancement of remediation endeavors. The availability of real-time data enables adaptive management, which ensures that remediation strategies are customized to suit the unique attributes of the contaminated site. Incorporating technology in agriculture, medicine, and environmental management has introduced a new period of inventive and effective practices. Implementing precision farming and crop enhancement techniques fundamentally transforms the agricultural sector, fostering sustainability and adaptability in response to climate change. Targeted drug delivery and personalized nutrition are emerging as practical treatment approaches in medicine and healthcare. These strategies enhance patient outcomes and prioritize individualized care. Technological advancements can tackle urgent ecological challenges by addressing nutrient management in water systems and soil remediation. These environmental applications promote the responsible stewardship of natural resources. As we further adopt and improve these technologies, it is crucial to consider the ethical ramifications, ecological sustainability, and equal availability. The potential of technology to bring about significant change is immense in tackling worldwide issues. Still, deploying it responsibly and ethically is essential to guarantee a beneficial societal influence. Through the utilization of technology in the fields of agriculture, medicine, and environmental management, we can establish a future that is both sustainable and resilient for future generations (Chojnacka & Chojnacki, 2024; Feng et al., 2023; Mazidi et al., 2022; Singh et al., 2024; Song et al., 2023; Wang et al., 2024b).

7.6 CHALLENGES AND FUTURE DIRECTIONS

In the ever-progressing realm of technology, ethical considerations have taken on a crucial role in ensuring that innovation is in harmony with human values, societal welfare, and environmental sustainability. This discourse explores the ethical aspects of technology, explicitly emphasizing its effects on the environment and potential risks to health and safety. In addition, we investigate forthcoming advancements, particularly the fusion of technology with AI and machine learning, and the significance of interdisciplinary partnerships in shaping an accountable technological environment.

7.6.1 ETHICAL CONSIDERATIONS

7.6.1.1 ENVIRONMENTAL IMPACT

A crucial ethical concern in the progress of technology is the ecological consequences of different innovations. Although technology has undeniably brought about beneficial transformations, it has also sparked apprehensions regarding the depletion of resources, pollution, and disruption of ecosystems. The production, utilization, and disposal of electronic devices contribute to generating electronic waste, which contains dangerous substances that present substantial environmental hazards. The manufacturing processes used for technology components, such as semiconductors and batteries, frequently employ resource-intensive methods that harm the environment. For example, the process of extracting rare earth metals leads to the destruction of habitats and the pollution of water. Moreover, the energy consumption of data centers and electronic devices prompts inquiries regarding the carbon footprint of the technology sector. Addressing the environmental consequences of technology necessitates a comprehensive and diverse strategy. Implementing sustainable design principles, such as prioritizing product longevity, recyclability, and eco-friendly materials, can effectively reduce the negative impact of electronic waste. Moreover, advocating for energy-efficient technologies and shifting to renewable energy sources for technology infrastructure is crucial to achieving a more sustainable technological ecosystem (Chakraborty et al., 2023; Dong et al., 2023; El-Husseiny et al., 2024; Gebresenbet et al., 2023; Logesh et al., 2021; Polaka et al., 2024).

7.6.1.2 OCCUPATIONAL HEALTH AND SAFETY ISSUES

The swift incorporation of technology into different facets of our lives has resulted in ethical concerns about health and safety. With our growing dependence on devices, networks, and AI, concerns arise about the possible health consequences of extended exposure to technology and the security of autonomous systems. The effects of radiation emitted by electronic devices, such as smartphones and laptops, on human health have generated discussions and controversies. Although research on the topic is inconclusive, ethical concerns necessitate clear and open communication regarding potential risks and ongoing investigation to gain a deeper understanding of the lasting impacts of technology usage. The emergence of autonomous systems, such as self-driving cars and UAVs, brings about safety concerns that require meticulous ethical analysis. To guarantee the safety of individuals and communities, it is necessary to have rigorous testing, regulatory frameworks, and ethical guidelines that govern the implementation of autonomous technologies. Achieving a harmonious equilibrium between technological advancement and the protection of public well-being remains a multifaceted dilemma that requires cooperation among various industry participants, policymakers, and experts in ethics (Dastghaib et al., 2020; Huang et al., 2024; Lu et al., 2024; Mazidi et al., 2022; Takeshita et al., 2023; Wang et al., 2024; Zhang et al., 2023).

7.6.2 PROSPECTIVE ADVANCEMENTS

7.6.2.1 INCORPORATION OF AI AND MACHINE LEARNING

The future of technology hinges on the seamless incorporation of AI and machine learning, offering unparalleled progress in various domains. As we explore the unexplored realms of AI-driven technologies, ethical considerations become increasingly important, given the immense potential of these innovations. The ethical ramifications of AI go beyond the stages of its development and implementation. The presence of bias in AI algorithms, which often mirrors societal prejudices found in the data used for training, can perpetuate and magnify preexisting inequalities. It is crucial to prioritize fairness, transparency, and accountability in AI systems to prevent inadvertent bias and promote inclusivity in technological applications. The widespread use of AI and machine learning raises significant privacy concerns. Given that these technologies handle extensive quantities of personal data to generate forecasts and make determinations, ensuring the protection of individual privacy emerges as a crucial ethical concern. To achieve a harmonious equilibrium between the advantages of data-driven insights and safeguarding user privacy, it is imperative to establish robust legal frameworks, ethical guidelines, and transparent practices by technology developers (Chen et al., 2024; Karimi et al., 2023; Liu et al., 2024; Moheb Afzali et al., 2024; Qiao et al., 2022; Shreya et al., 2023; Zehra et al., 2023). Tackling the ethical dilemmas posed by advancing technologies necessitates a cooperative and interdisciplinary methodology. As innovations grow more intricate, it becomes crucial to incorporate knowledge from various disciplines, such as ethics, law, sociology, and environmental science. Collaborations between different fields of study promote a comprehensive understanding of the ethical consequences of technology, facilitating well-informed decision-making. The ethical considerations in technology development go beyond the technical aspects and include the societal and cultural dimensions. Engaging in partnerships with social scientists and ethicists can assist technology developers in predicting and resolving the potential societal consequences of their innovations. This proactive strategy guarantees that technological progress is in harmony with human values and positively impacts societal welfare. Moreover, it is imperative to engage legal professionals in creating and implementing technology to establish regulatory structures that safeguard individual rights, guarantee responsibility, and tackle possible ethical transgressions. A legal and ethical framework guides responsible innovation and facilitates carefully managing the relationship between technological advancement and moral concerns. To effectively navigate the ethical boundaries of technology, addressing concerns related to environmental impact, health and safety, and promoting future innovations through a collaborative and intentional approach is necessary. Given the ongoing influence of technology on our society, it is imperative to incorporate ethical considerations throughout the entire process of technological development, spanning from its inception to its disposal. The environmental ramifications of technology necessitate implementing sustainable practices, conscientious resource management, and adopting environmentally friendly technologies. Transparent communication, ongoing research, and the implementation of ethical guidelines are essential to address health and safety concerns and safeguard the well-being of individuals and communities. When considering the future integration of AI and machine learning, addressing ethical concerns about fairness, transparency, and privacy is essential. Interdisciplinary collaborations are necessary for effectively managing the intricacies of technological ethics. These collaborations ensure that advancements align with societal values, uphold individual rights, and positively impact the well-being of the global community. In conclusion, the moral implications associated with technology emphasize the necessity for a deliberate and contemplative approach to advancement. By prioritizing ethical principles in technological progress, we can forge a future where innovations push the limits of what can be achieved and contribute to a fair, environmentally friendly, and impartial global society (Aliakbar Ahovan et al., 2022; Chojnacka & Chojnacki, 2024; Mazidi et al., 2022; Pandya et al., 2023; Raj et al., 2024; Singh et al., 2024; Song et al., 2023; Wang et al., 2024; Zhang et al., 2023).

7.7 CONCLUSION

To summarize, the study of intelligent nutrient delivery systems, their underlying mechanisms, and practical utilization highlights the significant impact of technology in improving nutrient management in different areas. Incorporating advanced systems, such as precision agriculture and personalized nutrition, signifies a fundamental change in our approach to providing nutrients to plants and humans. Advanced technologies, such as sensors, drones, and AI revolutionize agricultural nutrient delivery systems, provide unparalleled precision in managing fertilizers and water resources. These systems empower farmers to make data-based decisions, improving crop yields while reducing environmental impact. Customizing nutrient applications to match the precise requirements of crops enhances the efficient use of resources and supports sustainable farming methods in response to increasing global food needs. Personalized nutrition is becoming a fundamental aspect of human health and well-being, thanks to the advancement of intelligent nutrient delivery systems. By incorporating genetic information, dietary preferences, and health data, these systems offer personalized dietary plans that recognize the distinct nutritional needs of each individual. This personalized approach enhances the efficacy of dietary interventions and empowers individuals to make well-informed decisions for their overall health. The process of efficient nutrient delivery depends greatly on technology, including using nanotechnology for the precise administration of drugs and the application of sensors and data analytics in precision farming. These technological advancements have facilitated unprecedented accuracy and effectiveness, fundamentally transforming agriculture, healthcare, and environmental management. Acknowledging and dealing with ethical concerns when adopting these advanced systems and methods is essential. This will ensure that the advantages of intelligent nutrient delivery are available to all, fair, and environmentally sustainable. Achieving a harmonious equilibrium between technological advancement and ethical obligation will be crucial in fully utilizing these systems' capabilities to improve society. Intelligent nutrient delivery systems have the potential to enhance resource efficiency, promote better health outcomes, and contribute to a more sustainable and resilient future in the constantly changing field of technology. By persistently investigating, improving, and conscientiously implementing these technologies, we can create a future where nutrient distribution is intelligent and considerate of the broader ethical, social, and environmental consequences.

KEYWORDS

- **agriculture**
- **nutrient delivery**
- **nanotechnology**
- **sensing**
- **medicine**
- **healthcare**
- **environment**

REFERENCES

Abadi, B., Yazdanpanah, N., Nokhodchi, A., & Rezaei, N. (2021). Smart biomaterials to enhance the efficiency of immunotherapy in glioblastoma: State of the art and future perspectives. *Advanced Drug Delivery Reviews,* 179, 114035. https://doi.org/https://doi.org/10.1016/j.addr.2021.114035

Ahmed, F., Gill, S.S., Rao, T.N., Arshi, N., Kumar, S., & Prashanthi, Y. (2023). Chapter 13—Nanofertilizers: As smart nanoformulations in the agriculture industry. In N.S. Chauhan, & S. Gill, Eds., *Nanomaterial-Plant Interactions*. Academic Press, pp. 285–299. https://doi.org/https://doi.org/10.1016/B978-0-323-91703-2.00018-X

Aliakbar Ahovan, Z., Esmaeili, Z., Eftekhari, B.S., Khosravimelal, S., Alehosseini, M., Orive, G., Dolatshahi-Pirouz, A., Pal Singh Chauhan, N., Janmey, P.A., Hashemi, A., Kundu, S.C., & Gholipourmalekabadi, M. (2022). Antibacterial smart hydrogels: New hope for infectious wound management. *Materials Today Bio,* 17, 100499. https://doi.org/https://doi.org/10.1016/j.mtbio.2022.100499

Alipio, M.I., Dela Cruz, A.E.M., Doria, J.D.A., & Fruto, R.M.S. (2019). On the design of Nutrient Film Technique hydroponics farm for smart agriculture. *Engineering in Agriculture, Environment and Food,* 12(3). 315–324. https://doi.org/https://doi.org/10.1016/j.eaef.2019.02.008

Aminu, N., Ilyasu, S., Al-Kassim Hassan, M., Kurfi, F.S., Jatau, A.I., Chan, S.-Y., & Alfred-Ugbenbo, D. (2023). Applications of nanofibers drug delivery system in cancer therapy. *Journal of Drug Delivery Science and Technology,* 90, 105128. https://doi.org/https://doi.org/10.1016/j.jddst.2023.105128

Arif, Z.U., Khalid, M.Y., Zolfagharian, A., & Bodaghi, M. (2022). 4D bioprinting of smart polymers for biomedical applications: Recent progress, challenges, and future perspectives. *Reactive and Functional Polymers,* 179, 105374. https://doi.org/https://doi.org/10.1016/j.reactfunctpolym.2022.105374

Avşar, E., & Mowla, M.N. (2022). Wireless communication protocols in smart agriculture: A review on applications, challenges and future trends. *Ad Hoc Networks,* 136, 102982. https://doi.org/https://doi.org/10.1016/j.adhoc.2022.102982

Balu, S. kumar, Andra, S., Jeevanandam, J., Kulabhusan, P.K., Khamari, A., Vedarathinam, V., Hamimed, S., Chan, Y.S., & Danquah, M.K. (2023). Exploring the potential of metal oxide nanoparticles as fungicides and plant nutrient boosters. *Crop Protection,* 174, 106398. https://doi.org/https://doi.org/10.1016/j.cropro.2023.106398

Beheshtizadeh, N., Gharibshahian, M., Bayati, M., Maleki, R., Strachan, H., Doughty, S., & Tayebi, L. (2023). Vascular endothelial growth factor (VEGF) delivery approaches in regenerative medicine. *Biomedicine & Pharmacotherapy,* 166, 115301. https://doi.org/https://doi.org/10.1016/j.biopha.2023.115301

Bharathi, R., Ganesh, S.S., Harini, G., Vatsala, K., Anushikaa, R., Aravind, S., Abinaya, S., & Selvamurugan, N. (2022). Chitosan-based scaffolds as drug delivery systems in bone tissue engineering. *International Journal of Biological Macromolecules,* 222, 132–153. https://doi.org/https://doi.org/10.1016/j.ijbiomac.2022.09.058

Bhavsar, D., Limbasia, B., Mori, Y., Imtiyazali Aglodiya, M., & Shah, M. (2023). A comprehensive and systematic study in smart drip and sprinkler irrigation systems. *Smart Agricultural Technology,* 5, 100303. https://doi.org/https://doi.org/10.1016/j.atech.2023.100303

Biondo, F., Baldassarre, F., Vergaro, V., & Ciccarella, G. (2022). Chapter 3—Controlled biocide release from smart delivery systems: Materials engineering to tune release rate, biointeractions, and responsiveness. In: G.M. Balestra & P.D. Fortunati, Eds., *Micro and Nano Technologies*. Elsevier, pp. 31–147. https://doi.org/https://doi.org/10.1016/B978-0-12-823394-8.00010-X

Chakraborty, R., Mukhopadhyay, A., Paul, S., Sarkar, S., & Mukhopadhyay, R. (2023). Nanocomposite-based smart fertilizers: A boon to agricultural and environmental sustainability. *Science of the Total Environment,* 863, 160859. https://doi.org/https://doi.org/10.1016/j.scitotenv.2022.160859

Chen, D. (2020). Chapter 8—Nanocontainers for the encapsulation and delivery of antioxidants/nutrients to food. In: P. Nguyen-Tri, T.-O. Do, & S.N. Nguyen, Eds., *Micro and Nano Technologies*. Elsevier, pp. 119–136. https://doi.org/https://doi.org/10.1016/B978-0-12-816770-0.00008-3

Chen, J., Hu, S., Sun, M., Shi, J., Zhang, H., Yu, H., & Yang, Z. (2024). Recent advances and clinical translation of liposomal delivery systems in cancer therapy. *European Journal of Pharmaceutical Sciences,* 193, 106688. https://doi.org/https://doi.org/10.1016/j.ejps.2023.106688

Chen, L., Yang, Z., McClements, D.J., Jin, Z., & Miao, M. (2022). Chapter 20—Biological macromolecules for nutrients delivery. In: A.K. Nayak, A.K. Dhara, & D Pal, Eds., *Biological Macromolecules.* Academic Press, pp. 455–477. https://doi.org/https://doi.org/10.1016/B978-0-323-85759-8.00020-8

Chojnacka, K., & Chojnacki, M. (2024). Nutrient recovery from anaerobic digestate: Fertilizer informatics for circular economy. *Environmental Research,* 245, 117953. https://doi.org/https://doi.org/10.1016/j.envres.2023.117953

Dahiya, S., & Dahiya, R. (2023). Chapter 13—Smart drug delivery systems and their clinical potential. In: S.P. Vyas, U. Agrawal, & D.D. Sharma, Eds., *Smart Polymeric Nano-Constructs in Drug Delivery*. Academic Press, pp. 401–436. https://doi.org/https://doi.org/10.1016/B978-0-323-91248-8.00007-6

Dastghaib, S., Afkhami, A., Mokarram, P., Madrakian, T., & Ahmadi, M. (2020). Chapter 15—Smart nanocarriers in glucose transporters-targeted delivery of anticancer drugs. In: P. Nguyen-Tri, T.-O. Do, & T.A.B.T.-S.N. Nguyen, Eds., *Micro and Nano Technologies*. Elsevier, pp. 251–269. https://doi.org/https://doi.org/10.1016/B978-0-12-816770-0.00015-0

Daund, V., Chalke, S., Sherje, A.P., & Kale, P.P. (2021). ROS responsive mesoporous silica nanoparticles for smart drug delivery: A review. *Journal of Drug Delivery Science and Technology,* 64, 102599. https://doi.org/https://doi.org/10.1016/j.jddst.2021.102599

de León, E.H.-P., Valle-Pérez, A.U., Khan, Z.N., & Hauser, C.A.E. (2023). Intelligent and smart biomaterials for sustainable 3D printing applications. *Current Opinion in Biomedical Engineering,* 26, 100450. https://doi.org/https://doi.org/10.1016/j.cobme.2023.100450

Dong, J., Han, A., Zhao, Y., Li, H., Yang, Y., Yuan, B., Wang, Y., Liu, R., Yin, X., & Du, X. (2023). Smart, degradable, and eco-friendly carboxymethyl cellulose-CaII hydrogel-like networks gated MIL-101(FeIII) nanoherbicides for paraquat delivery. *Science of the Total Environment,* 903, 166424. https://doi.org/https://doi.org/10.1016/j.scitotenv.2023.166424

Eissa, N.G., Elsabahy, M., & Allam, A. (2021). Engineering of smart nanoconstructs for delivery of glucagon-like peptide-1 analogs. *International Journal of Pharmaceutics,* 597, 120317. https://doi.org/https://doi.org/10.1016/j.ijpharm.2021.120317

El-Husseiny, H.M., Mady, E.A., Doghish, A.S., Zewail, M.B., Abdelfatah, A.M., Noshy, M., Mohammed, O.A., & El-Dakroury, W.A. (2024). Smart/stimuli-responsive chitosan/gelatin and other polymeric macromolecules natural hydrogels vs. synthetic hydrogels systems for brain tissue engineering: A state-of-the-art review. *International Journal of Biological Macromolecules,* 260, 129323. https://doi.org/https://doi.org/10.1016/j.ijbiomac.2024.129323

El-Husseiny, H.M., Mady, E.A., El-Dakroury, W.A., Zewail, M.B., Noshy, M., Abdelfatah, A.M., & Doghish, A.S. (2022). Smart/stimuli-responsive hydrogels: State-of-the-art platforms for bone tissue engineering. *Applied Materials Today,* 29, 101560. https://doi.org/https://doi.org/10.1016/j.apmt.2022.101560

El-Husseiny, H.M., Mady, E.A., Hamabe, L., Abugomaa, A., Shimada, K., Yoshida, T., Tanaka, T., Yokoi, A., Elbadawy, M., & Tanaka, R. (2022). Smart/stimuli-responsive hydrogels: Cutting-edge platforms for tissue engineering and other biomedical applications. *Materials Today Bio,* 13, 100186. https://doi.org/https://doi.org/10.1016/j.mtbio.2021.100186

Fan, T., Meng, Z., Chen, X., Liang, Y., Zhao, M., Wu, Q., Cui, J., Xu, W., & Wang, J. (2023). Fabrication of stimuli-responsive nanoparticles for high-efficiency chlorantraniliprole delivery and smart control of *Spodoptera frugiperda. Industrial Crops and Products,* 205, 117427. https://doi.org/https://doi.org/10.1016/j.indcrop.2023.117427

Feng, Y., Mei, D., & Zhao, H. (2023). Auction-based deep learning-driven smart agricultural supply chain mechanism. *Applied Soft Computing,* 149, 111009. https://doi.org/https://doi.org/10.1016/j.asoc.2023.111009

Gebresenbet, G., Bosona, T., Patterson, D., Persson, H., Fischer, B., Mandaluniz, N., Chirici, G., Zacepins, A., Komasilovs, V., Pitulac, T., & Nasirahmadi, A. (2023). A concept for application of integrated digital technologies to enhance future smart agricultural systems. *Smart Agricultural Technology,* 5, 100255. https://doi.org/https://doi.org/10.1016/j.atech.2023.100255

Hani Shukri, A., Shawkat Mizzouri, N., Saifadeen, A., Saleem, D., & O'Malley, M. (2023). Assessing and improving the sanitation services in Kabarto 2 IDP camp using the Smart sanitation approach. *Ain Shams Engineering Journal,* 102593. https://doi.org/https://doi.org/10.1016/j.asej.2023.102593

He, J., Li, J., Gao, Y., He, X., & Hao, G. (2023). Nano-based smart formulations: A potential solution to the hazardous effects of pesticide on the environment. *Journal of Hazardous Materials*, 456, 131599. https://doi.org/https://doi.org/10.1016/j.jhazmat.2023.131599

Henn, J.G., Aguirre, T.A.S., Nugent, M., & Moura, D.J. (2024). Cancer nanomedicine: Recent developments in drug delivery systems and strategies to overcome eventual barriers to achieve a better outcome. *Journal of Drug Delivery Science and Technology,* 91, 105254. https://doi.org/https://doi.org/10.1016/j.jddst.2023.105254

Hossain, M.I., Zahid, M.S., Chowdhury, M.A., Hossain, M.M.M., Hossain, N., Islam, M.A., & Mobarak, M.H. (2024). Smart bandage: A device for wound monitoring and targeted treatment. *Results in Chemistry*, 7, 101292. https://doi.org/https://doi.org/10.1016/j.rechem.2023.101292

Huang, Q., Zeng, Y., Qiu, Y., Zou, J., Li, F., Liu, X., Nezamzadeh-Ejhieh, A., Song, H., & Liu, J. (2024). Applications and prospects of new transdermal drug delivery system based on metal-organic frameworks for diabetic wound healing. *Dyes and Pigments*, 222, 111865. https://doi.org/https://doi.org/10.1016/j.dyepig.2023.111865

Ismail, E.A., Devnarain, N., Govender, T., & Omolo, C.A. (2022). Stimuli-responsive and biomimetic delivery systems for sepsis and related complications. *Journal of Controlled Release*, 352, 1048–1070. https://doi.org/https://doi.org/10.1016/j.jconrel.2022.11.013

Jia, R., Li, T., Jiang, W., Wang, J., Li, X., Qu, Q., Dang, J., & Li, P. (2022). Smart Milli-capsules manipulated by nIR irradiation for controllable drug delivery in-vivo for renal cell carcinoma and neurodegenerative diseases. *Materials & Design*, 224, 111287. https://doi.org/https://doi.org/10.1016/j.matdes.2022.111287

Kamsani, N.H., Haris, M.S., Pandey, M., Taher, M., & Rullah, K. (2021). Biomedical application of responsive 'smart' electrospun nanofibers in drug delivery system: A minireview. *Arabian Journal of Chemistry*, 14(7), 103199. https://doi.org/https://doi.org/10.1016/j.arabjc.2021.103199

Karimi, K., Mojtabavi, S., Tehrany, P.M., Nejad, M.M., Rezaee, A., Mohtashamian, S., Hamedi, E., Yousefi, F., Salmani, F., Zandieh, M.A., Nabavi, N., Rabiee, N., Ertas, Y.N., Salimimoghadam, S., Rashidi, M., Rahmanian, P., Hushmandi, K., & Yu, W. (2023). Chitosan-based nanoscale delivery systems in hepatocellular carcinoma: Versatile bio-platform with theranostic application. *International Journal of Biological Macromolecules,* 242, 124935. https://doi.org/https://doi.org/10.1016/j.ijbiomac.2023.124935

Kaur, H., Hussain, S.J., Mir, R.A., Chandra Verma, V., Naik, B., Kumar, P., & Dubey, R.C. (2023). Nanofertilizers—Emerging smart fertilizers for modern and sustainable agriculture. *Biocatalysis and Agricultural Biotechnology*, 54, 102921. https://doi.org/https://doi.org/10.1016/j.bcab.2023.102921

Khan, M.U.A., Stojanović, G.M., Abdullah, M.F. Bin, Dolatshahi-Pirouz, A., Marei, H.E., Ashammakhi, N., & Hasan, A. (2024). Fundamental properties of smart hydrogels for tissue engineering applications: A review. *International Journal of Biological Macromolecules,* 254, 127882. https://doi.org/https://doi.org/10.1016/j.ijbiomac.2023.127882

Khanna, N., Chatterji, T., Singh, S., & Pandey, S. (2023). Application of stimuli responsive nanocomposites in drug delivery and theranostics to counter cancer proliferation and drug resistance. *Journal of Drug Delivery Science and Technology,* 88, 104958. https://doi.org/https://doi.org/10.1016/j.jddst.2023.104958

Liu, H., Fan, Y., Zhong, J., Malkoch, M., Cai, Z., & Wang, Z. (2023a). Advance in oral delivery of living material. *Biomedical Technology*, 3, 26–39. https://doi.org/https://doi.org/10.1016/j.bmt.2022.12.003

Liu, F., Meng, F., Yang, Z., Wang, H., Ren, Y., Cai, Y., & Zhang, X. (2023b). Exosome-biomimetic nanocarriers for oral drug delivery. *Chinese Chemical Letters*, 35, 109335. https://doi.org/https://doi.org/10.1016/j.cclet.2023.109335

Liu, L., Tang, H., & Wang, Y. (2023c). Polymeric biomaterials: Advanced drug delivery systems in osteoarthritis treatment. *Heliyon*, 9(11), e21544. https://doi.org/https://doi.org/10.1016/j.heliyon.2023.e21544

Liu, Y., Wu, J., Li, W., Li, J., Han, H., & Song, Z. (2023d). Responsive metal–organic framework nanocarrier delivery system: An effective solution against bacterial infection. *Coordination Chemistry Reviews*, 496, 215431. https://doi.org/https://doi.org/10.1016/j.ccr.2023.215431

Liu, F., Zheng, B., Zheng, N., Alfaiz, F.A., Ali, H.E., AL Garalleh, H., Assilzadeh, H., & Xia, S. (2024). Smart nano generation of transgenic algae expressing white spot syndrome virus in shrimps for inner ear-oral infection treatments using the spotted hyena optimizer (SHO)-Long short-term memory algorithm. *Environmental Research,* 243, 117519. https://doi.org/https://doi.org/10.1016/j.envres.2023.117519

Liu, S., Qamar, S.A., Qamar, M., Basharat, K., & Bilal, M. (2021). Engineered nanocellulose-based hydrogels for smart drug delivery applications. *International Journal of Biological Macromolecules,* 181, 275–290. https://doi.org/https://doi.org/10.1016/j.ijbiomac.2021.03.147

Logesh, D., Vallikkadan, M.S., Leena, M.M., Moses, J.A., & Anandharamakrishnan, C. (2021). Advances in microfluidic systems for the delivery of nutraceutical ingredients. *Trends in Food Science & Technology*, 116, 501–524. https://doi.org/https://doi.org/10.1016/j.tifs.2021.07.011

Lu, S., Zhang, C., Wang, J., Zhao, L., & Li, G. (2024). Research progress in nano-drug delivery systems based on the characteristics of the liver cancer microenvironment. *Biomedicine & Pharmacotherapy,* 170, 116059. https://doi.org/https://doi.org/10.1016/j.biopha.2023.116059

Machchhar, R.J., Toller, C.N.K., Bertoni, A., & Bertoni, M. (2022). Data-driven value creation in Smart Product-Service System design: State-of-the-art and research directions. *Computers in Industry,* 137, 103606. https://doi.org/https://doi.org/10.1016/j.compind.2022.103606

Mazidi, Z., Javanmardi, S., Naghib, S.M., & Mohammadpour, Z. (2022). Smart stimuli-responsive implantable drug delivery systems for programmed and on-demand cancer treatment: An overview on the emerging materials. *Chemical Engineering Journal,* 433, 134569. https://doi.org/https://doi.org/10.1016/j.cej.2022.134569

Moheb Afzali, A., Kheradmand, M.A., & Naghib, S.M. (2024). Bioreactor design-assisted bioprinting of stimuli-responsive materials for tissue engineering and drug delivery applications. *Bioprinting,* 37, e00325. https://doi.org/https://doi.org/10.1016/j.bprint.2023.e00325

Nandini, B., & Geetha, N. (2021). 15—Smart delivery mechanisms of nanofertilizers and nanocides in crop biotechology. In: S. Jogaiah, H.B. Singh, L.F. Fraceto, & R. de Lima, Eds., *Woodhead Publishing Series in Food Science, Technology and Nutrition.* Woodhead Publishing, pp. 385–414. https://doi.org/https://doi.org/10.1016/B978-0-12-820092-6.00015-X

Ni, X., Li, C., Zhou, J., Zhang, M., You, B., Wu, L., Li, W., & Guo, J. (2023). Design of chlorella-inspired and smart-responsive bio-microspheres and their application in eco-friendly anti-biofouling coatings. *Chemical Engineering Journal,* 478, 147426. https://doi.org/https://doi.org/10.1016/j.cej.2023.147426

Norouzi, Z., & Abdouss, M. (2023). Electrospun nanofibers using β-cyclodextrin grafted chitosan macromolecules loaded with indomethacin as an innovative drug delivery system. *International Journal of Biological Macromolecules,* 233, 123518. https://doi.org/https://doi.org/10.1016/j.ijbiomac.2023.123518

Pachuau, L. (2020). Chapter 5—Nanocellulose and nanohydrogel-mediated sustained drug delivery: Smart medical technology. In: F. Mohammad, H.A. Al-Lohedan, & M. Jawaid, Eds., *Micro and Nano Technologies.* Elsevier, pp. 115–130. https://doi.org/https://doi.org/10.1016/B978-0-12-816789-2.00005-5

Pandya, A.K., Vora, L.K., Umeyor, C., Surve, D., Patel, A., Biswas, S., Patel, K., & Patravale, V.B. (2023). Polymeric in situ forming depots for long-acting drug delivery systems. *Advanced Drug Delivery Reviews,* 200, 115003. https://doi.org/https://doi.org/10.1016/j.addr.2023.115003

Parekh, P.Y., Patel, V.I., Khimani, M.R., & Bahadur, P. (2023). Self-assembly of bile salts and their mixed aggregates as building blocks for smart aggregates. *Advances in Colloid and Interface Science,* 312, 102846. https://doi.org/https://doi.org/10.1016/j.cis.2023.102846

Paul, K., Chatterjee, S.S., Pai, P., Varshney, A., Juikar, S., Prasad, V., Bhadra, B., & Dasgupta, S. (2022). Viable smart sensors and their application in data driven agriculture. *Computers and Electronics in Agriculture,* 198, 107096. https://doi.org/https://doi.org/10.1016/j.compag.2022.107096

Polaka, S., Pawar, B., Vasdev, N., & Tekade, R.K. (2024). Development and biological evaluation of smart powder bandage for wound healing and dressing applications. *International Journal of Biological Macromolecules,* 258, 129044. https://doi.org/https://doi.org/10.1016/j.ijbiomac.2023.129044

Prajapati, V., & Roy, S. (2023). Chapter 14—Smart packaging for medicinal food supplements. In: T.A. Inamuddin, & J.N. Cruz, Eds., *Green Sustainable Process for*

Chemical and Environmental Engineering and Science. Elsevier, pp. 229–266. https://doi.org/https://doi.org/10.1016/B978-0-323-95644-4.00007-3

Prange, C.J., Hu, X., & Tang, L. (2023). Smart chemistry for traceless release of anticancer therapeutics. *Biomaterials*, 303, 122353. https://doi.org/https://doi.org/10.1016/j.biomaterials.2023.122353

Qiao, D., Li, J., Zhang, S., & Yang, X. (2022). Controlled release fertilizer with temperature-responsive behavior coated using polyether polyol (PPG)/polycaprolactone (PCL) blend-based polyurethane performs smart nutrient release. *Materials Today Chemistry*, 26, 101249. https://doi.org/https://doi.org/10.1016/j.mtchem.2022.101249

Raikwar, S., & Jain, S.K. (2022). Chapter 22—Stimuli-responsive nanovesicles for smart drug delivery. In: A.K. Nayak, M.S. Hasnain, T.M. Aminabhavi, & V.P. Torchilin, Eds., *Systems of Nanovesicular Drug Delivery*. Academic Press, pp. 353–365. https://doi.org/https://doi.org/10.1016/B978-0-323-91864-0.00013-9

Raj, V., Chun, K.-S., & Lee, S. (2024). State-of-the-art advancement in tara gum polysaccharide (*Caesalpinia spinosa*) modifications and their potential applications for drug delivery and the food industry. *Carbohydrate Polymers*, 323, 121440. https://doi.org/https://doi.org/10.1016/j.carbpol.2023.121440

Raja, H.N., Din, F. ud, Shabbir, K., Khan, S., Alamri, A.H., Al Awadh, A.A., Lahiq, A.A., & Alasiri, A. (2023). Sodium alginate-based smart gastro-retentive drug delivery system of revaprazan loaded SLNs; Formulation and characterization. *International Journal of Biological Macromolecules*, 253, 127402. https://doi.org/https://doi.org/10.1016/j.ijbiomac.2023.127402

Ramasamy, T., Ruttala, H.B., Gupta, B., Poudel, B.K., Choi, H.-G., Yong, C.S., & Kim, J.O. (2017). Smart chemistry-based nanosized drug delivery systems for systemic applications: A comprehensive review. *Journal of Controlled Release*, 258, 226–253. https://doi.org/https://doi.org/10.1016/j.jconrel.2017.04.043

Riseh, R.S., Vatankhah, M., Hassanisaadi, M., & Varma, R.S. (2024). A review of chitosan nanoparticles: Nature's gift for transforming agriculture through smart and effective delivery mechanisms. *International Journal of Biological Macromolecules*, 260, 129522. https://doi.org/https://doi.org/10.1016/j.ijbiomac.2024.129522

Saha, S., & Bharadwaj, A. (2023). A step towards smart agriculture using metallic nanostructures. *Plant Stress*, 10, 100216. https://doi.org/https://doi.org/10.1016/j.stress.2023.100216

Saif, S., Roy, P., Chowdhury, C., Biswas, S., & Maulik, U. (2022). Chapter 11 - Smart e-agriculture monitoring systems. In: A. Abraham, S. Dash, J.J.P.C. Rodrigues, B. Acharya, & S.K. Pani, Eds., *Intelligent Data-Centric Systems*. Academic Press, pp. 183–203. https://doi.org/https://doi.org/10.1016/B978-0-12-823694-9.00002-5

Saxena, A., Dutta, A., Kapoor, N., Kumar, A., & Tiwari, A. (2022). Envisaging marine diatom Thalassiosira weissflogii as a "SMART" drug delivery system for insoluble drugs. *Journal of Drug Delivery Science and Technology*, 68, 102983. https://doi.org/https://doi.org/10.1016/j.jddst.2021.102983

Shahbazi, M., Jäger, H., Ettelaie, R., Mohammadi, A., & Asghartabar Kashi, P. (2023). Multimaterial 3D printing of self-assembling smart thermo-responsive polymers into 4D printed objects: A review. *Additive Manufacturing*, 71, 103598. https://doi.org/https://doi.org/10.1016/j.addma.2023.103598

Shanmugavel, D., Rusyn, I., Solorza-Feria, O., & Kamaraj, S.-K. (2023). Sustainable SMART fertilizers in agriculture systems: A review on fundamentals to in-field applications. *Science of the Total Environment,* 904, 166729. https://doi.org/https://doi.org/10.1016/j.scitotenv.2023.166729

Sharma, R., Kumar, S., Malviya, R., Prajapati, B.G., Puri, D., Limmatvapirat, S., & Sriamornsak, P. (2024). Recent advances in biopolymer-based mucoadhesive drug delivery systems for oral application. *Journal of Drug Delivery Science and Technology,* 91, 105227. https://doi.org/https://doi.org/10.1016/j.jddst.2023.105227

Sharma, S., Kumar, A., Choudhary, A., Harish, B.M., Karmakar, P., Sharma, P., Singh, J., Pandey, V., & Mehta, S. (2022). Recent developments in smart nano-agrochemicals: A promise for revolutionizing present-day agriculture. *Materials Today: Proceedings*, 69, 530–534. https://doi.org/https://doi.org/10.1016/j.matpr.2022.09.306

Shi, Z., Zhou, Y., Fan, T., Lin, Y., Zhang, H., & Mei, L. (2020). Inorganic nano-carriers based smart drug delivery systems for tumor therapy. *Smart Materials in Medicine*, 1, 32–47. https://doi.org/https://doi.org/10.1016/j.smaim.2020.05.002

Shreya, S., Chatterjee, K., & Singh, A. (2023). BFSF: A secure IoT based framework for smart farming using blockchain. *Sustainable Computing: Informatics and Systems*, 40, 100917. https://doi.org/https://doi.org/10.1016/j.suscom.2023.100917

Singh, A.K., Pal, P., Pandey, B., Goksen, G., Sahoo, U.K., Lorenzo, J.M., & Sarangi, P.K. (2023). Development of "Smart Foods" for health by nanoencapsulation: Novel technologies and challenges. *Food Chemistry: X*, 20, 100910. https://doi.org/https://doi.org/10.1016/j.fochx.2023.100910

Singh, R.R., Mondal, I., Janjua, T., Popat, A., & Kulshreshtha, R. (2024). Engineered smart materials for RNA based molecular therapy to treat Glioblastoma. *Bioactive Materials*, 33, 396–423. https://doi.org/https://doi.org/10.1016/j.bioactmat.2023.11.007

Song, Y.H., De, R., & Lee, K.T. (2023). Emerging strategies to fabricate polymeric nanocarriers for enhanced drug delivery across blood-brain barrier: An overview. *Advances in Colloid and Interface Science,* 320, 103008. https://doi.org/https://doi.org/10.1016/j.cis.2023.103008

Srivastava, N., Srivastav, A.K., & Shanker, K. (2024). 11—Smart nanomaterials in food formulations and enhancing the bioavailability of nutrients/nutraceuticals. In: P. Singh, P. Khare, D. Mishra, M. Bilal, & M. Sillanpää, Eds., *Micro and Nano Technologies*. Elsevier, pp. 283–314. https://doi.org/https://doi.org/10.1016/B978-0-323-99682-2.00012-8

Takeshita, V., Campos, E.V.R., Rodrigues, J.S., & Fraceto, L.F. (2023). Opinion: Hybrid nanoparticle systems—Two-way delivery approach for agriculture. *Plant Nano Biology*, 6, 100053. https://doi.org/https://doi.org/10.1016/j.plana.2023.100053

Tan, M., Zhang, X., Sun, S., & Cui, G. (2023). Chapter Two—Nanostructured steady-state nanocarriers for nutrients preservation and delivery. In: F. Toldrá, Ed., *Advances in Food and Nutrition Research*. Academic Press, Vol. 106, pp. 31–93. https://doi.org/https://doi.org/10.1016/bs.afnr.2023.02.001

Teng, G., Chen, C., Jing, N., Chen, C., Duan, Y., Zhang, L., Wu, Z., & Zhang, J. (2023). Halloysite nanotubes-based composite material with acid/alkali dual pH response and foliar adhesion for smart delivery of hydrophobic pesticide. *Chemical Engineering Journal,* 451, 139052. https://doi.org/https://doi.org/10.1016/j.cej.2022.139052

Thananukul, K., Kaewsaneha, C., Opaprakasit, P., Lebaz, N., Errachid, A., & Elaissari, A. (2021). Smart gating porous particles as new carriers for drug delivery. *Advanced*

Drug Delivery Reviews, 174, 425–446. https://doi.org/https://doi.org/10.1016/j.addr.2021.04.023

Thorn, C.R., Howell, P.L., Wozniak, D.J., Prestidge, C.A., & Thomas, N. (2021). Enhancing the therapeutic use of biofilm-dispersing enzymes with smart drug delivery systems. *Advanced Drug Delivery Reviews,* 179, 113916. https://doi.org/https://doi.org/10.1016/j.addr.2021.113916

Tian, B., & Liu, J. (2023). Smart stimuli-responsive chitosan hydrogel for drug delivery: A review. *International Journal of Biological Macromolecules,* 235, 123902. https://doi.org/https://doi.org/10.1016/j.ijbiomac.2023.123902

Türk, S., Altınsoy, I., Efe, G.Ç., Ipek, M., Özacar, M., & Bindal, C. (2021). A novel multifunctional NCQDs-based injectable self-crosslinking and in situ forming hydrogel as an innovative stimuli responsive smart drug delivery system for cancer therapy. *Materials Science and Engineering: C*, 121, 111829. https://doi.org/https://doi.org/10.1016/j.msec.2020.111829

Wang, X., Li, C., Wang, Y., Chen, H., Zhang, X., Luo, C., Zhou, W., Li, L., Teng, L., Yu, H., & Wang, J. (2022). Smart drug delivery systems for precise cancer therapy. *Acta Pharmaceutica Sinica B*, 12(11), 4098–4121. https://doi.org/https://doi.org/10.1016/j.apsb.2022.08.013

Wang, P., Li, Z., Li, H., Zhang, D., Wang, W., Xu, X., Xie, Q., Duan, Z., Xia, X., Guo, G., Shaheen, A., Zhou, Y., Wang, D., Guo, S., Hu, Z., Galbraith, D.W., & Song, C.-P. (2023a). Smart crops. *New Crops*. 1, 100007. https://doi.org/https://doi.org/10.1016/j.ncrops.2023.12.002

Wang, Y., Li, J., Tang, M., Peng, C., Wang, G., Wang, J., Wang, X., Chang, X., Guo, J., & Gui, S. (2023b). Smart stimuli-responsive hydrogels for drug delivery in periodontitis treatment. *Biomedicine & Pharmacotherapy,* 162, 114688. https://doi.org/https://doi.org/10.1016/j.biopha.2023.114688

Wang, X., Li, R., & Zhao, H. (2024a). Enhancing angiogenesis: Innovative drug delivery systems to facilitate diabetic wound healing. *Biomedicine & Pharmacotherapy,* 170, 116035. https://doi.org/https://doi.org/10.1016/j.biopha.2023.116035

Wang, Z., Fu, S., Guo, Y., Han, Y., Ma, C., Li, R., & Yang, X. (2024b). Classification and design strategies of polysaccharide-based nano-nutrient delivery systems for enhanced bioactivity and targeted delivery: A review. *International Journal of Biological Macromolecules,* 256, 128440. https://doi.org/https://doi.org/10.1016/j.ijbiomac.2023.128440

Wanyama, J., Soddo, P., Nakawuka, P., Tumutegyereize, P., Bwambale, E., Oluk, I., Mutumba, W., & Komakech, A.J. (2023). Development of a solar powered smart irrigation control system Kit. *Smart Agricultural Technology,* 5, 100273. https://doi.org/https://doi.org/10.1016/j.atech.2023.100273

Wu, B., Gao, B., Xu, W., Wang, H., Yi, Y., & R, P. (2022). Sustainable food smart manufacturing technology. *Information Processing & Management*, 59(1), 102754. https://doi.org/https://doi.org/10.1016/j.ipm.2021.102754

Xu, C., Gantumur, M.-A., Sun, J., Guo, J., Ma, J., Jiang, Z., Wang, W., Zhang, J., Ma, Y., Hou, J., & McClements, D.J. (2024). Design of probiotic delivery systems for targeted release. *Food Hydrocolloids*, 149, 109588. https://doi.org/https://doi.org/10.1016/j.foodhyd.2023.109588

Zehra, S., Mobin, M., Aslam, R., & Bhat, S. ul I. (2023). Nanocontainers: A comprehensive review on their application in the stimuli-responsive smart functional coatings. *Progress in Organic Coatings*, 176, 107389. https://doi.org/https://doi.org/10.1016/j.porgcoat.2022.107389

Zhang, M., Han, Y., Li, D., Xu, S., & Huang, Y. (2023). Smart horticulture as an emerging interdisciplinary field combining novel solutions: Past development, current challenges, and future perspectives. *Horticultural Plant Journal*, 10(6), 1257–1273. https://doi.org/https://doi.org/10.1016/j.hpj.2023.03.015

Zhu, J., Zhou, H., Gerhard, E.M., Zhang, S., Parra Rodríguez, F.I., Pan, T., Yang, H., Lin, Y., Yang, J., & Cheng, H. (2023). Smart bioadhesives for wound healing and closure. *Bioactive Materials*, 19, 360–375. https://doi.org/https://doi.org/10.1016/j.bioactmat.2022.04.020

Zou, L.-B., Gong, J.-Y., Ju, X.-J., Liu, Z., Wang, W., Xie, R., & Chu, L.-Y. (2022). Smart membranes for biomedical applications. *Chinese Journal of Chemical Engineering*, 49, 34–45. https://doi.org/https://doi.org/10.1016/j.cjche.2022.06.006

CHAPTER 8

Nanochemicals Improvising Production Constraints in Horticulture

BHARAT TAYNATH[1], PRATAP KADAM[1], KHALANGRE ATUL[2*], ANIS MIRZA, and AHAMED SHABEER T. P.[1]

[1]*National Referral Laboratory, ICAR—National Research Centre for Grapes, Pune, Maharashtra, India*

[2]*Department of Horticulture, Lovely Professional University, Phagwara, Punjab, India*

*Corresponding author

ABSTRACT

This chapter explores the potential of nanochemicals in improving horticultural production constraints, focusing on the application of nanotechnology in the form of nanofertilizers, nanopesticides, and nanofilms. Nanofertilizers, engineered to release nutrients slowly and steadily, contribute to controlled release, improved nutrient absorption, and reduced environmental impact. These nanoscale fertilizers are designed to address the limitations of traditional fertilizers, offering a targeted and efficient approach to providing essential nutrients to plants. Nanopesticides, which utilize nanoparticles, exhibit target-specificity, environmental friendliness, and exceptional efficacy in the control of insect pests and diseases. The use of nanopesticides offers multiple advantages, including reduced pesticide application, lower environmental impact, and reduced risk of pesticide resistance. Nanofilms, composed of nanomaterials, are used as coatings on harvested products or packaging materials to enhance shelf-life, reduce spoilage, and maintain product quality. The nanofilms

Smart Horticulture Production: Integrating Advanced Technologies.
Rajni Rajan and Duniya Ram Singh (Eds.)

DOI: 10.1201/9781998511679-8

are designed to specifically target certain areas, gradually release their contents, and greatly enhance plant growth and productivity. The application of nanotechnology in horticulture production represents a significant advancement in agricultural sustainability, offering a targeted and efficient approach to providing essential nutrients to plants while minimizing environmental impact.

8.1 INTRODUCTION

Horticulture, a critical component of agriculture, plays a pivotal role in ensuring food security, economic prosperity, and environmental sustainability. However, the productivity of horticultural crops often faces significant challenges due to various factors, including nutrient limitations, pest and disease pressures, and adverse environmental conditions (Humbal & Pathak, 2023). In recent years, the application of nanotechnology in agriculture has emerged as a promising avenue to address these constraints. Specifically, the use of nanochemicals, which encompass nanomaterials and nanoformulations tailored for horticultural applications, has garnered substantial attention (Rai et al., 2022).

Nanochemicals are chemicals that have been intentionally modified or controlled at a very small scale, usually between 1 and 100 nm. Nanochemicals involve the deliberate control and manipulation of chemical structures and properties at the nanometer level. These materials exhibit unique properties and behaviors due to their small size, which can be significantly different from their bulk counterparts. Nanochemicals often display higher reactivity due to their increased surface area, allowing for more efficient chemical interactions and controlled release kinetics. In recent years, the use of nanotechnology has become a prominent method in the development of sustainable agricultural practices, particularly in the field of horticulture crop production. The different types of nanochemicals, such as nanoclays, nanotubes, and nanowires (Mittal et al., 2020), offer a unique set of chemical and physical properties which can be harnessed to improve the efficiency of plant growth, nutrient uptake and delivery, pest management, stress tolerance, and postharvest quality in horticultural crops (Table 8.1) (Khodakovskaya et al., 2012; Suriyaprabha et al., 2012; Li et al., 2018; Rai et al., 2022; Humbal & Pathak, 2023).

TABLE 8.1 Nanochemicals Used for Horticultural Crop Production

Nanochemicals	Description	Examples
Nanofertilizers	Improve plant nutrition by increasing nutrient uptake efficiency and reducing environmental stresses	Nano-encapsulated urea, nano-encapsulated phosphorus, and nano-encapsulated potassium
Nanoplant growth stimulators	Enhance plant growth and development by stimulating root growth, photosynthesis, and stress tolerance	Nano-encapsulated plant growth regulators, such as auxins, gibberellins, and cytokinins
Nanopesticides	Reduce pesticide applications and improve plant protection by targeting specific pests and diseases	Nano-encapsulated insecticides (e.g., imidacloprid, chlorantraniliprole, and spinosad). Nano-encapsulated fungicides (e.g., azoxystrobin, fludioxonil, and tebuconazole)
		Nano-encapsulated nematicides (e.g., aldicarb, oxamyl, and fenamiphos)
Nano-insecticides	Control insect pests by targeting specific insects and reducing the risk of resistance development	Nano-encapsulated insecticides (e.g., imidacloprid, chlorantraniliprole, and spinosad)
Nanofungicides	Control fungal diseases by targeting specific fungi and reducing the risk of resistance development	Nano-encapsulated fungicides (e.g., azoxystrobin, fludioxonil, and tebuconazole)
Nanonematicides	Control nematodes by targeting specific nematodes and reducing the risk of resistance development	Nano-encapsulated nematicides (e.g., aldicarb, oxamyl, and fenamiphos)
Nanofilms	Improve postharvest management by reducing water loss, maintaining freshness, and preventing spoilage	Nanocoated polyethylene films
Nanopackaging	Improve postharvest management by reducing water loss, maintaining freshness, and preventing spoilage	Nanocoated polyethylene films

Generally, pesticides are key important in crop management (Figure 8.1). However, the excess use and target delivery of agrochemicals are a critical issue in conventional agricultural practices. The inefficient delivery of agrochemicals can result in significant environmental impacts, reduced effectiveness, and potential risks to nontarget organisms (Babu et al., 2022). Fertilizers are required to increase crop productivity, but on the other hand, their overuse reduces soil fertility by upsetting the mineral balance in the soil (Mittal et al., 2020). Moreover, the exorbitant expenses associated with these pesticides and fertilizers necessitate regulation. The application of nanochemicals in the horticultural industry seeks to minimize excessive usage in order to enhance production; decrease the quantity of products required for plant protection, and enhance productivity while simultaneously enhancing quality (Usman et al., 2020).

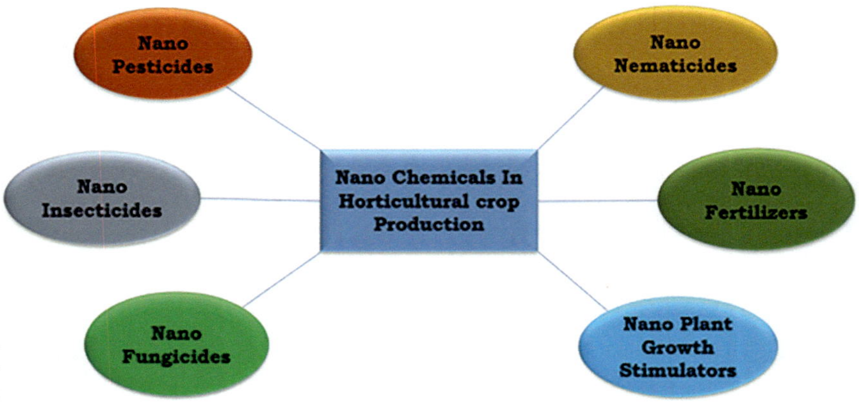

FIGURE 8.1 Utilization of nanotechnology in horticultural crop production.

Additionally, research has delved into the realm of postharvest handling and storage, where nanomaterials and coatings have demonstrated the ability to extend shelf life, reduce spoilage, and maintain quality during transportation (Babu et al., 2022). These advancements are poised to revolutionize the logistics and economics of horticultural supply chains. In addition, the smart nanomaterials, designed to improve water retention in soil, offer a sustainable solution to mitigate water scarcity issues in horticulture (Hu et al., 2019). This innovation not only reduces irrigation requirements but also aids in preserving vital water resources for agricultural use. In conclusion, the integration of nanochemicals into horticultural practices holds great promise for surmounting production constraints and elevating crop yield, quality, and resilience (Li et al., 2018; Rai et al., 2022; Humbal & Pathak, 2023).

This study seeks to offer a thorough summary of the present state of research on the use of nanochemicals in horticulture. It will combine findings from important studies and clarify the potential of nanotechnology in improving the productivity and quality of horticultural crops, with the intention of encouraging further investigation in this field.

8.2 NANOCHEMICALS IN HORTICULTURAL CROP PRODUCTION

8.2.1 NANOPESTICIDES

Nanotechnology has emerged as an innovative approach, enabling the design, formulation, and delivery mechanisms of various types of nanoscale molecules with district properties collectively referred to as "nanopesticides." Pesticides are often administered to crops through methods of spraying and/or broadcasting in traditional practices. Consequently, only a little quantity of pesticides effectively reaches the intended areas of crops, while a significant portion of the pesticide mixture is squandered by building up on unintended areas (Rana et al., 2021; Babu et al., 2022). However, nanopesticides have the potential to have more targeted and precise action subsequently that reduces the overall amount of pesticide needed while also minimizing collateral damage to nontarget organisms (Shang et al., 2019). The small size of nanoparticles can enhance the bioavailability and efficacy of pesticides, potentially resulting in greater pest control with lower doses (Chaud et al., 2021; Rana et al., 2021). Although nanopesticides might have lower environmental persistence and reduced off-target effects compared to some conventional pesticides. Significantly, nanoscale pesticides exhibit a 31.5% greater effectiveness against target organisms compared to traditional pesticides. Nevertheless, the toxicity of nanopesticides to nontarget organisms is 43.1% lower (Wang et al., 2022).

Nanostructured matrix systems, such as nanocapsules, nanospheres, and nanovesicles, have been created and altered by employing responsive polymer material to enhance the delivery of pesticides (Chaud et al., 2021). Carbon nanotubes, both single-walled and multiwalled, have been researched for their potential in the precise and directed release of insecticides (Adeel et al., 2020). Several nanomaterials have shown significant potential in enhancing the growth and development of horticultural crops.

8.2.2 NANO-INSECTICIDES

In the past, many researchers clearly investigated the use of nano-insecticides in crop improvement. In a soil-pot study, cypermethrin incorporated with Ca-alginate nanomaterial outperformed better results than direct application to plants (Patel et al., 2018). Ebadollahi et al. (2022) conducted a study on the insecticide acetamiprid, which belongs to the neonicotinoid group. They loaded this insecticide with carriers such as sodium alginate and polyethylene glycol. The study focused on its effectiveness against the elm leaf beetle (*Xanthogaleruca luteola*). The results showed that this formulation significantly enhanced the insecticidal activity of acetamiprid, both in terms of lethal and sublethal toxicity. However, the insecticidal impact of nanosilica on *Spodoptera littoralis* in tomato fields was examined. The study revealed enhanced plant resistance through the reduction of insect longevity, nymph production, and the reproductive potential of female insects. Consequently, there was a decrease in insect populations, alleviating crop damage and mitigating yield losses (El-Bendary et al., 2013). Zinc oxide nanoparticles were evaluated for insecticidal effectiveness on *Galleria mellonella* larvae by integrating them into the artificial rearing feed, resulting in significant larval mortality (Mekewi et al., 2012).

8.2.3 NANOFUNGICIDES

Hexaconazole, a fungicide, was efficiently intercalated into zinc/aluminum-layered double hydroxide (ZALDH), increasing the basal spacing and enhancing the thermal stability of intercalated hexaconazole. As a result, intercalated of ZALDH with hexaconazole exhibited superior inhibition of *Ganoderma boninense* growth compared to free hexaconazole against the basal stem rot disease in oil palm (Mustafa et al., 2018). In a study conducted by Le et al. (2019), silver–chitosan nanocomposites (Ag@CS) were initially synthesized using chitosan (CS) as both a reducing and stabilizing component. Subsequently, the synthesized nanomaterials were combined with the antifungal Antracol (An). The antifungal activity of the resultant Ag@CS/An composite against *Phytophthora capsici*, the causative agent of *Phytophthora blight* in pepper, was then tested. The intercomposite Ag@CS/An material exhibits better antifungal potential against the fungus *Phytophthora capsica*.

The previous study demonstrated that the synthesized graphene oxide (GO)–Fe_3O_4 nanomaterial effectively suppresses the formation of fungal sporangia and hinders the growth of downy mildew. The utilization of (GO)–Fe_3O_4 (250 µg/mL) intercalate nanomaterial in grapevine fields resulted in a notable decrease in the intensity of downy mildew disease. Furthermore, these discoveries have the potential to greatly advance the development of innovative ways for managing plant diseases (Wang et al., 2017).

8.2.4 NANONEMATICIDES

Nanotechnology offers innovative solutions for enhancing horticultural production, including the use of nanonematicides. These advanced nematicides are designed to address the challenges associated with traditional pesticides by providing targeted delivery, controlled release, and enhanced efficacy in combating pests and diseases. Research has shown that nanonematicides can be tailored to specifically target pests and diseases, minimizing harm to nontarget organisms and the environment (Pandey et al., 2021). Their controlled release mechanism ensures prolonged effectiveness, reducing the frequency of applications and minimizing environmental impact. Furthermore, the enhanced efficacy of nanonematicides can lead to reduced pesticide use, lowering the risk of resistance development. The application of nanonematicides in horticulture holds great promise for revolutionizing agricultural practices, contributing to sustainable horticultural development, and ensuring food security (Rai et al., 2022). The use of these advanced nanomaterials represents a significant step toward improving pest management strategies in horticulture, with the potential to minimize environmental impact and enhance overall productivity (Ghosh et al., 2017; Rana et al., 2021; El-Ashry et al., 2022).

8.2.5 NANOFERTILIZERS

Nanofertilizers (NFs) are referred to fertilizers that have been engineered at the nanoscale, typically using nanoparticles or nanomaterials which can potentially lead to more efficient nutrient delivery to plants. NFs can be formulated with carbon-based materials, metal oxides, different other types of nanomaterial components specified upon their distinct characteristics and properties (Liu et al., 2015).

In recent years, NFs play vital role in reducing the usage of chemical fertilizers and their associated adverse environmental effects. As well, NFs contribute to reducing application rates, boosting nutrient efficiency, mitigating adverse effects on crops, and reducing the chemical impact on soil compared to conventional fertilizers. Consequently, NFs enhance product quality, elevate nutrient accessibility for plants, and increase soil fertility (Singh et al., 2017; Zahra et al., 2022).

One novel example of an NF is nanohydroxyapatite (nano-HAP). Currently, researchers are exploring nano-HAP NF as a viable substitute for conventional water-soluble phosphorus (P) fertilizers. Traditional P fertilizers rapidly dissolve in soil, leading to easily soluble phosphates that further run off into surface water resources. On the other hand, nano-HA, with its low solubility, can provide nutrient P similar to conventional fertilizers while reducing contamination risks, such as eutrophication (Maghsoodi et al., 2020).

8.2.6 NANOPLANT GROWTH STIMULATORS

Nanotechnology has arisen as a promising domain for the advancement of plant growth stimulants in horticulture production. More precisely, researchers have studied NFs to determine their ability to improve the development, antioxidant defense, and production of fruit-bearing plants (Muhammad et al., 2024). This innovative approach offers controlled release of fertilizers, improved targeted delivery, and enhanced nutrient absorption by plants, thereby revolutionizing agricultural practices and contributing to sustainable horticultural development (Pandey et al., 2021; Ghidan et al., 2020). Nanobionics, another nanotechnology-based application, has shown potential in improving various plant parameters, including seed germination, root growth, biomass, and fruit production, further highlighting the significant role of nanotechnology in promoting sustainable crop production (Bhaskar et al., 2024). The use of nanotechnology in horticulture presents opportunities for addressing the challenges associated with traditional plant growth stimulants and fertilizers, ultimately contributing to enhanced agricultural productivity and food security.

NFs have attracted considerable interest in horticulture for their ability to enhance plant growth, antioxidant defense, and production in fruit-bearing plants. Nanotechnology-based fertilizers provide regulated nutrient delivery, ensuring plants acquire the necessary elements over a prolonged duration. The use of this regulated discharge mechanism reduces the amount of nutrients that are wasted and leached, effectively resolving the environmental issues linked to conventional fertilizers. Additionally, NFs can be tailored to enhance nutrient absorption by plants, leading to improved growth and productivity. The application of NFs in horticulture holds promise for sustainable agricultural practices, as it offers a targeted and efficient approach to providing essential nutrients to fruit-bearing plants, ultimately contributing to enhanced crop yield and quality (Muhammad et al., 2024).

In the context of sustainable horticulture development, nanotechnology presents opportunities for overcoming the limitations of traditional plant growth stimulants and fertilizers. The controlled release and targeted delivery of nutrients by NFs contribute to reduced environmental impact and improved resource utilization. By providing actual concentration and controlled release of fertilizers, NFs address the challenges associated with nutrient wastage and leaching, offering a more sustainable approach to enhancing plant growth and productivity. Furthermore, the ability of NFs to improve antioxidant defense in plants is of particular significance in horticulture, as it can contribute to enhanced stress tolerance and overall plant health. Therefore, the application of nanotechnology in the form of NFs represents a significant advancement in promoting sustainable horticultural development (Rai et al., 2022; Pandey et al., 2021).

Nanotechnology in horticulture also encompasses the field of nanobionics, which focuses on the development of nanomaterials to improve plant parameters and crop production. Nanobionics has demonstrated the potential to enhance various aspects of plant growth, including seed germination, root growth, biomass, and fruit production. The capacity of nanoparticles (NMs) to infiltrate the tough outer layer of seeds and facilitate water absorption is the determining element for enhanced development and vitality. Furthermore, the utilization of nanotechnology in seed priming is a highly promising approach that enhances the potential of achieving high agricultural yields by treating seeds prior to planting (Acharya et al., 2020; Anand et al., 2020).

The application of nanotechnology in horticulture, particularly in the form of NFs and nanobionics, presents a transformative opportunity to enhance plant growth and productivity while addressing the challenges of traditional plant growth stimulants and fertilizers. The controlled release, targeted delivery, and improved nutrient absorption offered by NFs contribute to reduced environmental impact and enhanced agricultural sustainability. Similarly, the development of nanobionics holds promise for improving various plant parameters, ultimately contributing to sustainable crop production. As research in this field continues to advance, the integration of nanotechnology into horticulture production is expected to play a significant role in ensuring food security and promoting environmentally sustainable agricultural practices. Metal nanoparticles, like silver or copper nanoparticles, have been explored for their antimicrobial properties in pesticide formulations. Eimer et al. (2016) reported that nanoparticles (AlO, CuO, FeO, MnO, NiO, and ZnO) were sprayed on infested (*Fusarium* wilt) soilless medium tomatoes, which reduced disease estimates by 31%, 28%, or 28%, respectively.

8.3 NANOCHEMICALS IN POSTHARVEST MANAGEMENT OF HORTICULTURAL CROPS

Nanotechnology offers promising applications in the postharvest management of horticultural crops, aiming to improve product quality, reduce spoilage, and enhance shelf-life. Nanomaterials, such as nanofilms, have demonstrated potential in this field. These thin films, typically composed of nanomaterials, are used as coatings on harvested products or packaging materials to achieve these objectives. Research and academic literature highlight the potential of nanofilms in improving the postharvest management of horticultural crops. For instance, nanofilms composed of antimicrobial nanomaterials have been identified as suitable coatings for harvested products and packaging materials, offering enhanced storage and transportation capabilities. These nanomaterial-based films contribute to reducing postharvest damage, minimizing microbial contamination, and extending the shelf-life of horticultural products (Rana et al., 2021; Yadollahi et al., 2009).

Furthermore, nanofilms have been recognized for their ability to increase the vitality of cut flowers, thereby contributing to the preservation of floral products and the maintenance of their quality during storage and transportation. The use of nanotechnology, including nanofilms, in postharvest management aligns with the goals of sustainable agriculture and food security, offering advanced solutions for maintaining the quality and safety of horticultural crops throughout the supply chain (Yadollahi et al., 2009; Islam, 2021).

In addition, nanofilms have been explored as part of nanobiotechnology-mediated sustainable agriculture, emphasizing their potential as intelligent delivery systems for crop protection and postharvest management. The purpose of these films, which are made from nanomaterials, is to release active compounds in a specific and controlled manner. This helps to decrease postharvest losses and maintain the quality of agricultural goods (Babu et al., 2022).

Overall, the application of nanofilms in the postharvest management of horticultural crops represents a significant advancement in agricultural sustainability, offering innovative solutions for addressing postharvest challenges and contributing to the preservation of product quality, reduction of waste, and enhancement of food safety. As research in this field continues to advance, the integration of nanotechnology, including nanofilms, into postharvest management practices is expected to play a significant role in ensuring food security and promoting environmentally sustainable agricultural practices (Yadollahi et al., 2009; Babu et al., 2022).

8.3.1 NANOFILMS

Nanotechnology-based approaches, including the use of nanofilms and nano-packaging, have shown promise in the postharvest management of horticultural crops. These thin films, typically composed of nanomaterials, are used as coatings on harvested products or packaging materials to enhance shelf-life, reduce spoilage, and maintain product quality. The role of nanofilms and nanopackaging in horticulture crop postharvest management is significant, offering innovative solutions for addressing challenges related to storage, transportation, and preservation of fruits, vegetables, and cut flowers.

Research and academic literature highlight the potential of nanofilms and nanopackaging in improving the postharvest management of horticultural crops. For instance, nanofilms composed of antimicrobial nanomaterials have been identified as suitable coatings for harvested products and packaging materials, offering enhanced storage and transportation capabilities. These nanomaterial-based films contribute to reducing postharvest damage, minimizing microbial contamination, and extending the shelf-life of horticultural products (Upadhyay et al., 2021; Rana et al., 2021).

Furthermore, nanofilms and nanopackaging have been recognized for their ability to increase the vitality of cut flowers, thereby contributing to the preservation of floral products and the maintenance of their quality during storage and transportation. Nanotechnology, namely the application of nanofilms and nanopackaging, supports the objectives of sustainable agriculture and food security by providing sophisticated methods to preserve the quality and safety of horticultural crops at every stage of the supply chain (Rana et al., 2021).

In addition, nanofilms and nanopackaging have been explored as part of nanobiotechnology-mediated sustainable agriculture, emphasizing their potential as intelligent delivery systems for crop protection and postharvest management. These nanomaterial-based films are designed to provide targeted and controlled release of active compounds, contributing to the reduction of postharvest losses and the preservation of horticultural products (Babu et al., 2022).

Overall, the application of nanofilms and nanopackaging in the postharvest management of horticultural crops represents a significant advancement in agricultural sustainability, offering innovative solutions for addressing postharvest challenges and contributing to the preservation of product quality, reduction of waste, and enhancement of food safety. As research in this field continues to advance, the integration of nanotechnology, including nanofilms and nanopackaging, into postharvest management practices is expected to play a significant role in ensuring food security and promoting environmentally sustainable agricultural practices (Rana et al., 2021; Babu et al., 2022).

8.3.2 SILVER NANOPARTICLES

Silver nanoparticles (AgNPs) have shown potential for postharvest management of horticultural crops. They have been investigated for their antimicrobial properties, which can help reduce spoilage and extend the shelf-life of fruits, vegetables, and cut flowers. AgNPs have been used in nanofilms and nanopackaging to create protective barriers that minimize microbial contamination and reduce postharvest damage (Manzoor et al., 2023; Jasrotia et al., 2022). Research has demonstrated the effectiveness of plant-based silver nanoparticles in repelling insect pests, as well as the ability of green silver nanoparticles to control pests and reduce postharvest losses. Additionally, AgNPs have been found to enhance the vitality of cut flowers, contributing to the preservation of their quality during storage and transportation. These findings suggest that the use of silver nanoparticles in nanocoatings holds promise for improving the postharvest management of horticultural crops by reducing spoilage, controlling pests, and extending the shelf-life of produce (Rana et al., 2021; Ruff et al., 2019).

8.4 CONCLUSION

The implications of nanochemicals in addressing production constraints in horticulture are significant and offer promising solutions for sustainable agricultural practices. The application of NFs, nanopesticides, and nanofilms has demonstrated the potential to improve nutrient management, pest control, and postharvest management, thereby contributing to enhanced agricultural productivity and reduced environmental impact. Embracing nanotechnology in horticulture presents an opportunity to capitalize on improved product visibility, food safety, and reduced chemical usage, ultimately contributing to the development of a well-organized, sustainable agricultural industry. As research in this field continues to advance, the integration of nanotechnology into horticulture production is expected to play a significant role in ensuring food security and promoting environmentally sustainable agricultural practices. Therefore, the utilization of nanochemicals in horticulture holds great promise for addressing production constraints and advancing the transition toward a more sustainable and efficient agricultural sector.

KEYWORDS

- **environmental stewardship**
- **food security**
- **horticulture**
- **nanotechnology**
- **nanofertilizers**
- **nanopesticides**
- **nanofilms**
- **sustainable agriculture**

REFERENCES

Acharya, P., Jayaprakasha, G.K., Crosby, K.M., Jifon, J.L., & Patil, B.S. (2020). Nanoparticle-mediated seed priming improves germination, growth, yield, and quality of watermelons (Citrullus lanatus) at multi-locations in Texas. *Scientific Reports,* 10, 5037.

Adeel, M., Rahman, L., Bilal, M., Tahir, M., & Raza, A. (2020). Synthesis of chitosan–multi-walled carbon nanotubes nanocomposites and their potential applications in agriculture. *International Journal of Biological Macromolecules,* 164, 2898–2912.

Anand, K.V., Anugraga, A.R., Kannan, M., Singaravelu, G., & Govindaraju, K. (2020). Bio-engineered magnesium oxide nanoparticles as nano-priming agent for enhancing seed germination and seedling vigor of green gram (*Vigna radiata* L.). Materials Letters 8, 127792.

Babu, P.J., Saranya, S., Longchar, B., & Rajasekhar, A. (2022). Nanobiotechnology-mediated sustainable agriculture and post-harvest management. *Current Research in Biotechnology,* 4, 326–336.

Bhaskar, R., Pandey, S.P., Kumar, U., Kim, H., Jayakodi, S.K., Gupta, M.K., & Han, S.S. (2024). Nanobionics for sustainable crop production: Recent development to regulate plant growth and protection strategies from pests. *OpenNano*, 15, 100198.

Chaud, M., Souto, E.B., Zielinska, A., Severino, P., Batain, F., Oliveira-Junior, J., & Alves, T. (2021). Nanopesticides in agriculture: Benefits and challenge in agricultural productivity, toxicological risks to human health and environment. *Toxics*, 9(6), 131. https://doi.org/10.3390/toxics9060131.

Ebadollahi, A., Valizadeh, B., Panahandeh, S., Mirhosseini, H., Zolfaghari, M., & Changbunjong, T. (2022). Nanoencapsulation of acetamiprid by sodium alginate and polyethylene glycol enhanced its insecticidal efficiency. *Nanomaterials*, 12(17), 2971.

El-Ashry, R.M., El-Saadony, M.T., El-Sobki, A.E., El-Tahan, A.M., Al-Otaibi, S., El-Shehawi, A.M., Saad, A.M., & Elshaer, N. (2022). Biological silicon nanoparticles maximize the efficiency of nematicides against biotic stress induced by Meloidogyne incognita in eggplant. *Saudi Journal of Biological Sciences*, 29(2), 920–932.

El-Bendary, H.M., & El-Helaly, A.A. (2013). First record nanotechnology in agricultural: Silica nano-particles a potential new insecticide for pest control. *Applied Science Report*, 4(3), 241–246.

Elmer, W.H., & White, J.C. (2016). The use of metallic oxide nanoparticles to enhance growth of tomatoes and eggplants in disease infested soil or soilless medium. *Environmental Science: Nano*, 3(5), 1072–1079.

Ghosh, A., Dey, K., Bauri, F.K., Mani, A., Dey, A.N., & Das, B.C. (2017). Implication of nano technology in horticulture. In *Hi-tech Horticulture—Vol. 7: Advance Techniques*, New India Publishing Agency, pp. 141–150.

Hu, J., et al. (2019). Advances in nanotechnology for sustainable agriculture: A comprehensive review. *Sustainability*, 11(11), 3056.

Humbal, A., & Pathak, B. (2023). Application of nanotechnology in plant growth and diseases management: Tool for sustainable agriculture. In: *Agricultural and Environmental Nanotechnology: Novel Technologies and their Ecological Impact.* Singapore: Springer Nature Singapore, pp. 145–168.

Islam, T. (2021). Nanotechnology in horticultural crops. In: *Encyclopedia.* https://encyclopedia.pub/entry/14649.

Jasrotia, P., Nagpal, M., Mishra, C.N., Sharma, A.K., Kumar, S., Kamble, U., Bhardwaj, A.K., Kashyap, P.L., Kumar, S., & Singh, G.P. (2022). Nanomaterials for postharvest management of insect pests: Current state and future perspectives. *Frontiers in Nanotechnology*, 3, 811056.

Khodakovskaya, M., et al. (2012). Carbon nanotubes are able to penetrate plant seed coat and dramatically affect seed germination and plant growth. *ACS Nano*, 6(8), 6548–6552.

Le, V.T., Bach, L.G., Pham, T.T., Le, N.T.T., Ngoc, U.T.P., Tran, D.H.N., & Nguyen, D.H. (2019). Synthesis and antifungal activity of chitosan-silver nanocomposite synergize fungicide against *Phytophthora capsici. Journal of Macromolecular Science, Part A*, 56(6), 522–528.

Li, Y., et al. (2018). Nanotechnology promotes the R&D in fertilizers: Current status and future perspectives. *Journal of Nanoscience and Nanotechnology*, 18(9), 5866–5874.

Liu, R., & Lal, R. (2015). Potentials of engineered nanoparticles as fertilizers for increasing agronomic productions. *Science of the Total Environment*, 514, 131–139.

Manzoor, M.A., Xu, Y., Xu, J., Wang, Y., Sun, W., Liu, X., Wang, L., Usman, M., Wang, J., Liu, R., Whiting, M.D., Jiu, S., Zhang, C., & Zhang, C. (2023). Nanotechnology-based approaches for promoting horticulture crop growth, antioxidant response and abiotic stresses tolerance: an overview. *Plant Stress*, 11, 100337.

Mekewi, M., Shebl, A., Imam, A.I., Amin, M.S., & Albert, T. (2012). Screening the insecticidal efficacy of nano-ZnO synthesized via in-situ polymerization of crosslinked polyacrylic acid as a template. *Journal of Materials Science & Technology*, 28(11), 961–968.

Mittal, D., Kaur, G., Singh, P., Yadav, K., & Ali, S.A. (2020). Nanoparticle-based sustainable agriculture and food science: Recent advances and future outlook. *Frontiers in Nanotechnology*, 2, 579954.

Muhammad, A.M., Xu, Y., lv, Z., Xu, J., Wang, Y., Sun, W., Liu, X., Wang, L., Usman, M., Wang, J., Liu, R., Whiting, M.D. Jiu, S., & Zhang, C. (2024). Nanotechnology-based approaches for promoting horticulture crop growth, antioxidant response and abiotic stresses tolerance. *Plant Stress*, 11, 100337.

Mustafa, I.F., Hussein, M.Z., Saifullah, B., Idris, A.S., Hilmi, N.H.Z., & Fakurazi, S. (2018). Synthesis of (hexaconazole-zinc/aluminum-layered double hydroxide nanocomposite) fungicide nanodelivery system for controlling Ganoderma disease in oil palm. *Journal of Agricultural and Food Chemistry*, 66(4), 806–813.

Pandey, V., & Singh, S., Alam, K., & Keshari, D. (2021). Nanotechnology in horticulture.

Patel, S., Bajpai, J., Saini, R., Bajpai, A.K., & Acharya, S. (2018). Sustained release of pesticide (Cypermethrin) from nanocarriers: An effective technique for environmental and crop protection. *Process Safety and Environmental Protection*. 117, 315–325.

Rai, R., Nalini, P., & Singh, Y.P. (2022). Nanotechnology for sustainable horticulture development: Opportunities and challenges. *Innovative Approaches for Sustainable Development: Theories and Practices in Agriculture*, 191–210.

Ghidan, A.Y., & Al Antary, T.M. (2020). Applications of nanotechnology in agriculture. In: *Applications of Nanobiotechnology*. https://doi.org/10.5772/intechopen.88390

Rana, R.A., Siddiqui, M.N., Skalicky, M., Brestic, M., Hossain, A., Kayesh, E., Popov, M., Hejnak, V., Gupta, D.R., Mahmud, N.U., & Islam, T. (2021). Prospects of nanotechnology in improving the productivity and quality of horticultural crops. *Horticulturae* 7(10), 332.

Shang, Y., Hasan, M.K., Ahammed, G.J., Li, M., Yin, H., & Zhou, J. (2019). Applications of nanotechnology in plant growth and crop protection: A review. *Molecules*, 24(14), 2558.

Singh, M.D., Kumar, B. Bio efficacy of nano zinc sulphide (ZnS) on growth and yield of sunflower (*Helianthus annuus* L.) and nutrient status in the soil. *International Journal of Agricultural Science*, 2017(9), 0975–3710.

Suriyaprabha, R., et al. (2012). Zinc oxide nano particle induced growth enhancement in first generation of soybean (Glycine max L.). *The Plant Growth Regulation*, 68(2), 165–174.

Upadhyay, T., Kumar, V., Sharangi, A., Upadhye, V., Khan, F., Pandey, P, Kamal, M., Baba, A., & Hakeem, K. (2021). Nanotechnology-based advancements in postharvest management of horticultural crops. *Phyton*. 91.

Usman, M., Farooq, M., Wakeel, A., Nawaz, A., Cheema, S.A., Rehman, H.U., Ashraf, I., Sanaullah, M. (2020). Nanotechnology in agriculture: Current status, challenges and future opportunities. *Science of the Total Environment*, 721:137778.

Wang, D., Saleh, N.B., Byro, A., Zepp, R., Sahle-Demessie, E., Luxton, T.P., Ho, K.T., Burgess, R.M., Flury, M., White, J.C., & Su, C. (2022). Nano-enabled pesticides for sustainable agriculture and global food security. *Nature Nanotechnology*, 17(4), 347–360. https://doi.org/10.1038/s41565-022-01082-8.

Wang, X., Cai, A., Wen, X., Jing, D., Qi, H., & Yuan, H. (2017). Graphene oxide-Fe3O4 nanocomposites as high-performance antifungal agents against *Plasmopara viticola*. *Science China Materials*, 60(3), 258–268.

Yadollahi, A., Arzani, K., & Khoshghalb, H. (2009). The role of nanotechnology in horticultural crops postharvest management. In: *Southeast Asia Symposium on Quality and Safety of Fresh and Fresh-Cut Produce*, 875, pp. 49–56.

Zahra, Z., Habib, Z., Hyun, H., & Shahzad, H.M.A. (2022). Overview on recent developments in the design, application, and impacts of nanofertilizers in agriculture. *Sustainability*, 14(15), 9397.

CHAPTER 9

Application of Drones and Robotics in Horticultural Crop Production

SUSHIL KUMAR[1], VANDANA THAKUR[2*], and SOURABH MISHRA[3]

[1]*ICAR-Central Potato Research Institute, Regional Station, Jalandhar, Punjab, India*

[2]*Department of Horticulture, School of Agriculture, Lovely Professional University, Phagwara, Punjab, India*

[3]*Hohal University, Xi Kang Lu, Gulou, Nanjing, Jiangsu, China*

*Corresponding author

ABSTRACT

Modern agriculture has to increase production efficiency given the growing global population and climate change. The cultivation of horticultural crops affects the environment significantly and is essential for human nourishment. The agricultural industry must modernize and make use of cutting-edge technology, such as drones, to cut down on resource use, boost output, and enhance quality. Because of their adaptability and agility, these gadgets, known as unmanned aerial vehicles, or UAVs, are essential for operations monitoring and spraying. The reviewed articles provide an overview of how these instruments fit into the subject of precision farming, having been categorized according to the sorts of drone uses in vegetable crop production. The contributions of machine learning, deep learning, and artificial intelligence were then examined as technological advancements in these instruments and information processing techniques were investigated. In order to fully achieve drones' potential in agricultural precision and crop production, their practical application and future

Smart Horticulture Production: Integrating Advanced Technologies.
Rajni Rajan and Duniya Ram Singh (Eds.)

DOI: 10.1201/9781998511679-9

technical and scientific issues were considered. Drones can reduce inputs like water, pesticides, fertilizers, and herbicides, and avoid damage by identifying stress types early. Though input reductions may benefit the environment, these technologies may enhance vegetable crop ecological viability.

9.1 INTRODUCTION

Horticulture stands out as one of the most rapidly expanding and valuable sectors in agriculture globally. It covers the growth of fruits, vegetables, ornamentals, plantation crops, aromatic/medicinal plants, and spices. Given the growing global population and the shift toward healthier eating habits, horticulture plays a vital role in ensuring food and nutritional security. Horticultural crops provide essential antioxidants, fiber, vitamins, minerals, and phytonutrients and are vital for human health. Robots and drones are sophisticated, intelligent machines that can perform complex tasks, play a big part in agriculture, and are becoming a necessary component of scientific and technical advancement. Farm robotics applications lead to significant cost savings, improve production capacity, and reduce the deployment of human resources. Drones are at the forefront of technological advancements in the field of precision farming, making significant contributions to the scientific and productive sectors. Nevertheless, their origins can be traced back to different areas of application. Additionally, the drone was initially developed as a device for military use to protect anthropological personnel during survey and surveillance tasks. Over time, their uses have expanded far beyond the military realm, being utilized in sectors such as entertainment, security, transportation, photography, and exploration of the environment. In recent years, the farming industry has embraced robotic systems, leading to significant changes in agricultural practices. Advancements in technology are creating new opportunities for improving resource management, reducing labor costs, and boosting efficiency and sustainability in the farming industry. Automating cultivation, feeding, and harvesting procedures is increasingly seen as a solution to address global challenges like growing populations, depletion of natural resources, as well as safety issues related to the environment. Regardless of the significant upfront costs and the need for skilled workers, incorporating robotic systems bids assists in lowering operational expenses and improving production efficiency. Innovative advancements in robotics, like automatic calf feeding systems, mobile feed mixers, and grassland management systems, are transforming cultivation and establishing higher benchmarks in farm management. This chapter depicts the general introduction of agricultural drones and robotics with application in horticultural crops.

9.2 NEED FOR DRONES AND ROBOTICS

The world's main source of food is the agriculture sector (Friha et al., 2021). However, it has been confronted with several difficult issues, including the growing population, the need for food products, concerns around food safety and security, and calls for ecological preservation, water preservation, and sustainability (Inoue, 2020). Furthermore, future environmental issues may result from the expansion of farming practices combined with rising fertilizer and pesticide consumption. In a similar vein, the amount of arable terrestrial is finite and the worldwide farmer population is deteriorating. These issues highlight the mandate for inventive and sustainable farming approaches (Tzounis et al., 2017; Elijah et al., 2018; Inoue, 2020; Friha et al., 2021). Somewhat crop grows through three basic vital phases in their life cycle: the stages of cultivation, monitoring, and harvesting (Wakchaure et al., 2023). Selecting which crops to plant, planning the land, preparing the soil for irrigation, preparing the seeds, and seed sowing are all part of the cultivation phase. After the cultivation stage, keeping an eye on and managing crop growth, including strategic crop health monitoring, utilization of fertilizer, disease and weed identification, and application of pesticides are time-sensitive during the monitoring stage. The harvesting stage, which comprises tasks including crop harvesting, sorting, curing, storing, and market sale, is the ultimate and furthermost significant phase of the agricultural cycle. As mentioned above, numerous tasks are involved in each phase of the crop life cycle and each task needs proper attention from the grower. Out of these tasks, most of the tasks are done manually or semiautomatic, which takes a lot of time, labor, cost, and effort.

It has been recognized that incorporating innovative technology could be a successful strategy to address these challenges. As a consequence of these discussions, precision farming (Khanna & Kaur, 2019; Feng et al.,

2019) and smart farming have been developed (Brewster et al., 2017; Tang et al., 2021). To overcome all of these problems, the agricultural expansion revolt occurred from 1.0 to 4.0 (today). The systematic utilization of an AI-based approach has replaced the conventional farming scheme, where these machines resolve the farming-related problem in real-time. In a recent period, researchers have made great efforts to combine agricultural practices with engineering to make agriculture, modest, economical, intelligent, time efficient, highly productive, sustainable, and also pay to a healthy and wealthy society.

Currently, there is a great deal of effort being made by young scientists and engineers to make agriculture simple, intelligent, economical, highly productive, time-efficient, sustainable, and contribute to a wealthy and healthy society.

9.3 ADVANCEMENT OF TECHNOLOGY IN THE AGRICULTURE SECTOR

The design and development of precisely novel agricultural tools (automatic systems), mobile robots (aerial and ground), and related equipment, such as human–machine interfaces, communications equipment, fleet management, machine vision, safety systems, and location equipment) as well as their all-inclusive amalgamation are currently of interest to young scholars and agricultural scientists. These systems are intended to carry out the principal tasks of observation, decision-making, and execution on croplands to diminish the usage of pesticides, though improving effectiveness. Prominent technologies in this sector that have wedged the interest of academics are big data (Gill et al., 2017), computation technology (Zamora-Izquierdo et al., 2019; Jinbo et al., 2019; Hsu et al., 2020), artificial intelligence techniques such as deep and machine learning (Liakos et al., 2018; Shadrin et al., 2019; Parsaeian et al., 2020), and wireless sensor networks (Zhou et al., 2016; Zheng & Yang, 2018). Drones and robotics have an inclusive assortment of applications in agriculture because they might be combined with computation power, innovative technologies, and sensors to provision disaster relief, early warning systems, forestry and wildlife conservation, and crop management under the abovementioned three phases of the crop life cycle (such as cultivation, monitoring, and harvesting phase). Drone robots can be used for several agricultural tasks, such as seedling transplanting, crop and growth monitoring, precision irrigation, removal of weeds, insect–pest and disease detection, harvesting, yield estimation, and evaluation of various stresses (water, temperature, and salt stress) and deficiencies (nutritional) in crop plants (Panday et al., 2020; Inoue, 2020).

9.4 AGRICULTURAL ROBOTS

9.4.1 *AERIAL ROBOTS*

Aerial robots are prevalently identified as unmanned aerial vehicles, that is, UAVs. A UAV is a radio-controlled aircraft that is capable of taking off and landing without the need for a human pilot. Overall, UAVs are equipped with sensors and cameras for crop monitoring and sprayers for pesticide spraying (Roldán et al., 2018). "Yamaha" created the first UAV (RMAX) prototype specifically for crop monitoring and pest management in farming. In precision agriculture, a technical investigation of UAVs involves evaluating their suitability for various agricultural tasks such as crop monitoring (for measuring stress, yield prediction, disease detection, fungicide, and fertilizer requirement), soil and field examination, crop height evaluation, and pesticide spraying (Zarco-Tejada et al., 2012; Gago et al., 2015).

9.4.2 *GROUND ROBOTS*

Ground robots are known as unmanned ground vehicles (UGVs) prepared to apply chemical or physical treatments for weed and pest management-based agricultural tools. They are mobile robots with crawler and caterpillar platforms and wheeled robots allowing them to follow crop rows (Chatzimichali et al., 2009). The robots were equipped with combined approaches alike odometry, GPS, line guidance, and manned approaches as navigation plans and path plans (Sakai et al., 2008; De-An et al., 2011).

9.4.3 *SPECIAL ROBOTS*

The primary purpose of special robots is to overcome the restrictions of conventional ground robots, which are associated with wheeled, caterpillar motion, as well as the limitations of aerial robots, which are incapable of performing tasks on surfaces or indoors (Zhan et al., 2008). Sphere robots, whose movements are caused by instability, are employed in a variety of precision agriculture applications and scenarios, including exploration, surveillance, soil moisture, and temperature measurement (Hernández et al., 2013a, 2013b). Compared to wheeled robots, sphere robots are substantially stronger and lighter. Several robots are created using bio-inspiration, a design approach that aims to mimic biological evolution. Harvard University's RoboBees are micro-aerial robots designed to resemble bees and are used to monitor the environment and aid in crop pollination (Wood, 2015).

9.5 RECENT TECHNOLOGIES BEING UTILIZED IN HORTICULTURAL CROPS

Advanced robotics, drone technologies, and integrated intervention mechanisms for specific horticultural crop applications developed till now and being operative in advanced and developing countries are mentioned below.

9.5.1 AUTOMATIC SEED MAPPING

The conventional approach for seeding is the labor-intensive one or by tractor-mounted seeder. However, it takes longer, and there is a persistent lack of manpower. In the agriculture industry, smart robots reduced reliance on human labor and reduced seed wastage. Recent years have seen the development of highly precise pneumatic planters for various crops that can handle a wide range of seed sizes, resulting in a uniform distribution of seeds. To decrease human labor and increase yield, Abdulrahman et al. (2017) designed a robot to automate the seed-sowing process in crops, for example, baby corn, sunflower, groundnut, and vegetables, such as okra, beans, pumpkin, and also pulses, namely, green gram, black gram, and so on. DC motor is used to plant seeds automatically, and a microcontroller is used to control and vary the spacing between each seed. Indian Institute of Horticultural Research, Bangalore, developed "Arka Automatic Dibbler Cum Seeder for Vegetable Nursery." The automatic machine is designed to sow seeds in 98 protray cells which is commonly used to raise vegetable nurseries with a seed sowing efficiency of 150 protrays per hour as compared to the roughly eight man hours needed to sow 10 trays every hour.

9.5.2 AUTOMATIC WEED MAPPING

The weed control area has been an actual active area for study for a long time. Weeds are harmful or undesirable plants in the crop field and strive with the main produce for nutrients, light, and water. So, their control not only results in higher yield but also high-quality produce. The techniques to control the weeds depend upon the planted crop, planting method, and type of weed. Conventionally, weeds were managed by manual and automated means; nevertheless, currently, several sprays of herbicides are applied to manage weed development. These chemicals negatively affect soil health, microorganisms, and also human health. As a result, new expert systems for weed management are proposed in precision agriculture. This system only sprays on the desired spot by identifying weeds and crop plants and lessens the chance of contaminating crops, people, animals, and water sources (Sabzi et al., 2018). The robotic system deployed on the field can proficiently distinguish between a weed and a crop plant, which eradicates human engagement in the distribution of chemicals. By way of precision farming is implemented, reducing the number of pesticides applied to the crop, making it healthier and more cost-effective than the control of weeds by traditional chemical application (Rani et al., 2022). An automatic weed system, based on computer vision embattled to perceive the weeds by using features such as morphology (shape, curvature, area, perimeter of leaves, and veins), spectral property, visual texture, and spatial context of the leaf produced the precision of 92% in detecting weed plants (Nejati et al., 2008; Rani et al., 2022). Natural external surrounding conditions present a wide range of external factors that can affect the accuracy of the visual system. Thus, it is a well-thought-out feasible scientific challenge and a worthwhile figure from an agricultural perspective to detect more than 90% of weeds (Gonzalez-de-Santos et al., 2017). Kargar & Shirzadifar (2013) developed a smart herbicide sprayer robot that identifies the weeds using image analysis and applies spray only on the desired space with an accuracy of 95.89%. Thermal/Mechanical equipment detects 90% of the weeds and destroys them with liquefied petroleum gas making it feasible and alluring from an agricultural perspective.

9.5.3 WATER MANAGEMENT THROUGH DRONES IN VEGETABLE CROPS

To guarantee agricultural production's quantity and quality, as well as to adapt it to changes in the environment to make it more sustainable, water resource management is crucial (Pereira et al., 2002). Water resources are trending downward and becoming less of a good quality, particularly in Mediterranean environments. Therefore, it is imperative to implement strategies that maximize irrigation efficiency in Mediterranean regions. The utilization of drones is especially beneficial in this situation. These tools enable the adoption of various strategies aimed at enhancing the efficiency of water usage. For instance, crop water status can be determined using optical, multispectral, as well as thermal sensors installed on drones. Coulombe et al. (2020) used thermal and multispectral sensors to estimate the crop water status of chili crops. While Mwinuka et al. (2021) used UAV-based multispectral sensors in conjunction with handheld thermal sensors in eggplant, Stutsel et al. (2021) verified it utilizing RGB as well as thermal sensors in tomatoes. The most popular and widely utilized index for measuring water stress is the Crop Water Stress Index (CWSI), which can be calculated using thermal sensors among these other types of sensors. Three CWSI models (CWSIe—empirical, CWSIt—theoretical, and CWSIh hybrid) were compared by Ekinzog et al. (2022) for crop water stress detection in potato cultivation. With CWSIh, these indices produced the best results as accurate forecasters of soil volumetric water content (θ) (R = 0.57–0.63). Multispectral sensors have been used in other studies to assess crop water status. For instance, the estimation of stem water potential in tomatoes has been made possible by the use of photochemical reflectance and Normalized Difference Vegetation Index (Tang et al., 2023). However, the best correlations between the Normalized Difference Water Index and surface soil moisture were discovered by Casamitjana et al. (2020) when studying potatoes. Other methods have included mapping crop evapotranspiration (ET) using energy flux models, which may be utilized for calculating irrigation needs for water. Chandel et al. (2020) used this method on several crops, including potatoes, substituting landsat data for UAV-based optical and thermal data in an already-existing model. The outcomes demonstrated that the UAV-based data provided more accuracy and were consistent with the model. A two-source energy balance model (TSEB) was applied in a bell pepper field by Tunca et al. (2022). The UAV platform showed a lot of promise in this instance as well for mapping ET, particularly in small plots where observations at the plant level are possible. Additionally, this method allowed for the circumvention of the constraints imposed by satellites, which provide a temporal resolution of roughly fifteen days (Ihuoma et al., 2021). A new energy flux modeling framework based on TSEB was proposed by Peng et al. (2023) for high spatial resolution thermal and multispectral UAV data. The model performed exceptionally well in recovering ET dynamics, as demonstrated by the results, which validated the accuracy and dependability of these instruments.

9.5.4 MORPHOLOGICAL AND GEOMETRICAL FEATURE EXTRACTION THROUGH DRONES IN VEGETABLE CROPS

The process of assessing a plant's morphological, physical, and phenological stage is known as phenotyping. Drones can take detailed pictures of crops at high definition, which may then be processed to provide details like their dimensions, form, architecture, and phenological phases. Zhang et al. (2022) used a multispectral camera-equipped drone at a height of 10 m with a spatial resolution of 5 cm to map the dimensions of cabbage, calculating its length and breadth. Din et al. (2021) used a combination of multispectral data and deep learning approaches to identify the stage of growth of onions. In the meanwhile, Kim et al. (2018) estimated the biophysical characteristics of radish and cabbage using UAV-RGB-based imaging at a height of 20 m and a spatial resolution of 0.64 cm. Certain crop characteristics may be computed using machine learning, deep learning algorithms, photogrammetry, and image analysis methods characteristics as a result of their exposed, upward-oriented head form and plant structure. Numerous authors' investigations have verified that this feature makes it easier to quickly and accurately monitor broad regions, giving precise information about plant phenology (Grenzd¨orffer, 2019; Zhou et al., 2022). The capacity to quickly estimate crop biomass is one of the direct uses that may be obtained from possessing biometric characteristics like plant area, height, and volume. For instance, Zheng et al. (2022) evaluated this parameter in the cultivation of strawberries, while Ballesteros et al. (2018) assessed it in the cultivation of onions. Certain tubers, such as potatoes, cannot be directly characterized since they grow underground. Because of the potential agronomic insights, calculating plant geometry as well as aboveground biomass is thus quite interesting (Yang et al., 2023).

The right spraying volume may be estimated by integrating measures such as canopy coverage and plant height (Xie et al., 2022). Additionally, quick phenotyping might be a very useful technique for breeders: To show how point cloud data from inexpensive UAV photos may be used to generate 3D surface models of plant canopies on potatoes, De Jesus Colwell et al. (2021) devised a technique. All of the geometric and morphological traits may be inferred from these models, opening the door for a broadening of upcoming genotype-phenotype association research. Drone-based phenotyping has been effectively used in tomatoes, where it was used to evaluate the efficacy of particular biostimulants (Antonucci et al., 2023). To assess inbred lines in lettuce with different carotenoid contents, picture phenotyping was used (Mascarenhas Maciel et al., 2019). It is possible to distinguish certain phenological phases by identifying the physical traits of plants. By estimating plant height observed by drone, it was feasible to identify the commencement of tuberization in potatoes, an important phenological stage for modifying irrigation management (Martins et al., 2023). For cucurbits, where the flower is one of the end products, it might be very critical to identify certain phenological phases, such as blossoming. For this reason, Mithra & Nagamalleswari (2023) created a transfer-learning algorithm that can identify the genus and gender of nine cucurbit species, such as cucumber, muskmelon, squash, pumpkin, and watermelon, using the CuCuFlower dataset. For crops like lettuce, where the blooming process is essential to genetic improvement initiatives, flowering detection may also be important (Han et al., 2021).

9.5.5 CROP MONITORING

UAVs can observe crops with several indices. In a single trip, the UAVs can cover up to hectares of fields. Thermal and multispectral cameras are attached to record the reflectance of the crop canopy for record of observation. The camera records one frame per second, can store it in memory, and transmits it via telemetry to the ground station.

9.5.6 DISEASE IDENTIFICATION

Incidence of diseases and pests are the important cause of yield reduction in crops and timely optimization of disease will improve the quality as well as extent of produce. In the past few years, several disease optimization models were evaluated using AI, deep learning, robotics, and drones (listed in Table 9.1). In horticultural crops, various fungi, bacteria, nematodes, viruses, and physiological disorders are accountable for disease development. Meticulous acknowledgment of disease is a multiclass classification problem. Chaudhary et al. (2020) predicted the infestation of various diseases such as anthracnose, black rot, blackleg, bacterial wilt, club root, chili-mosaic virus, early blight, downy mildew, gray mold, Fusarium wilt, leaf-spot, late blight, onion-smut, rust, powdery mildew, Septoria leaf spot, yellow vein mosaic, and Verticillium wilt in vegetable crops using machine learning algorithms. Zou et al. (2020) recognized three communal tea diseases, namely *Colletotrichum camelliae*, *Phyllosticta fheaefolia* Hara, and *Phyllosticta theicola* Petch using image recognition technology. As an alternative to human monitoring, drones might be utilized to identify leaf spot disease in wide areas of apple plantations (Anam, 2020). To identify disease, the image taken by drone must be processed by separating the healthy part from the affected parts.

TABLE 9.1 Application of Robots and Drones

Cultivation Phase	Monitoring Phase	Harvesting Phase
Preparing land	Manage crop growth	Harvesting
Nursery raising	Crop health	Sorting and grading
Irrigation for the soil	Fertilizer use	Storing
Seed wowing	Peat Disease Identification	
	Weed identification	
	Pesticide application	
	Application of Drones and Robotics	
Robots (UGVs)	Drone (UAVs)	Automatic harvester
Automated tractor	and	Automatic grading
Automatic seed mapping	Robots (UGVs)	Automatic temperature control system
Automatic irrigation system		

9.5.7 DISEASE AND PEST CONTROL

The amalgamation of UAVs with a sprayer system values in a probable platform for vector control and pest management. For large-area crop fields, an accurate site-specific application is possible. The pulse width modulation controller on the spraying system attached to the UAV improves its efficiency while spraying pesticides. Generally, the sprinkling mechanism is often mounted on the lower part of the UAV and functions as a nozzle beneath the pesticide tank to sprinkle the chemical downstream crop plants. The sprinkling system comprises a spraying nozzle, which is stimulated by a controller and the substance to be sprayed, like fertilizer or pesticides.

9.6 HARVESTING

The efficiency and acceptance of harvesting robots are taken into consideration based on four factors: a controlled planting environment, recognition and placement algorithms, picking effectiveness, and being economically viable (Wang et al., 2022). Identification and harvesting of ripe fruits without any injury to plants and fruit is one of the challenges that are faced by robots. Cameras and sensors in the robot arms have been utilized to assess the data for fruit harvesting. The harvesting method can be expanded and improved with the use of robots that take real-time pictures and touch-based data that allows data to be gathered in a variety of ways (Sabzi et al., 2020). In the current era, numerous fruit and vegetable-picking robots have been planned by investigators worldwide, such as apple-picking robots, strawberry-picking robots, kiwi fruit-picking robots, and tomato-picking robots (Table 9.2). Image processing method, in comparison to manual methods, offers information that is practical, straightforward, quick, and accurate. Apart from the aforementioned technologies, remote sensing has been recognized as a technical instrument possessing noteworthy potential to enhance intelligent and focused agriculture. Prevalent remote-sensing technologies comprise drones, human-piloted planes, and satellites (Tsouros et al., 2019).

Selecting which crops to plant, planning the land, preparing the soil for irrigation, preparing the seeds, and sowing the seeds are all part of the cultivation phase. After the cultivation stage, keeping an eye on and managing crop growth, including prearranged crop health monitoring, fertilizer use, disease, and weed identification, as well as pesticide application is time-sensitive during the monitoring phase. The harvesting phase, which comprises tasks including crop harvesting, sorting, curing, storing, and market sale, is the final and most important stage of the agricultural cycle. As mentioned above, numerous tasks are involved in each phase of the crop life cycle and individual task needs proper attention from the grower. Out of these tasks, most of the tasks are done manually or semiautomatic, which takes a lot of time, labor, cost, and effort.

TABLE 9.2 Advancement in Horticultural Crops Via Utilization of Drones and Robotics

S. No.	Work	Operation	References
1.	Greenhouse automation	Intelligent control technique to optimally control the relative humidity	Morimoto and Hashimoto (1998)
		Control of heating, ventilating, fogging, watering, fertilizing, lighting processes by the automation system	
2.	Vision system	Classify crops and weeds	Noguchi et al. (1998)
3.	Crop monitoring	Evaluation and the judgment of the seedling using image processing incarnation	Fujiwara (1991)
4.	Disease identification	Dissection of leaf spots disease in apple via drone	Anam (2020)
		Identification of diseases in beet, brinjal, celery, cabbage, chili, French bean, onion, okra, potato, turnip, pepper, and tomato	Chaudhary et al. (2020)
		Recognition of tea diseases	Zou et al. (2020)
5.	Pest control	Robots (UGVs and UAVs) for environmentally safe pest control	Gonzalez-de-Santos et al. (2017)
6.	Weed control	Automated detection system in crop plants	Rani et al. (2022)
		Classify crops and weeds	Noguchi et al. (1998b)
		Automated weed detection system and smart herbicide sprayer robot	Kargar and Shirzadifar (2013)
		Weed identification system in potato crop	Sabzi et al. (2018)
7.	Harvesting	Hybrid robot system (heavy produce harvesting)	Kim et al. (2008)
		Ambiguous controller for automating crop harvesting	Hagras et al. (1998)
		Automated fruit detection tool using deep learning for plucking of fruits and packing by robots	Munir et al. (2020)
		Autonomous robot for detecting maturity of greenhouse strawberries	Habaragamuwa et al. (2018)
		Automatic fruit-picking robots: apple, strawberry, cherry, citrus, kiwi, plum, wolfberry, tomato, cucumber, bell pepper, green pepper, eggplant, mushroom	Onishi et al. (2019); Ge et al. (2019); Ma et al. (2021); Yida (2020); Sabzi et al. (2020); Mehta and Burks (2014); Wang et al. (2022); Ji et al. (2019, 2020)

TABLE 9.2 *(Continued)*

S. No.	Work	Operation	References
		Vision-based estimation and control system for robotic citrus fruit harvesting	Mehta and Burks (2014)
		Estimating the yield of a vegetable by considering temperature, humidity, and moisture of the soil	Upadhya and Mathew (2019)
		Robot for grading and sorting root-trimmed garlic	Thuyet et al. (2020)
		Robotic harvesting system for eggplant	Hayashi et al. (2002)
		Robot for harvesting lettuce	Cho et al. (2002)
		Grading of tomatoes based on fruit color, size, and hardness	Nassiri et al. (2022)
		UAV for mapping plant density in potato	Mhango et al. (2021)
		Genetic algorithm approach for classification of potatoes based on their external defects and diseases	Dacal-Nieto et al. (2009)

9.7 CHALLENGES

The commercialization of robotic technology is still much behind the present state of affairs. Robotic performance issues, knowledge gaps, and installation costs may cause the noncommercialization of robots. Zhou et al. (2022) systematically examined the challenges behind the inadequate performance of harvesting robots. The major challenges of robotics in agriculture are fruit visibility, obstacle interference, light variation, low contrast, wind disturbance, undesirable robot contact, fruit fragility, plant stem/spur fragility, ripening pattern, diseases/pests, fruit quality variation, adverse weather, and so on. In most horticultural crops, mainly fruit, stem, leaves, and flowers are the economic part and they are very perishable. These economic parts can sustain damage via squeezing, clamping, or grabbing actions that involve pulling, bending, or twisting motions by robots while working. Drones are also not without their restrictions for use in farms. Among these include the involvement of pilots, engine power, stability and reliability, quality of the sensors because of shipment weight restrictions, execution costs, and aviation regulations which are of major concern. The assessment highlights the importance of using drones for horticultural crops and the enormous potential these instruments have to improve agricultural efficiency. The amount of literature on drone uses in horticulture is growing annually, as seen by the increasing number of specialized publications on the topic. Agronomists, agrotechnicians, and specialized advisors in precision agriculture will use the scientific understanding of this subject along with the wealth of information that drones can supply. These experts will be able to provide farmers with more precise and knowledgeable operational advice, which will help to optimize agricultural management techniques and produce positive effects on the economy and environment. Drones can offer two benefits from an economic perspective. Through early identification of various stress types, their use not only permits the prevention of damage but also the decrease of input amounts like water, fertilizer, pesticides, and herbicides. These technologies are positioned as feasible options for the ecological viability of flower, fruit, and vegetable crops since they can also produce environmental advantages through reduced input costs. But to meet technological obstacles while rendering these instruments more accessible and useful for the agriculture industry, where tradition is deeply ingrained and innovations are progressively embraced and adopted research and development must continue.

KEYWORDS

- **robotics**
- **agriculture**
- **drone sensors**
- **artificial intelligence**
- **horticultural crops**

REFERENCES

Abdulrahman, M., Koli, U., & Kori, A. (2017). Seed sowing robot. *International Journal of Computer Science Trends and Technology (IJCST)*, 5(2), 131–143.

Anam, S. (2020). Segmentation of leaf spots disease in apple plants using particle swarm optimization and K-means Algorithm. *Journal of Physics: Conference Series,* 1562(1), 012011. https://doi.org/10.1088/1742-6596/1562/1/012011

Antonucci, G., Impollonia, G., Croci, M., Potenza, E., Marcone, A., & Amaducci, S. (2023). Evaluating biostimulants via high-throughput field phenotyping: Biophysical traits retrieval through PROSAIL inversion. *Smart Agricultural Technology,* 3, 100067, https://doi.org/10.1016/j.atech.2022.100067.

Brewster, C., Roussaki, I., Kalatzis, N., Doolin, K., & Ellis, K. (2017). IoT in agriculture: Designing a Europe-wide large-scale pilot. *IEEE Communications Magazine,* 55(9), 26–33.

Casamitjana, M., Torres-Madro˜nero, M.C., Bernal-Riobo, J., & Varga, D. (2020). Soil moisture analysis by means of multispectral images according to land use and spatial resolution on andosols in the Colombian Andes. *Applied Sciences,* 10, 5540, https://doi.org/10.3390/app10165540.

Chandel, A.K., Molaei, B., Khot, L.R., Peters, R.T., & St¨ockle, C.O. (2020). High resolution geospatial evapotranspiration mapping of irrigated field crops using multispectral and thermal infrared imagery with METRIC energy balance model. *Drones*, 4, 52. https://doi.org/10.3390/drones4030052.

Chatzimichali, A.P., Georgilas, I.P., & Tourassis, V.D. (2009). Design of an advanced prototype robot for white asparagus harvesting. In: *2009 IEEE/ASME International Conference on Advanced Intelligent Mechatronics*, 14–17 July 2009; Singapore. IEEE; pp. 887–892

Chaudhary, A., Thakur, R., Kolhe, S., & Kamal, R. (2020). A particle swarm optimization-based ensemble for vegetable crop disease recognition. *Computers and Electronics in Agriculture*, 178, 105747.

Cho, S.I., Chang, S.J., Kim, Y.Y., & An, K.J. (2002). Developing a three-degrees-of-freedom robot for harvesting lettuce using machine vision and fuzzy logic control. *Biosystems Engineering*, 82(2), 143–149. https://doi.org/10.1006/bioe.2002.0061

Coulombe, J., Brown, P., White, S., Xu, C., & Koech, R. (2020). Detection of crop water status using UAV mounted sensor. *Acta Horticulturae*, 271–278, https://doi.org/10.17660/ActaHortic.2020.1279.39.

Dacal-Nieto, A., Vázquez-Fernández, E., Formella, A., Martin, F., Torres-Guijarro, S., & González-Jorge, H. (2009). A genetic algorithm approach for feature selection in potatoes classification by computer vision. *2009 35th Annual Conference of IEEE Industrial Electronics*, Porto, Portugal, 2009, pp. 1955–1960, doi: 10.1109/IECON.2009.5414871.

de Jesus Colwell, F., Souter, J., Bryan, G.J., Compton, L.J., Boonham, N., & Prashar, A. (2021). Development and validation of methodology for estimating potato canopy structure for field crop phenotyping and improved breeding. *Frontiers in Plant Science*, 12, https://www.frontiersin.org/articles/10.3389/fpls.2021.612843.

De-An, Z., Jidong, L., Wei, J., Ying, Z., & Yu, C. (2011). Design and control of an apple harvesting robot. *Biosystems Engineering*, 110(2), 112–122

Din, N.U., Naz, B., Zai, S., & Bakhtawer, W.A. (2021). Onion crop monitoring with multispectral imagery using deep neural network. *International Journal of Advanced Computer Science and Applications, (IJACSA)*, 12, https://doi.org/10.14569/IJACSA.2021.0120537.

Ekinzog, E.K., Schlerf, M., Kraft, M., Werner, F., Riedel, A., Rock, G., & Mallick, K. (2022) Revisiting crop water stress index based on potato field experiments in Northern Germany. *Agricultural Water Management*, 269, 107664, https://doi.org/10.1016/j.agwat.2022.107664.

Elijah, O., Rahman, T.A., Orikumhi, I., Leow, C.Y., & Hindia, M.N. (2018). An overview of the Internet of Things (IoT) and data analytics in agriculture: Benefits and challenges. *IEEE Internet Things Journal*, 5(5), 3758–3773.

Feng, X., Yan, F., & Liu, X. (2019). Study of wireless communication technologies on Internet of Things for precision agriculture. *Wireless Personal Communications*, 108(3), 1785–1802.

Friha, O., Ferrag, M.A., Shu, L., Maglaras, L.A., & Wang, X. (2021). Internet of Things for the future of smart agriculture: A comprehensive survey of emerging technologies. *IEEE/CAA Journal of Automatica Sinica*, 8(4), 718–752.

Fujiwara, H. (1991). Discriminating robot system for carnation seedlings with fuzzy logic. *IFAC Proceedings Volumes*, 24(11), 231–235.

Gago, J., Douthe, C., Coopman, R.E., Gallego, P.P., Ribas-Carbo, M., Flexas, J., Escalona, J., & Medrano, H. (2015). UAVs challenge to assess water stress for sustainable agriculture. *Agricultural Water Management*, 153, 9–19

Ge, Y., Xiong, Y., Tenorio, G.L., & From, P.J. (2019). Fruit localization and environment perception for strawberry harvesting robots. *IEEE Access*, 7, 147642–147652.

Gill, S.S., Chana, I., & Buyya, R. (2017). IoT-based agriculture as a cloud and big data service: The beginning of digital India. *Journal of Organizational and End User Computing (JOEUC)*, 29(4), 1–23.

Gonzalez-de-Santos, P., Ribeiro, A., Fernandez-Quintanilla, C., Lopez-Granados, F., Brandstoetter, M., Tomic, S., Pedrazzi, S., Peruzzi, A., Pajares, G., Kaplanis, G., Perez-Ruiz, M., Valero, C., del Cerro, J., Vieri, M., Rabatel, G., & Debilde, B. (2017). Fleets of robots for environmentally-safe pest control in agriculture. *Precision Agriculture*, 18(4), 574–614. https://doi.org/10.1007/s11119-016-9476-3

Grenzd¨orffer, G.J. (2019). Automatic generation of geometric parameters of individual cauliflower plants for rapid phenotyping using drone images. *The International Archives of the Photogrammetry, Remote Sensing and Spatial Information Sciences*, XLII-2/W13, 329–335, https://doi.org/10.5194/isprs-archives-XLII-2-W13- 329-2019. XLII-2-W13.

Habaragamuwa, H., Ogawa, Y., Suzuki, T., Shiigi, T., Ono, M., & Kondo, N. (2018). Detecting greenhouse strawberries (mature and immature), using a deep convolutional neural network. *Engineering in Agriculture, Environment, and Food*, 11(3), 127–138. https://doi.org/10.1016/j.eaef.2018.03.001

Hagras, H., Callaghan, V., Colley, M., & Carr-West, M. (1998). Developing a fuzzy logic controlled agricultural vehicle. *Third International Conference of Fuzzy Systems and Soft Computing (FSC'88)*, pp. 9–16, Wiesbaden , Germany.

Han, R., Wong, A.J.Y., Tang, Z., Truco, M.J., Lavelle, D.O., Kozik, A., Jin, Y., & Michelmore, R.W. (2021). Drone phenotyping, and machine learning enable discovery of loci regulating daily floral opening in lettuce. *Journal of Experimental Botany*, 72, 2979–2994, https://doi.org/10.1093/jxb/erab081.

Hayashi, S., Ganno, K., Ishii, Y., & Tanaka, I. (2002). Robotic harvesting system for eggplants. *Japan Agricultural Research Quarterly*, 36(3), 163–168.

Hernández, J.D., Barrientos, J., del Cerro, J., Barrientos, A., & Sanz, D. (2013b). Moisture measurement in crops using spherical robots. *Industrial Robot*, 40(1), 59–66

Hernández, J.D., Sanz, D., Rodríguez-Canosa, G.R., Barrientos, J., del Cerro, J., & Barrientos, A. (2013a). Sensorized robotic sphere for large exterior critical infrastructures supervision. *Journal of Applied Remote Sensing*, 7(1), 073522–073522

Hsu, T.-C., Yang, H., Chung, Y.-C., & Hsu, C.-H. (2020). A creative IoT agriculture platform for cloud fog computing. *Sustainable Computer Information Systems*, 28, 100285.

Ihuoma, S.O., Madramootoo, C.A., & Kalacska, M. (2021). Integration of satellite imagery and in situ soil moisture data for estimating irrigation water requirements. *International Journal of Applied Earth Observation*, 102, 102396, https://doi.org/10.1016/j.jag.2021.102396.

Inoue, Y. (2020). Satellite- and drone-based remote sensing of crops and soils for smart farming–a review. *Soil Science and Plant Nutrition*, 66(6), 798–810. https://doi.org/10.1080/00380768.2020.1738899.

Ji, W., Chen, G., Xu, B., Meng, X., & Zhao, D. (2019). Recognition method of green pepper in greenhouse based on least-squares support vector machine optimized by the improved. *IEEE Access*, 7, 119742–119754. doi: 10.1109/ACCESS.2019.2937326.

Jia, B., Yang, S., & Yu, T. (2020). Research on three picking arm avoidance algorithms for agaricus mushroom picking robot. In: *Proceedings of 2020 IEEE International Conference on Advances in Electrical Engineering and Computer Applications, AEECA*, 2020, pp. 325–328. https://doi.org/10.1109/AEECA49918.2020.9213575

Jinbo, C., Xiangliang, C., Han-Chi, F., & Lam, A. (2019). Agricultural product monitoring system supported by cloud computing. *Cluster Computing*, 22(4), 8929–8938.

Kargar, A.H., & Shirzadifar, A. (2013). Automatic weed detection system and smart herbicide sprayer robot for corn fields. In: *2013 First RSI/ISM International Conference on Robotics and Mechatronics (ICRoM)*, pp. 468–473.

Khanna, A., & Kaur, S. (2019). Evolution of the Internet of Things (IoT) and its significant impact in the field of precision agriculture. *Computers and Electronics in Agriculture*, 157, 218–231.

Kim, D.W., Yun, H.S., Jeong, S.J., Kwon, Y.S., Kim, S.G., Lee, W.S., & Kim, H.J. (2018). Modeling and testing of growth status for Chinese cabbage and white radish with UAV-based RGB imagery. *Remote Sensing*, 10, 563. https://doi.org /10.3390/rs10040563.

Kim, Y.Y., Hwang, H., & Cho, S.I. (2008). A hybrid robotic system for harvesting heavy produce. *Engineering in Agriculture, Environment and Food*, 1(1), 18–23. https://doi.org/10.1016/S1881-8366(08)80009-2

Liakos, K.G., Busato, P., Moshou, D., Pearson, S., & Bochtis, D. (2018). Machine learning in agriculture: A review. *Sensors*, 18(8), 2674.

Ma, Y., Zhang, W., Qureshi, W.S., Gao, C., Zhang, C., & Li, W. (2021). Autonomous navigation for a wolfberry-picking robot using visual cues and fuzzy control. *Information Processing in Agriculture*, 8(1), 15–26. https://doi.org/10.1016/j.inpa.2020.04.005

Martins, S., Lhissou, R., Chokmani, K., & Cambouris, A. (2023). Determining the beginning of potato Tuberization period using plant height detected by drone for irrigation purposes. *Agronomy*, 13, 492, https://doi.org/10.3390/agronomy13020492.

Mascarenhas Maciel, G., de A. Gallis, R.B., Barbosa, R.L., Pereira, L.M., Siquieroli, A.C.S., & Vit´ oria Miranda Peixoto, J. (2019). Image phenotyping of inbred red lettuce lines with genetic diversity regarding carotenoid levels. *International Journal of Applied Earth Observation*, 81, 154–160, https://doi.org/10.1016/j.jag.2019.05.016.

Mehta, S.S., & Burks, T.F. (2014). Vision-based control of robotic manipulator for citrus harvesting. *Computers and Electronics in Agriculture*, 102, 146–158. https://doi.org/10.1016/j.compag.2014.01.003

Mhango, J.K., Harris, E.W., Green, R., & Monaghan, J.M. (2021). Mapping potato plant density variation using aerial imagery and deep learning techniques for precision agriculture. *Remote Sensing*, 13(14), 2705. https://doi.org/10.3390/rs13142705

Mithra, S., & Nagamalleswari, T.Y.J. (2023). Cucurbitaceous family flower inferencing using deep transfer learning approaches: CuCuFlower UAV imagery data. *Soft Computing*, 27,8345–8356. https://doi.org/10.1007/s00500-023-08186-w.

Morimoto, T., & Hashimoto, Y. (1998). An intelligent control technique based on fuzzy controls, neural networks, and genetic algorithms for greenhouse automation. *IFAC Proceedings Volumes*, 31(5), 61–66.

Munir, K., Umar, A.I., & Yousaf, W. (2020). Automatic fruits classification system based on deep neural network. *NUST Journal of Engineering Sciences*, 13(1), 37–44. https://doi.org/10.24949/njes.v13i1.501

Mwinuka, P.R., Mbilinyi, B.P., Mbungu, W.B., Mourice, S.K., Mahoo, H.F., & Schmitter, P. (2021). The feasibility of hand-held thermal and UAV-based multispectral imaging for canopy water status assessment and yield prediction of irrigated African eggplant (*Solanum aethopicum* L). *Agricultural Water Management*, 245, 106584, https://doi.org/10.1016/j.agwat.2020.106584.

Nassiri, S.M., Tahavoor, A., & Jafari, A. (2022). Fuzzy logic classification of mature tomatoes based on physical properties fusion. *Information Processing in Agriculture*, 9(4), 547–555. https://doi.org/10.1016/j.inpa.2021.09.001

Nejati, H., Azimifar, Z., & Zamani, M. (2008, October). Using fast Fourier transform for weed detection in corn fields. In: *2008 IEEE International Conference on Systems, Man, and Cybernetics*, pp. 1215–1219. IEEE.

Noguchi, N., Reid, J.F., Zhang, Q., & Tian, L.F. (1998). Vision intelligence for precision farming using fuzzy logic optimized genetic algorithm and artificial neural network. *ASAE Paper* 983034. St. Joseph, MI

Onishi, Y., Yoshida, T., Kurita, H., Fukao, T., Arihara, H., & Iwai, A. (2019). An automated fruit harvesting robot by using deep learning. *ROBOMECH Journal*, 6(1), 13. https://doi.org/10.1186/s40648-019-0141-2

Panday, U.S., Pratihast, A.K., Aryal, J., & Kayastha, R.B. (2020). A review on drone-based data solutions for cereal crops. *Drones*, 4(3), 1–29. https://doi.org/10.3390/drones4030041.

Parsaeian, M., Shahabi, M., & Hassanpour, H. (2020). Estimating oil and protein content of sesame seeds using image processing and artificial neural network. *Journal of the American Oil Chemists' Society*, 97(7), 691–702.

Peng, J., Nieto, H., Neumann Andersen, M., Kørup, K., Larsen, R., Morel, J., Parsons, D., Zhou, Z., & Manevski, K. (2023). Accurate estimates of land surface energy fluxes and irrigation requirements from UAV-based thermal and multispectral sensors. *ISPRS Journal of Photogrammetry and Remote Sensing*, 198, 238–254, https://doi.org/10.1016/j.isprsjprs.2023.03.009.

Pereira, L.S., Oweis, T., & Zairi, A. (2002). Irrigation management under water scarcity, *Agricultural Water Management*, 57, 175–206, https://doi.org/10.1016/S0378-3774(02)00075-6.

Rani, S.V.J., Kumar, P.S., Priyadharsini, R., Srividya, S.J., & Harshana, S. (2022). Automated weed detection system in smart farming for developing sustainable agriculture. *International Journal of Environmental Science and Technology*, 19(9), 9083–9094. https://doi.org/10.1007/s13762-021-03606-6

Roldán, J.J., del Cerro, J., Garzón-Ramos, D., Garcia-Aunon, P., Garzón, M., de León, J., & Barrientos, A. (2018). *Robots in Agriculture: State of Art and Practical Experiences.* InTech. doi: 10.5772/intechopen.69874

Sabzi, S., Abbaspour-Gilandeh, Y., & García-Mateos, G. (2018). A fast and accurate expert system for weed identification in potato crops using metaheuristic algorithms. *Computers in Industry*, 98, 80–89. https://doi.org/10.1016/j.compind.2018.03.001

Sabzi, S., Pourdarbani, R., Kalantari, D., & Panagopoulos, T. (2020). Designing a fruit identification algorithm in orchard conditions to develop robots using video processing and majority voting based on a hybrid artificial neural network. *Applied Sciences (Switzerland)*, 10(1), 383. https://doi.org/10.3390/app10010383

Sakai, S., Iida, M., Osuka, K., & Umeda, M. (2008). Design and control of a heavy material handling manipulator for agricultural robots. *Autonomous Robots*, 25(3), 189–204

Shadrin, D., Menshchikov, A., Somov, A., Bornemann, G., Hauslage, J., & Fedorov, M. (2019). Enabling precision agriculture through embedded sensing with artificial intelligence. *IEEE Transactions on Instrumentation and Measurement*, 69(7), 4103–4113.

Stutsel, B., Johansen, K., Malb′, Y.M. eteau, & McCabe, M.F. (2021). Detecting plant stress using thermal and optical imagery from an unoccupied aerial vehicle. *Frontiers in Plant Science*, 12, 734944. https://www.frontiersin.org/articles/10.3389/fpls.

Tang, Y., Dananjayan, S., Hou, C., Guo, Q., Luo, S., & He, Y. (2021). A survey on the 5G network and its impact on agriculture: Challenges and opportunities. *Computers and Electronics in Agriculture*, 180, 105895 https://doi.org/10.1016/j.compag.2020.105895.

Tang, Z., Jin, Y., Brown, P.H., & Park, M. (2023). Estimation of tomato water status with photochemical reflectance index and machine learning: Assessment from proximal sensors and UAV imagery. *Frontiers in Plant Science*, 14, 1057733. https://www.frontiersin.org/articles/10.3389/fpls.2023.1057733.

Thuyet, D.Q., Kobayashi, Y., & Matsuo, M. (2020). A robot system equipped with a deep convolutional neural network for autonomous grading and sorting of root-trimmed garlic. *Computers and Electronics in Agriculture*, 178, 105727. https://doi.org/10.1016/j.compag.2020.105727

Tsouros, D.C., Bibi, S., & Sarigiannidis, P.G. (2019). A review on UAV-based applications for precision agriculture. *Information*, 10(11), 349.

Tunca, E., K¨ oksal, E.S., Torres-Rua, A.F., Kustas, W.P., & Nieto, H.S. (2022). Estimation of bell pepper evapotranspiration using two-source energy balance model based on high- resolution thermal and visible imagery from unmanned aerial vehicles. *JARS*, 16, 022204, https://doi.org/10.1117/1.JRS.16.022204.

Tzounis, A., Katsoulas, N., Bartzanas, T., & Kittas, C. (2017). Internet of Things in agriculture, recent advances and future challenges. *Biosystems Engineering*, 164, 31–48. https://doi.org/10.1016/j.biosystemseng.2017.09.007

Upadhya, S.M., & Mathew, S. (2019). Implementation of fuzzy logic in estimating the yield of a vegetable crop. *Journal of Physics, Conference Series*, pp. 1–12. https://doi.org/10.1088/1742-6596/1427/1/012013

Wakchaure, M., Patel, B.K., & Mahindrakar, A.K. (2023). Application of AI techniques and robotics in agriculture: A review. In: *Artificial Intelligence in the Life Sciences (Vol. 3)*. Elsevier B.V. https://doi.org/10.1016/j.ailsci.2023.100057

Wang, Z., Xun, Y., Wang, Y., & Yang, Q. (2022). Review of smart robots for fruit and vegetable picking in agriculture. *International Journal of Agricultural and Biological Engineering*, 15(1), 33–54. https://doi.org/10.25165/j.ijabe.20221501.7232

Wood, R. *RoboBees Project*. Cambridge, MA, USA: Harvard University, 2015.

Xie, J., Zhou, Z., Zhang, H., Zhang, L., & Li, M. (2022). Combining canopy coverage and plant height from UAV-Based RGB images to estimate spraying volume on potato. *Sustainability*, 14, 6473, https://doi.org/10.3390/su14116473.

Yang, S., Feng, Q., Yang, W., & Gao, X. (2023). Simple, low-cost estimation of potato above-ground biomass using improved canopy leaf detection method. *American Journal of Potato Research*, 100, 143–162, https://doi.org/10.1007/s12230-022-09897-w.

Yida, N.I.U. (2020). Research state and trend of fruit picking robot manipulator structure. *International Journal of Plant Engineering and Management*, 25(1), 36–50

Zamora-Izquierdo, M.A., Santa, J., Martínez, J.A., Martínez, V., & Skarmeta, A.F. (2019). Smart farming IoT platform based on edge and cloud computing. *Biosystems Engineering*, 177, 4–17.

Zarco-Tejada, P.J., González-Dugo, V., & Berni, J.A.J. (2012). Fluorescence, temperature, and narrow-band indices were acquired from a UAV platform for water stress detection using a micro-hyperspectral imager and a thermal camera. *Remote Sensing of Environment*, 117, 322–337

Zhan, Q., Cai, Y., & Liu, Z. (2008). Near-optimal trajectory planning of a spherical mobile robot for environment exploration. In: *IEEE Conference on Robotics, Automation and Mechatronics*, 21–24 Sept. 2008; Chengdu, China. IEEE, pp. 84–89.

Zhang, J., Wang, X., Liu, J., Zhang, D., Lu, Y., Zhou, Y., Sun, L., Hou, S., Fan, X., Shen, S., & Zhao, J. (2022). *Multispectral Drone Imagery and SRGAN for Rapid Phenotypic Mapping of Individual Chinese Cabbage Plants, Plant Phenomics*, 2022, 0007, https://doi.org/10.34133/plantphenomics.0007.

Zheng, J., & Yang, W. (2018). Design of a precision agriculture leakage seeding system based on wireless sensors. *International Journal of Online Engineering*, 14(05), 184. Z

Zhou, C., Ye, H., Sun, D., Yue, J., Yang, G., & Hu, J. (2022a). An automated, high-performance approach for detecting and characterizing broccoli based on UAV remote-sensing and transformers: A case study from Haining, China. *International Journal of Applied Earth Observation and Geoinformation,* 114, 103055, https://doi.org/10.1016/j.jag.2022.103055.

Zhou, H., Wang, X., Au, W., Kang, H., & Chen, C. (2022b). Intelligent robots for fruit harvesting: Recent developments and future challenges. In: *Precision Agriculture*. Springer, Vol.23(5), pp. 1856–1907. https://doi.org/10.1007/s11119-022-09913-3

Zhou, Y., Xie, Y., & Shao, L. (2016). Simulation of the core technology of a greenhouse monitoring system based on a wireless sensor network. *International Journal of Online Engineering*, 12(05), 43.

Zou, X., Zhang, J., Huang, S., Wang, J., Yao, H., & Song, Y. (2020). Recognition of tea diseases under natural background based on particle swarm optimization algorithm optimized support vector machine. In: *IEEE International Conference on Industrial Informatics (INDIN)*, 2020 July, pp. 547–552. https://doi.org/10.1109/INDIN45582.2020.9442092

CHAPTER 10

Smart Water, Soil, and Environment Management System: An Approach Toward Sustainable Farming

SMITA RANI SWAIN* and RAKESH KUMAR GUPTA

Department of Agricultural and Food Engineering, Indian Institute of Technology, Kharagpur, West Bengal, India

*Corresponding author

ABSTRACT

The growing global population has intensified the demand for food production, placing 28 significant pressure on agricultural systems. This chapter introduces a smart water, soil, and 29 environment management system designed to promote sustainable farming by leveraging 30 advanced technologies. The system integrates Internet of Things (IoT) devices, sensors, and 31 data analytics to provide a comprehensive framework for precision agriculture. Real-time data 32 collected from sensors deployed in agricultural fields enables informed decision-making for 33 irrigation, fertilization, and environmental management. Smart irrigation, guided by soil 34 moisture levels and weather forecasts, ensures efficient water use, reduces over-irrigation, and 35 mitigates soil degradation. The soil management component analyzes nutrient levels and 36 composition, delivering targeted fertilization recommendations that minimize synthetic input 37 use and reduce nutrient runoff. Environmental monitoring enhances crop management by 38 tracking temperature, humidity, and air quality, while also enabling early detection of pests and 39 diseases. This proactive approach supports timely interventions and reduces dependence on 40 chemical treatments.

Smart Horticulture Production: Integrating Advanced Technologies.
Rajni Rajan and Duniya Ram Singh (Eds.)

DOI: 10.1201/9781998511679-10

By optimizing resource utilization and improving crop productivity, the 41 proposed system offers a scalable, data-driven solution to the environmental challenges of 42 modern agriculture. It contributes to the resilience and long-term sustainability of farming 43 practices, addressing both current demands and future uncertainties related to climate change 44.

10.1 INTRODUCTION

Modern agriculture faces numerous challenges, including water scarcity, soil degradation, and environmental concerns. To address these issues, a holistic approach integrating smart technologies has emerged, transforming traditional farming into a sustainable and efficient system. A system known as "smart agriculture" makes use of cutting-edge technology to rationalize the use of natural resources, particularly freshwater and soil, in order to produce food in a sustainable and clean manner (Hadid & Ahmed, 2024). Food and clean water make agriculture a valuable asset to humanity, as it uses water to provide us with food. Environmental destruction and rapid population growth have had a massive effect on agriculture, a detrimental impact on the world's water supplies, and are crucial for sustained development. To resolve the issue, we need to implement the intelligent irrigation method using automated and Internet of Things (IoT) technologies (Parvathi Sangeetha et al., 2022). The smart management of water for precision irrigation in agriculture is essential for increasing crop yield and decreasing costs, while at the same time contributing to environmental sustainability (Kamienski et al., 2019). The environmental degradation that has come along with the use of technology in the last few decades has to be reversed and farming be made sustainable (Patil Shirish & Bhalerao, 2013). This chapter aims to explore the integration of smart technologies in water, soil, and environment management for sustainable farming. It also discusses the benefits, challenges, and future prospects of implementing such systems.

10.2 SMART WATER MANAGEMENT

10.2.1 IOT-BASED WATER MONITORING

The IoT is the natural choice for smart water management applications, even though the integration of different technologies required for making it work seamlessly in practice is still not fully accomplished. A number of factors have come together to cause the IoT to emerge, including low-cost devices, low-power wireless technologies, availability of cloud data centers for processing and storing data, management frameworks for handling unstructured data from social networks, high-performance computing resources in commodity platforms, and computational intelligence algorithms to handle the massive amount of data (Kamienski et al., 2019). The primary goal of the IoT application in agriculture is to use wireless network connectivity to connect physical objects (things) like sensors, cameras, flow meters, actuators, and robots to the Internet in order to measure various parameters like soil moisture, temperature, humidity, plant images, and other weather conditions. As a result, it would be beneficial to use an IoT system platform to improve irrigation operation monitoring and control and track plant response to water stress and growth (Baruah et al., 2024).

10.2.2 PRECISION IRRIGATION SYSTEM

Considering the global population boom, water availability for irrigation will be a significant challenge. Precision irrigation is a sustainable agricultural approach that gives farmers the ability to use irrigation water more effectively and efficiently. This smart irrigation system optimizes water usage by delivering the right amount of water to crops based on their specific needs. This not only conserves water but also improves crop yield and quality without polluting the environment and helps save energy by decreasing the cost of cultivation using low-cost wireless sensors. Sensors and devices provide real-time data on soil moisture levels, weather conditions, and crop water requirements (CWRs). These data are then transmitted to a central cloud-based platform, where it is analyzed using machine learning (ML) algorithms to generate actionable insights for precise irrigation scheduling. It also helps the farmers by providing stability over all uncertainties associated with cultivation by predicting the climate and its consequent management (Baruah et al., 2024).

10.2.3 DATA-DRIVEN DECISION-MAKING

Sustainable agriculture demands a lot of knowledge to be attained. The human brain's limited ability to notice, evaluate, and remember the myriad interactions and synergies that can sustain biological systems was the foundation of traditional agriculture. More of those interactions can be observed, recorded, and understood by farmers, businesses, and researchers thanks to data-driven technology than is possible with human brain capacity alone (Rozenstein et al., 2023). Analyzing data from various sources allows farmers to make informed decisions about water usage for crop cultivation. For example, artificial intelligence (AI) technologies enable farmers to apply the right amount of inputs for each crop, increasing yields while lowering water consumption and greenhouse gas emissions. They do this by monitoring circumstances and optimizing output (Linaza et al., 2021). Similarly, ML algorithms predict water demand, helping farmers adjust irrigation plans for maximum efficiency. Recently, the crop water supply (CWS) was compared to the CWR by Huang et al. (2023). Both parameters are made up of multiple different data sources. Pumping subterranean water and precipitation are included in the CWS. The CWR takes runoff and evaporation into account.

10.2.4 STRATEGIES OF A SMART WATER MANAGEMENT SYSTEM

A smart water management system for sustainable farming integrates various technologies and practices to optimize water usage, reduce wastage, and improve crop yield. Here is a breakdown of components and strategies for such a system:

Soil moisture sensors: Deploy sensors in the soil to monitor moisture levels. These data help farmers understand when and how much to irrigate, preventing overwatering or underwatering.

Weather forecast integration: Integrate weather forecasts into the system to anticipate rainfall and adjust irrigation schedules accordingly, minimizing water wastage during rainy periods.

Drip irrigation systems: Utilize drip irrigation systems that deliver water directly to the roots of plants, reducing evaporation and ensuring efficient water usage.

Remote monitoring and control: Implement remote monitoring and control systems that allow farmers to manage irrigation schedules and monitor water usage from anywhere using mobile apps or web interfaces.

Water recycling and reuse: Set up systems to collect and treat runoff water for reuse in irrigation, reducing the dependency on freshwater sources.

Crop selection and rotation: Opt for drought-resistant crops and practice crop rotation to minimize water consumption and maintain soil health.

Data analytics and AI: Employ data analytics and AI algorithms to analyze historical data, predict water requirements, and optimize irrigation schedules for maximum efficiency.

Variable rate irrigation (VRI): Use VRI technology to adjust irrigation rates based on soil moisture levels, topography, and crop requirements, ensuring uniform water distribution across the field.

Regenerative agriculture practices: Implement regenerative agriculture practices such as cover cropping, mulching, and no-till farming to improve soil structure and water retention, reducing the need for irrigation.

Education and training: Provide farmers with education and training on sustainable water management practices to enhance awareness and adoption of efficient irrigation techniques.

Government support and policies: Encourage government support and policies that promote sustainable water management practices, such as subsidies for implementing water-saving technologies and incentives for adopting conservation practices.

Community engagement: Foster collaboration among farmers, researchers, and stakeholders to share knowledge and best practices for sustainable water management in agriculture.

By integrating these components and strategies, a smart water management system can help farmers optimize water usage, conserve resources, and ensure the long-term sustainability of agricultural operations.

10.3 SMART SOIL MANAGEMENT

10.3.1 SOIL HEALTH MONITORING

Sensors and imaging technologies assess soil health by measuring parameters such as nutrient levels, pH, and organic matter content. This information guides farmers in implementing soil conservation practices.

10.3.2 PRECISION FARMING TECHNIQUES

Precision farming nowadays is an emerging area in agriculture that has a basic goal to maximize output (i.e., crop yield) while minimizing input (i.e., fertilizer, pesticide, herbicide, etc.). Precision farming can be defined as "application of a holistic management strategy that uses information technology to bring data from multiple source to make decision associated with agricultural production, marketing, finance and personnel" (Patil Shirish & Bhalerao, 2013). The smart technologies enable precision farming, allowing farmers to apply fertilizers and pesticides with precision. This reduces environmental impact, minimizes input costs, and enhances overall soil health. By comparing the soil erosion potential of different soil types and the effectiveness of different conservation measures, researchers can develop recommendations for improving soil health and productivity. By analyzing changes in soil properties over time and their effects on crop productivity, researchers can develop strategies to adapt to the changing climate and mitigate its impact on horticultural production (Vista, 2023).

10.3.3 AUTOMATED CROP ROTATION PLANNING

Algorithms analyze historical data and recommend optimal crop rotation plans to prevent soil exhaustion. This promotes sustainable farming practices and enhances soil fertility over time.

10.3.4 STRATEGIES OF A SMART SOIL MANAGEMENT SYSTEM

A smart soil management system for sustainable farming involves the integration of technology and data-driven approaches to optimize soil health, conserve resources, and enhance overall agricultural sustainability. Here are key components and features of such a system:

10.3.4.1 SOIL SENSORS

Deploy soil sensors to measure key parameters such as moisture content, temperature, nutrient levels, and pH. Use advanced sensors to collect real-time data and provide insights into soil health.

10.3.4.2 IOT CONNECTIVITY

Connect sensors and other devices through IoT to enable seamless data transfer and monitoring. IoT connectivity allows for remote monitoring and control of soil conditions.

10.3.4.3 DATA ANALYTICS AND AI

Utilize data analytics and AI to analyze the collected data. Develop algorithms that can provide actionable insights, such as optimal planting times, irrigation schedules, and nutrient application.

10.3.4.4 PRECISION AGRICULTURE

Implement precision agriculture techniques to apply inputs (water, fertilizers, and pesticides) precisely where and when they are needed. Use GPS technology for precise mapping and application of inputs.

10.3.4.4 WEATHER INTEGRATION

Integrate weather forecasts and real-time weather data into the system to adjust farming practices based on upcoming weather conditions. Predictive analytics can help farmers plan for extreme weather events.

Remote monitoring and control: Enable farmers to monitor and control soil conditions remotely through a user-friendly interface, such as a mobile app or web portal. Automated alerts can notify farmers of any critical changes in soil parameters.

10.3.4.5 DRONES AND SATELLITE IMAGING

Use drones and satellite imaging for high-resolution mapping of fields. These technologies can provide valuable insights into overall crop health, identify areas of concern, and monitor changes over time.

10.3.4.6 DECISION SUPPORT SYSTEM

Develop a decision support system that provides farmers with actionable recommendations based on the analyzed data. The system could suggest optimal planting times, irrigation amounts, and fertilizer applications.

10.3.4.7 FARM MANAGEMENT SOFTWARE

Integrate the smart soil management system with overall farm management software for a comprehensive approach. This allows farmers to manage multiple aspects of their operation from a centralized platform.

10.3.4.8 EDUCATION AND TRAINING

Provide training and educational resources to farmers on how to effectively use and interpret data from the smart soil management system. Encourage the adoption of sustainable farming practices based on the insights provided.

By incorporating these components, a smart soil management system can help farmers make informed decisions, reduce input waste, and contribute to long-term soil health and sustainable agriculture.

10.4 SMART ENVIRONMENT MANAGEMENT

10.4.1 CLIMATE MONITORING

Farmers can monitor climatic patterns with the use of satellite data and advanced weather stations. This knowledge helps to provide resilience and sustainability in farming methods by adjusting them to changing climatic circumstances. The notion of sustainable agriculture has gained popularity due to the belief held by research and farming groups that productivity must be carefully balanced with consideration for the environment and society (Vishnoi & Goel, 2024).

10.4.2 PEST AND DISEASE PREDICTION

Applications powered by AI have the capacity to analyze a wide range of data inputs, such as crop health, meteorological conditions, soil quality, and market trends. According to Vishnoi & Goel (2024), having access to real-time data can help make informed decisions on crop preferences, the best times to plant, irrigation, fertilization, and pest control. Based on environmental factors, ML models forecast outbreaks of diseases and pests. Farmers can take preventive action thanks to early detection, which lessens the need for chemical interventions.

10.4.3 BIODIVERSITY CONSERVATION

Smart farming practices promote biodiversity by preserving natural habitats and implementing agroecological approaches. Ecology applied to agriculture farms, or agroecology, is defined as the research, planning, and administration of a sustainable agroecosystem using ecological science. With the use of agroecological resource management systems, the natural equilibrium of all available resources can be managed (Gangwar et al., 2019). This fosters a healthier environment for crops, beneficial organisms, and wildlife.

10.4.4 STRATEGIES OF A SMART ENVIRONMENT MANAGEMENT SYSTEM

A smart environment management system for sustainable farming integrates various technologies to optimize resource usage, improve crop yields, and minimize environmental impact. Here is an outline of components and features such a system might include the following:

Sensor networks: Deploy sensors to monitor various environmental parameters such as soil moisture, temperature, humidity, light intensity, pH levels, and nutrient levels.

Data collection and analysis: Collect data from sensors in real time and analyze it to gain insights into environmental conditions and crop health. ML algorithms can be employed to predict trends and provide actionable recommendations.

Automated irrigation systems: Use sensor data to automate irrigation systems, ensuring that crops receive the right amount of water at the right time. This prevents overwatering or underwatering, conserving water resources, and promoting plant health.

Precision agriculture techniques: Implement precision agriculture techniques such as variable rate application of fertilizers and pesticides based on real-time soil and crop data. This minimizes input usage and reduces environmental pollution.

Climate control systems: Utilize climate control systems such as greenhouse automation to regulate temperature, humidity, and ventilation, creating optimal growing conditions for crops regardless of external weather conditions.

Energy management: Integrate renewable energy sources such as solar panels or wind turbines to power farm operations, reducing reliance on fossil fuels and lowering carbon footprint.

Crop monitoring and management: Use drones, satellite imagery, or unmanned ground vehicles equipped with cameras to monitor crop health, detect pests and diseases early, and optimize harvesting schedules.

Resource optimization and planning: Utilize predictive analytics to optimize resource allocation, crop rotation planning, and land usage, maximizing yield while minimizing resource wastage.

Integrated farm management platform: Develop a centralized platform or dashboard where farmers can access all data, insights, and control systems from a single interface. This enhances ease of use and decision-making.

Environmental monitoring and compliance: Implement systems to monitor environmental parameters beyond the farm, such as water quality in nearby water bodies, to ensure compliance with environmental regulations and minimize negative impacts on surrounding ecosystems.

Community engagement and education: Involve local communities and stakeholders by providing educational resources and demonstrations on sustainable farming practices, fostering awareness and support for environmentally friendly agriculture.

Feedback loop and continuous improvement: Establish a feedback loop where data collected from the system's operation is analyzed to identify areas for improvement, leading to iterative enhancements and optimization of the smart farming system over time.

By integrating these components and features, a smart environment management system for sustainable farming can help farmers increase productivity, reduce costs, and mitigate environmental impacts, contributing to long-term agricultural sustainability.

10.5 CHALLENGES AND FUTURE DIRECTIONS

10.5.1 DATA SECURITY AND PRIVACY CONCERNS

The integration of smart technologies raises issues related to data security and privacy. Despite the fact that a data-driven approach to agriculture may contribute to the development of sustainable food and fiber production systems, it is hampered by social, legal, and technological obstacles as well as economic and practical limitations. These challenges together make it more difficult to share data and reap its wider benefits (Rozenstein et al., 2023). So, addressing these concerns is crucial for widespread adoption and acceptance among farmers.

10.5.2 ACCESSIBILITY AND AFFORDABILITY

Ensuring that smart farming technologies are accessible and affordable to all farmers, regardless of the scale of their operation, is essential for widespread adoption and global sustainability. First, we need to bring back the farmers, who have left the farming profession because of various reasons by encouraging them to adopt smart technologies for sustainable farming. It should be the priority of international organizations, regional government, and policymakers to support small and marginal farmers and encourage family farming so that future generations do not shy away from adopting this profession (Vishnoi & Goel, 2024).

10.5.3 CONTINUOUS TECHNOLOGICAL ADVANCEMENTS

As technology evolves, ongoing research and development are necessary to keep smart farming systems up-to-date. Collaboration between researchers, technology developers, and farmers is crucial for successful implementation. As a trending technological advancement, digital soil maps (DSM) can be used as an evidence-based decision-making tool for improving soil, managing inputs and resources for horticultural crops. DSM has the tremendous potential to transform our understanding and management of soil resources, contributing to sustainable agriculture, land-use and environmental management (Vista, 2023). Much less growth has been made in the highly valued crops in dry and semiarid regions (where the crop is primarily rainfed) because of the shortage of water and the expense of irrigation. As a result, these areas could be improved with the use of precision irrigation systems by enabling cultivation and producing greater yields with less water input (Baruah et al., 2024). However, there are still several obstacles in the way of the widespread application of IoT for precision irrigation. First off, there is still work to be done in fully automating software development for IoT-based smart applications like agricultural irrigation. Second, in order to automate a portion of the process and integrate various technologies including big data analytics, cloud computing, fog computing, and the IoTs for the deployment of pilot applications for smart water management, sophisticated IoT software platforms are still lacking. Third, appropriate standards and information models are needed for the integration of sophisticated and diverse sensors (Kamienski et al., 2019). The synergy between technology and agriculture has the potential to create more sustainable and efficient agricultural practices through the recycling of resources (Vishnoi & Goel, 2024).

10.6 CONCLUSION

The integration of smart water, soil, and environment management systems marks a significant step toward achieving sustainable agriculture by optimizing resource utilization, enhancing crop yield, and reducing the environmental footprint of agriculture. This innovative solution not only meets the challenges posed by a growing global population but also contributes to the long-term resilience and viability of agriculture in the face of climate change and resource scarcity. By harnessing the power of technology, farmers can mitigate environmental impact, conserve resources, and improve overall productivity. Governments, researchers, and industry stakeholders must collaborate to promote the adoption of smart farming practices. Investments in research, education, and infrastructure are essential to create an enabling environment for sustainable agriculture.

This chapter provides a comprehensive overview of the current state, challenges, and future directions of smart water, soil, and environment management systems in the context of sustainable farming.

KEYWORDS

- **agriculture**
- **sustainable farming**
- **Internet of Things**
- **precision irrigation**

REFERENCES

Baruah, V.J., Begum, M., Sarmah, B., Deka, B., Bhagawati, R., Paul, S., & Dutta, M. (2024). Chapter 11. Precision irrigation management: A step toward sustainable agriculture. In: *Remote Sensing in Precision Agriculture*. Remote Sensing in Precision Agriculture. https://doi.org/10.1016/B978-0-323-91068-2.00011-4

Gangwar, D.S., Tyagi, S., & Soni, S.K. (2019). A conceptual framework of agroecological resource management system for climate-smart agriculture. *International Journal of Environmental Science and Technology*, 16(8), 4123–4102. https://doi.org/10.1007/s10762-018-1827-3

Hadid, M., & Ahmed, S.M. (2024). Role of smart agriculture on food security in Saudi Arabia. In: A.E. Ahmed, J.M. Al-Khayri, A.A. Elbushra, Eds., *Food and Nutrition Security in the Kingdom of Saudi Arabia*. Cham: Springer, Vol. 1. https://doi.org/10.1007/978-3-031-46716-5_10

Huang, Y., Srivastava, R., Ngo, C., Gao, J., Wu, J., & Chiao, S. (2023). Data-driven soil analysis and evaluation for smart farming using machine learning approaches. *Agriculture (Switzerland)*, *13*(9), 1777. https://doi.org/10.3390/agriculture10091777

Kamienski, C., Soininen, J.P., Taumberger, M., Dantas, R., Toscano, A., Cinotti, T.S., Maia, R.F., & Neto, A.T. (2019). Smart water management platform: IoT-based precision irrigation for agriculture. *Sensors (Switzerland)*, 19(2) 276. https://doi.org/10.3390/s19020276.

Linaza, M.T., Posada, J., Bund, J., Eisert, P., Quartulli, M., Döllner, J., Pagani, A., Olaizola, I.G., Barriguinha, A., Moysiadis, T., & Lucat, L. (2021). Data-driven artificial intelligence applications for sustainable precision agriculture. *Agronomy*, 11(6) 1227. https://doi.org/10.3390/agronomy11061227

Patil Shirish, S. & Bhalerao, S.A. (2013). Precision farming: The most scientific and modern approach to sustainable agriculture. *International Research Journal of Science and Engineering*, 1(2), 21–30.

Parvathi Sangeetha, B., Kumar, N., Ambalgi, A.P., Abdul Haleem, S.L., Thilagam, K., & Vijayakumar, P. (2022). IOT based smart irrigation management system for environmental sustainability in India. *Sustainable Energy Technologies and Assessments*, 52(PA), 101973. https://doi.org/10.1016/j.seta.2022.101973

Rozenstein, O., Cohen, Y., Alchanatis, V., Behrendt, K., Bonfil, D.J., Eshel, G., Harari, A., Harris, W.E., Klapp, I., Laor, Y., Linker, R., Paz-Kagan, T., Peets, S., Rutter, S.M., Salzer, Y., & Lowenberg-DeBoer, J. (2023). Data-driven agriculture and sustainable farming: friends or foes? *Precision Agriculture*, 25(1), 520–531. https://doi.org/10.1007/s11119-023-10061-5

Vishnoi, S., & Goel, R.K. (2024). Climate smart agriculture for sustainable productivity and healthy landscapes. *Environmental Science and Policy*, 151(March 2022), 103600. https://doi.org/10.1016/j.envsci.2023.103600

Vista, S.P. (2023). Digital soil maps: Its use for managing horticultural crops. In: *Digital Soil Maps: Its Use for Managing Horticultural Crops*. November.

CHAPTER 11

An Approach to Machine Learning for Smart Horticultural Production Systems

AMBUJ*, RAJENDRA MACHAVARAM, and GAJENDRA SINGH

Indian Institute of Technology Kharagpur, Kharagpur, West Bengal,India

*Corresponding author

ABSTRACT

In recent years, deep learning and machine vision have significantly advanced horticultural research, transforming practices in disease management, growth monitoring, genetic prediction, and market forecasting. Convolutional neural networks and other deep learning models have achieved high accuracy in detecting and classifying plant diseases across crops like tomatoes, cucumbers, and bananas, enabling timely interventions. These technologies also support precise monitoring of crop growth stages, maturity, and health in fruits such as apples, mangoes, and lettuce, improving yield quality and management. In genomics, deep learning has shown promise in predicting complex genetic traits in polyploid crops like blueberries and strawberries, offering insights into fruit size, yield, and firmness by capturing intricate gene interactions. Additionally, neural networks, including recurrent models, have been used for market forecasting, accurately predicting price trends of produce such as cucumbers and tomatoes using web-sourced data. Collectively, these applications highlight the transformative role of deep learning in enhancing efficiency, sustainability, and resilience in horticulture.

Smart Horticulture Production: Integrating Advanced Technologies.
Rajni Rajan and Duniya Ram Singh (Eds.)

DOI: 10.1201/9781998511679-11

11.1 INTRODUCTION

The cultivation and utilization of horticultural crops, encompassing fruits, vegetables, materials for beverages, herbal medicine, ornamental plants, and their diverse applications, have transcended their mere economic significance to become integral aspects of human culture, landscape aesthetics, and lifestyle influences. However, as societal changes unfold, the pivotal role of horticultural crops has expanded, urging workers to strive for increased variety and higher quality products. Unfortunately, intricate tasks in horticulture often rely heavily on manual labor, a challenge exacerbated by the declining engagement of younger generations in gardening practices (Nturambirwe & Opara, 2020; Chen et al., 2019). Yet, propelled by technological advancements, the introduction of advanced and automated instruments in horticulture presents a promising shift toward enhanced efficiency and reduced reliance on manual labor.

This evolution in horticultural practices underscores the need for researchers to redirect their focus toward innovative technologies. The primary goal is to revolutionize orchard management and maximize horticultural productivity by employing cutting-edge technologies, equipment, and intelligent systems (Colaço et al., 2018; Edwards & Moghadam, 2020). Intelligent horticulture strives to produce superior-quality fruits, vegetables, and ornamental crops while minimizing reliance on human labor and amplifying efficiency. Crucially, the rapid, precise, and automatic identification of horticultural crops, coupled with the acquisition of comprehensive planting and postharvest data, stands as a fundamental direction in this pursuit of intelligent horticulture (Edwards & Moghadam, 2020). Advancements in computer technologies and data analysis methods have paved the way for the development of intelligent systems, offering unprecedented opportunities to refine cultivation practices and crop management. In recent years, the orchestration of various sensor technologies has facilitated the collection of a myriad of digital information from orchards and groves. Remote sensing techniques utilizing unmanned aerial vehicles (UAVs), planes, satellites, coupled with internet of things devices, thermal and near-infrared cameras, and X-ray imaging technology have become commonplace (Kamilaris & Prenafeta-Boldú, 2018). This array of technologies captures diverse digital data from horticultural crops, forming the basis for models applied in practical production processes. By monitoring crop growth, established models enable informed management decisions to optimize the growth trajectory. However, the exponential growth in imaging sensor capabilities has resulted in a deluge of digital horticultural data (Edwards & Moghadam, 2020; Singh et al., 2018). This inundation poses challenges for the horticultural science community, necessitating the development of analytical technologies capable of sifting through this vast pool of data cluttered with irrelevant and redundant information. Addressing this challenge demands urgent attention to devise suitable analytical methodologies to distil useful features from unprocessed data and effectively translate these technologies into real-world applications.

In response to this data complexity, a plethora of data analysis approaches, including various machine-learning methodologies, have emerged. Techniques such as partial least squares (PLS) (Wendel et al., 2018), artificial neural networks (ANNs) (Pan et al., 2016), support vector machines (SVMs) (Leiva-Valenzuela & Aguilera, 2013), random forests (RFs) (Che et al., 2018), and k-nearest neighbors (KNNs) (van Dael et al., 2016; Wrzesień et al., 2019) have been instrumental in handling heterogeneous data. Deep learning, a subset of machine-learning approaches, has garnered significant attention for its ability to assimilate vast and diverse data sets. This versatile tool, with its deep network structures and hierarchical data representations, outperforms traditional machine-learning techniques, especially in processing data collected through digital cameras and spectroscopy.

Notably, the integration of deep learning into horticultural sectors for analyzing RGB and spectral images from horticultural crops has shown promising results. Despite its potential, understanding the principles and practical applications of deep learning remains a challenge for researchers and workers in horticultural sectors. To bridge this gap, this chapter aims to comprehensively present the latest research advancements in the application of deep learning to horticultural sciences. By providing an overview and guidance, this chapter endeavors to assist researchers in navigating the complexities of deep learning applications within the realm of horticulture (Figure 11.1).

11.2 BRIEF OVERVIEW OF MACHINE LEARNING AND DEEP LEARNING TECHNIQUES

Machine learning techniques serve as a powerful arsenal within horticulture, revolutionizing how we understand, manage, and optimize crop

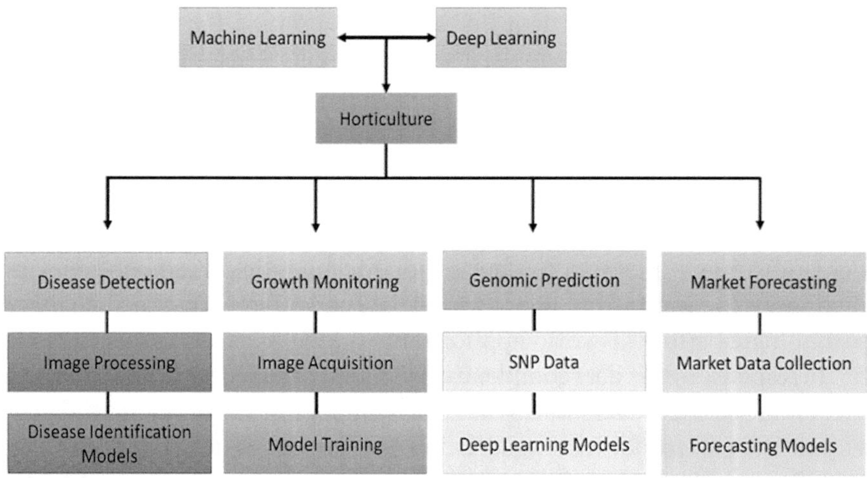

FIGURE 11.1 Deep learning applications in smart horticulture systems.

production (Singh et al., 2016; Colmer et al., 2020). In the realm of supervised learning, regression models stand as stalwarts for predicting continuous variables. Within horticulture, these models are deployed for forecasting crop yields based on historical data, environmental variables like temperature, humidity, and soil nutrients. Additionally, they aid in estimating growth patterns and optimizing resource allocation, ensuring efficient utilization of resources while maximizing yield (Kamilaris & Prenafeta-Boldú, 2018). Classification methods, another facet of supervised learning, play a pivotal role in categorizing horticultural data. These methods are essential for disease identification, crop quality assessment, and pest detection. By categorizing plant-related data into distinct classes, these techniques empower decision-making processes critical for maintaining healthy crops and enhancing overall productivity. Unsupervised learning techniques, such as clustering and dimensionality reduction, offer insights without the need for predefined outcomes. Clustering algorithms segment datasets, facilitating the grouping of plants exhibiting similar growth behaviors or identifying inherent patterns in crop behavior. Dimensionality reduction methods like principal component analysis (PCA) or t-distributed stochastic neighbor embedding (t-SNE) assist in simplifying complex data, aiding visualization and interpretation while retaining critical information.

Transfer learning, a technique leveraging pretrained models on extensive datasets, adapts them to specific horticultural tasks. This adaptation enables efficient utilization of deep learning methodologies in diverse horticultural contexts with limited labeled data, enhancing the precision and applicability of models. Ensemble methods such as RFs and Gradient Boosting amalgamate models or construct iterative models to improve prediction accuracy. Their applications span from yield prediction and disease diagnosis to precision agriculture, integrating diverse datasets for comprehensive insights and decision-making (de Ridder et al., 2017).

Reinforcement learning represents a frontier in horticulture, empowering autonomous decision-making in crop management. This approach allows robotic systems to execute tasks such as harvesting or weeding based on real-time feedback from the environment. Hybrid approaches, often combining multiple machine learning techniques, offer nuanced solutions to complex horticultural challenges. By integrating methodologies such as image analysis with classification techniques, these hybrids address intricacies in disease diagnosis, thereby enhancing overall crop health and productivity. Collectively, these machine learning methods usher in a new era of precision agriculture, where data-driven insights and intelligent systems optimize crop management, disease prevention, and resource utilization, contributing to sustainable and efficient horticultural practices.

Deep learning, a subset of machine learning, has evolved as a potent paradigm in diverse fields, including horticulture, revolutionizing the analysis and interpretation of complex datasets (Koirala et al., 2019). At the forefront of this revolution are various deep learning methodologies, each contributing unique capabilities to the understanding and management of crops. Convolutional neural networks (CNNs), the cornerstone of image analysis, stand prominent in horticulture. These networks excel in discerning intricate patterns within images, enabling precise analysis of plant health and growth. In the realm of horticulture, CNNs serve as crucial tools for disease identification, assessing fruit quality through image processing, and performing leaf-level classifications for species recognition. By decoding detailed features from images, CNNs offer valuable insights into plant diseases, growth anomalies, and overall crop health.

RNNs specialize in processing sequential data and retaining memory from past inputs. Within horticulture, they are pivotal in monitoring crop growth patterns over time, forecasting weather conditions, and predicting changes in plant behavior based on temporal data. In practical applications, RNNs forecast crop yields, anticipate disease outbreaks, and optimize irrigation schedules by considering fluctuating environmental factors. Their capacity to capture temporal dependencies renders them indispensable for analyzing dynamic agricultural data. Transfer learning leverages pretrained models on extensive datasets and adapts them to specific horticultural tasks, enhancing the efficiency and applicability of deep learning methodologies. In horticulture, transfer learning expedites the adaptation of powerful models trained on vast image datasets, enabling quick identification of diseases or assessment of plant health across different crop varieties with limited labeled data. Deep reinforcement learning merges deep learning with reinforcement learning principles, showing promise in autonomous decision-making for tasks like robotic harvesting or automated crop management based on real-time environmental feedback. Although in its nascent stages in horticulture, its potential lies in fostering smart and adaptable agricultural systems. Generative adversarial networks (GANs), although relatively new in horticulture, offer prospects in creating synthetic datasets for training models or generating simulated images. This synthetic data generation supplements limited datasets, potentially enhancing the robustness of models trained on smaller or scarce data sources.

Long short-term memory (LSTM) networks, a variant of RNNs, specialize in processing sequences, particularly in time-series data analysis. In agriculture, LSTM models forecast crop yields, predict disease outbreaks, and optimize irrigation schedules based on changing environmental factors. Their ability to capture long-term dependencies makes them invaluable for understanding and predicting temporal patterns in crop behavior. Additional methodologies like autoencoders, attention mechanisms, capsule networks, and hybrid architectures bring further depth to deep learning in horticulture. Autoencoders aid in feature extraction, attention mechanisms enhance models' focus on relevant data portions, capsule networks improve pattern recognition, and hybrid architectures integrate diverse models for comprehensive insights. Collectively, these diverse deep learning methodologies empower horticulturists to unravel intricate patterns in plant behavior, predict diseases, optimize resource allocation, and revolutionize precision agriculture for sustainable crop management (Altaheri et al., 2019). Their integration offers a promising pathway toward enhancing agricultural productivity and resilience in the face of evolving challenges.

11.3 DESCRIPTION OF MACHINE LEARNING TERMINOLOGIES

This section deals with the key terminologies that are widely used in the domain of machine learning. The key terms are as follows:

1. Supervised Learning:

   Imagine a student learning from a labeled dataset (think of a textbook with practice problems and solutions). Supervised learning works similarly. The model, acting as the student, ingests labeled data (input–output pairs) to learn patterns and make predictions for unseen data. For instance, a supervised learning algorithm trained on labeled images of cats and dogs could predict the category of a new, unseen image.

2. Unsupervised Learning:

   Unlike supervised learning, unsupervised learning does not have a predefined answer sheet. The model sifts through unlabeled data, uncovering hidden patterns and structures. It is like exploring a new world and identifying natural groupings or relationships within the data. For example, an unsupervised learning algorithm might discover customer segments within a large dataset based on their purchase history.

3. Regression:

   Regression techniques excel at predicting continuous values, like housing prices or stock market trends. They build models that map the relationship between independent variables (factors influencing the outcome) and a dependent variable (the value we want to predict). Imagine a model that considers factors like square footage and location to predict a house's price.

4. Classification:

   Classification algorithms act like sorting machines. They categorize data points into predefined classes based on features learned from the training data. This is akin to sorting emails into spam and inbox folders. For example, a spam filter might classify emails based on keywords or sender information.

5. Clustering:

   Clustering algorithms are like detectives uncovering hidden connections within data. They group data points together based on their inherent characteristics or patterns, identifying natural clusters without predefined categories. Imagine grouping customers with similar purchase behaviors to personalize marketing campaigns.

6. Dimensionality Reduction:

   When dealing with a multitude of features (dimensions), visualization and analysis can become cumbersome. Dimensionality reduction techniques come to the rescue! They condense the number of features while preserving essential information. Think of it like summarizing a complex document into its key points for easier comprehension.

7. Ensemble Methods:

   Ensemble methods leverage the wisdom of crowds. They combine predictions from multiple models (like RFs or gradient boosting) to achieve superior accuracy. Imagine consulting multiple experts for their insights before making a final decision.

8. Reinforcement Learning:

   Reinforcement learning mimics how we learn through experience. An agent interacts with an environment, taking actions and receiving rewards (or penalties) for its choices. Through this iterative process, the agent learns to make optimal decisions to maximize rewards. Think of training an AI agent to play a game by rewarding successful moves.

9. Deep Learning:

   Deep learning is a cutting-edge subfield of machine learning inspired by the structure and function of the human brain. It utilizes ANNs with multiple layers to extract complex patterns from data. This technique excels in tasks like image recognition and natural language processing.

10. Artificial Neural Networks (ANNs):

    ANNs are computational models loosely inspired by the biological neural networks in our brains. They consist of interconnected nodes (artificial neurons) arranged in layers. These nodes process information, learn from patterns, and transmit signals through weighted connections.

11. Transfer Learning:

    Transfer learning capitalizes on the power of knowledge transfer. Imagine a student excelling in math applying those skills to solve physics problems. Similarly, transfer learning utilizes a pretrained model on a large dataset for a new, related task with limited data. This approach saves time and resources by leveraging existing knowledge.

12. Overfitting and Underfitting:

    Overfitting is akin to memorizing every detail on a practice test but failing a real examination. It occurs when a model learns the training data "too well," including noise, leading to poor performance on unseen data. Underfitting, on the other hand, is like skipping practice altogether. The model remains too simple to capture the underlying patterns, resulting in low accuracy on both training and test data.

13. Hyperparameters versus Parameters:

    Imagine the knobs and dials on a complex machine. Hyperparameters are like those controls—they are set before training and influence how the model learns. Examples include learning rates or the number of layers in a neural network. In contrast, parameters are the internal tunings learned during training. They define the model's structure and are optimized to minimize error.

14. Bias-Variance Trade-Off:

    The bias-variance trade-off is a fundamental concept in machine learning, akin to finding the perfect temperature—not too hot, not too cold. It revolves around the interplay between a model's complexity (variance) and its underlying assumptions (bias), both of which significantly impact its ability to generalize to unseen data. Imagine a model like a thermometer. High bias is like a faulty thermometer with a fixed offset—it consistently underestimates or overestimates the true temperature (prediction) regardless of the actual data. On the other hand, high variance is like a very sensitive thermometer that fluctuates wildly based on minor variations—it might perfectly capture the training data's idiosyncrasies but fail to generalize to new situations. The ideal scenario lies in the sweet spot—a model with low bias that can capture the essential patterns in the data and low variance that avoids overfitting to noise or specific characteristics of the training data. Techniques like regularization can help us achieve this balance, essentially fine-tuning the model's complexity to prevent overfitting while retaining its ability to learn from the data.

15. Feature Engineering:

    Feature engineering is the cornerstone of successful machine learning projects. It is the art of transforming raw data into meaningful features (inputs) that a model can effectively learn from. This process involves three key steps: (1) Feature Selection: Choosing the most relevant features from the available data. Irrelevant or redundant features can add noise and hinder the model's learning process. (2) Feature Transformation: Transforming the existing features to improve their quality and facilitate learning. This might involve scaling numerical features, encoding categorical data, or creating new features through combinations of existing ones. Imagine transforming raw pixel values of an image into edges or shapes, making them more meaningful for image recognition tasks. (3) Feature Creation: Deriving entirely new features from the data. For example, in a customer transaction dataset, you might create a new feature representing the total purchase amount per month to understand buying behavior. Effective feature engineering can significantly improve a model's performance and interpretability. By carefully crafting the data the model sees, you essentially guide it toward learning the right patterns and relationships for accurate predictions.

11.4 DESCRIPTION OF DEEP LEARNING TERMINOLOGIES

This section explores advanced concepts within deep learning, venturing beyond the fundamentals to unveil the specialized architectures and techniques that empower neural networks to tackle complex tasks.

1. Convolutional Neural Networks (CNNs):

   CNNs reign supreme in the realm of image recognition and analysis. They excel at processing grid-like data (think images!) by employing convolutional layers. These layers act like filters, scanning the input data and extracting features at different levels of abstraction. Imagine progressively identifying edges, shapes, and ultimately entire objects within an image. This hierarchical feature extraction allows CNNs to effectively perform image classification, object detection, and pattern recognition.

2. Recurrent Neural Networks (RNNs):

   Unlike CNNs built for spatial data, RNNs are tailor-made for sequential data like language or time series. They incorporate a unique architecture with feedback loops, enabling the network to remember past inputs and make predictions based on the context. This memory is crucial for tasks like natural language processing, speech recognition, and sentiment analysis.

3. Long Short-Term Memory (LSTM) Networks:

   While RNNs are powerful, they can struggle with long sequences due to the vanishing or exploding gradient problem. LSTMs, a specific type of RNN, address this challenge. They possess specialized units (cells) that can selectively retain information for extended periods, making them ideal for tasks requiring analysis of long-term dependencies like machine translation or video captioning.

4. Autoencoders:

   Autoencoders are a fascinating breed of neural networks that operate in the unsupervised learning domain. They function by learning to compress input data into a lower-dimensional representation (encoding) and then reconstructing the original data from this compressed version (decoding). This process unveils the underlying structure of the data and has applications in data compression, anomaly detection, and feature extraction for dimensionality reduction.

5. Generative Adversarial Networks (GANs):

   i. GANs introduce a captivating concept—a two-player game between a generator and a discriminator network. The generator, like a skilled artist, learns to create new, synthetic data that mimics the real data distribution. The discriminator, acting as the art critic, tries to distinguish between real and generated data. This adversarial training pushes both networks to improve, enabling GANs to generate incredibly realistic images, translate languages more creatively, and even create entirely new data for specific applications.

6. Attention Mechanisms:

   Attention mechanisms are like spotlights within a neural network, enabling them to selectively focus on crucial parts of the input data when making predictions. Imagine translating a sentence, where the attention mechanism might focus on specific words or phrases to grasp the meaning more accurately. This selective attention empowers neural networks in tasks like machine translation, image captioning, and improving overall performance by prioritizing relevant information.

7. Beyond the Basics:

   The world of deep learning is constantly evolving, and this section merely scratches the surface. Here are some additional noteworthy techniques:

   i. Dropout and Batch Normalization: Regularization techniques that combat overfitting and improve training stability.
   ii. Activation Functions: The hidden language within neural networks, introducing nonlinearity for complex relationship learning.
   iii. Word Embeddings: Capturing semantic relationships between words by representing them as vectors in a high-dimensional space.
   iv. Recurrent Neural Network Variants: Alternative architectures like GRUs (gated recurrent units) offer efficient ways to handle sequential data.
   v. Self-Attention Mechanism: A powerful tool in transformer models, enabling the network to focus on different parts of a sequence for improved natural language understanding.
   vi. Neural Architecture Search (NAS): Automating the discovery of optimal neural network architectures for specific tasks.

vii. Adversarial Training: Enhancing model robustness against adversarial attacks by exposing them to manipulated data.
viii. Knowledge Distillation: Transferring knowledge from complex models to smaller, faster models, enabling efficient deployment.

11.5 SYNERGIZING MACHINE LEARNING AND DEEP LEARNING FOR ADVANCING HORTICULTURE CROP APPLICATIONS

11.5.1 IDENTIFYING AND CATEGORIZING HORTICULTURAL CROPS THROUGH RECOGNITION AND CLASSIFICATION TECHNIQUES

Deep learning has emerged as a game-changer in horticulture, particularly in the recognition and classification of diverse crops (Majeed et al., 2020). Within horticultural landscapes boasting over 91 fruit-bearing plant species, 200 vegetable varieties, and a staggering 6000 ornamental cultivars, the challenge of accurate classification becomes complex due to extensive intraclass variations. Manual classification, though feasible, is time-consuming and prone to errors, especially when dealing with a large number of samples.

The integration of CNNs has significantly bolstered image classification accuracy in horticulture (Prasad et al., 2018). For instance, utilizing ImageNet, a DCNN (AlexNet model) achieved remarkable accuracy in classifying various vegetable categories, outperforming traditional methods like SVM classifiers and back propagation (BP) neural networks. In flower species classification, researchers employed combined models to extract features, showcasing notable accuracy rates surpassing other advanced classifiers. Researchers have leveraged open-access image datasets like ImageNet and OUFD, providing rich data for training deep learning models in horticultural crop recognition. However, the field remains in its developmental phase, lacking establishment on a larger scale. Researchers often prefer using their collected image sets, indicating the ongoing evolution of horticultural crop recognition systems.

Deep learning is making waves in the agricultural sector, particularly with the potential to automate fruit harvesting (Noda et al., 2015; Ng et al., 2015; Yang & Sun, 2019). One of the initial hurdles is developing reliable fruit detection systems. Unlike lab settings, orchards are complex environments with ever-changing lighting conditions, leaves obscuring fruits, and the fruits themselves transforming in appearance as they ripen. Deep learning models need to be robust enough to identify fruits bathed in sunlight, shrouded in shadows, or partially hidden by leaves, all while accounting for the various stages of fruit maturity. Researchers are tackling these challenges by creating deep learning models that can handle the inherent variability of orchards. One technique involves image augmentation, where synthetic variations of real-world images are used to train the model on a wider range of scenarios, improving its ability to generalize.

Moving beyond just detecting the presence of fruit, truly autonomous harvesting systems require even more precise information. This is where instance segmentation comes into play. Instance segmentation delves deeper than simply identifying a fruit—it aims to precisely outline the boundaries of each individual fruit within an image (LeCun et al., 2015). This detailed information is essential for robots to accurately locate and pick fruits, especially when they are clustered together or obscured by leaves. By combining the power of fruit detection with instance segmentation, deep learning paves the way for a future where robots take over the labor-intensive task of fruit harvesting in orchards. This automation has the potential to increase efficiency and reduce costs for farmers, revolutionizing the agricultural landscape.

Researchers have developed advanced deep neural network models, such as DaSNet-V2 (LeCun & Bengio, 1998), specifically designed for tasks like fruit recognition and instance segmentation (Pan & Yang, 2009). These models excel in accurately delineating fruit shapes and sizes within images, facilitating the identification of fruit poses, particularly in situations where fruits overlap or are partially shaded. DaSNet-V2, leveraging the ResNet-101 architecture, showcases notable efficiency and robustness in visual sensing applications, particularly in automating harvesting processes within orchards.

Moreover, studies have explored techniques like Mask R-CNN combined with structure-from-motion (SfM) photogrammetry to detect and locate apples in both two-dimensional (2D) and three-dimensional (3D) spaces (Yu et al., 2018). This approach enhances accuracy in identifying apples and precisely locating them within the environment. Acknowledging the diverse shapes and colors of horticultural produce, researchers have broadened their scope beyond fruits, experimenting with models tailored to identify and classify various agricultural products. These efforts aim to refine the existing models to encompass a wider array of horticultural items, adapting them to recognize the unique characteristics of different agricultural products beyond fruits.

Horticultural crop cultivation demands unique techniques compared to other common crops. Regular pruning and thinning of flowers and fruits are essential practices to ensure robust tree growth and enhance yield quality. These practices play a pivotal role in maintaining tree strength and promoting the production of high-quality fruits.

Automating horticultural cultivation heavily relies on segmenting various components like branches, trunks, flowers, and fruits (Figure 11.2). Accurately delineating these segments stands as a fundamental step in the development of automated technologies for horticultural cultivation. Segmenting these elements enables precise management strategies and interventions, facilitating efficient and targeted agricultural practices within horticulture.

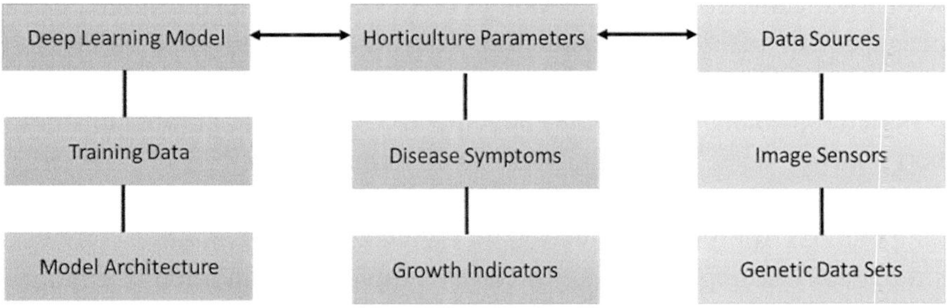

FIGURE 11.2 Components of horticulture deep learning models.

11.5.2 PREDICTIVE MODELING FOR YIELD ESTIMATION IN HORTICULTURAL CROPS USING ADVANCED TECHNIQUES

In horticulture, accurately estimating yields before harvesting is vital for producers and agricultural companies. It guides crucial decisions in planning harvests and marketing strategies. This estimation process, akin to object counting in computer vision, has shown promise in accurately counting various items, including plants and crowds. Recent studies have applied these counting methods to estimate yields in horticultural crops (Mai et al., 2020). They also employed neural networks like LSTM to estimate fruit yields. Their findings highlighted the effectiveness of these methods, showcasing their ability to accurately estimate yields with minimal errors when compared to manual counting. This indicates the potential of these computational models as reliable tools for yield estimation in horticulture.

Researchers have been applying various advanced techniques to estimate preharvest yields for horticultural products, showcasing promising outcomes. For instance, Rebortera et al. (Rebortera & Fajardo, 2019) developed a multifaceted system employing LSTM layers to forecast banana harvest yields. Their enhanced model exhibited substantial improvements in performance metrics compared to earlier models, demonstrating its efficacy in yield prediction.

In another study, Neupane et al. (2019) utilized a deep-learning algorithm (Faster R-CNN Inception-V2 model) to accurately detect and count bananas in high-resolution images acquired via UAV. This method achieved high accuracy rates in detecting bananas at different altitudes, showcasing its potential for accurate fruit detection in horticulture.

Additionally, techniques involving video-tracking systems combined with advanced algorithms like MangoYOLO and Kalman filters have been proposed for mango fruit detection, offering improved yield estimation by capturing additional imaging perspectives. In the domain of vegetable production, Zhang et al. (2022) developed a method for automatic strawberry flower detection using small UAVs equipped with RGB cameras. Their system demonstrated high accuracy in flower detection, indicating its potential for yield prediction in strawberry cultivation. Moreover, simulated learning methods, such as employing modified Inception-ResNet models for synthetic tomato images, and the utilization of Mask R-CNN models for tomato detection in greenhouse environments, exhibited promising results in estimating crop yields with high precision and recall scores. These applications in estimating preharvest yields for horticultural products indicate the potential of these techniques to guide planting, harvesting, and marketing decisions in real production settings. With ongoing scientific efforts, these methods are expected to play pivotal roles in shaping future agricultural practices (Figure 11.3).

11.5.3 ENHANCING QUALITY ASSESSMENT IN HORTICULTURAL CROPS THROUGH ADVANCED DETECTION METHODS

Improving living standards has heightened expectations for the quality of daily fruits and vegetables. Yet, these products remain vulnerable

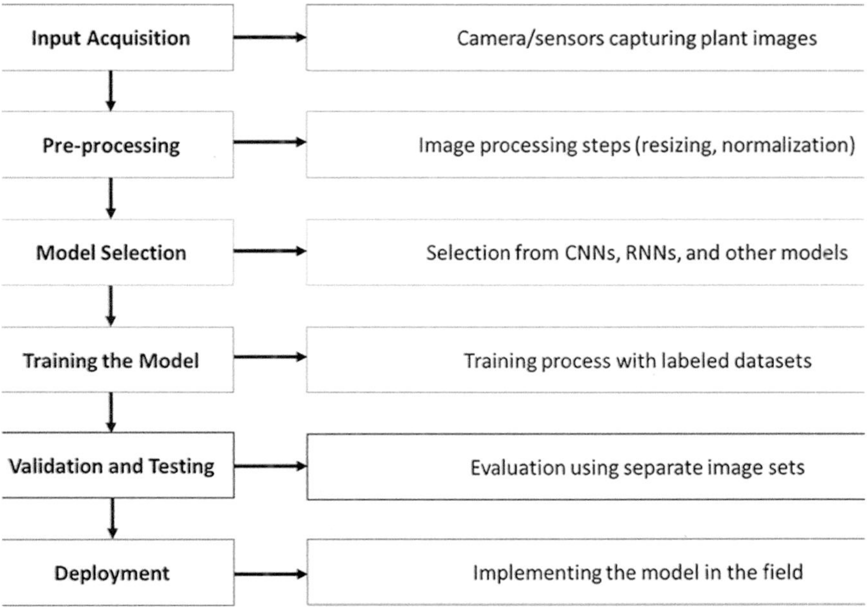

FIGURE 11.3 Growth monitoring in horticulture.

to diseases, pests, damage, and mishandling across various stages—production, storage, transportation, and marketing. Such issues can significantly diminish both their nutritional value and economic worth. Hence, the detection of horticultural product quality, especially in fruits and vegetables, presents a pressing challenge.

In recent research, the integration of deep learning techniques with RGB images or spectrographic methods has emerged as a powerful and noninvasive approach for assessing horticultural product quality (Nturambirwe & Opara, 2020). This innovative method addresses practical concerns such as postharvest grading, maturity assessment, bruise identification, and nutrient content prediction (Okere, 2023; Pandiselvam et al., 2022; Walsh et al., 2020; Hewett, 2012). Compared to conventional machine-learning methods, deep learning excels in analyzing and processing image, spectral, or sound data, demonstrating superior classification performance in these applications.

Several studies have delved into utilizing deep-learning approaches to distinguish between defective and healthy fruits based on visual characteristics. For instance, researchers applied advanced architectures like visual geometry group (VGG)-16 to differentiate defective dates from healthy ones with high accuracy (Tulbure et al., 2022; Pérez-Pérez et al., 2021; Yang & Xu, 2021). Another study explored employing deep learning for classifying different grades of bananas, achieving promising accuracy rates above 90% (Selvaraj et al., 2020). While visible surface defects significantly impact fruit grading, identifying internal or subdermal damage remains a challenge using visual inspection alone. To address this limitation, nondestructive testing technologies such as near-infrared spectrographs, hyperspectral imagers, and thermal imagers have emerged as valuable tools. These technologies allow for assessing internal conditions without causing harm to the produce.

Deep learning has played a crucial role in processing complex data obtained from nondestructive testing technologies. Various studies have demonstrated the efficacy of deep-learning methods in detecting internal mechanical damage, bruises, and defects in fruits using spectral data captured by advanced imaging techniques (Okere, 2023; He et al., 2022; Mei & Li, 2023). These studies have showcased the potential of deep learning in accurately analyzing spectral information to identify defects in fruits and vegetables without causing any harm to the produce. These applications highlight the promise of deep learning in quality assessment and defect identification in fruits and vegetables, especially in nondestructive testing methodologies applied in agriculture.

The quality of horticultural products, such as fruits, is often evaluated based on properties like their sweetness (soluble solid content [SSC]) and firmness, which greatly impact consumer satisfaction. Traditionally, assessing these attributes required destructive testing. However, nondestructive methods like Vis/near infrared (NIR) hyperspectral imaging have emerged (López-Maestresalas et al., 2016). These techniques analyze the reflected or transmitted light from fruits to establish connections between their spectral patterns and internal characteristics, like SSC and firmness.

Researchers have explored the potential of these methods combined with advanced technologies, such as deep learning, to predict these crucial attributes. Studies have shown promising results in predicting sweetness and firmness in fruits like apples and pears using deep-learning models coupled with spectral imaging techniques. Moreover, optical techniques like laser backscattering imaging have been employed to identify specific regions or defects in fruits, achieving higher accuracy compared to traditional machine-learning approaches. In addition to optical methods, acoustic sensing has proven effective in distinguishing between different apple types. Combining acoustic sensing with deep-learning models showcased superior performance in identifying specific apple qualities, demonstrating the potential of these combined techniques in nondestructive quality assessment of horticultural produce (Nturambirwe et al., 2021).

In commercial orchards, closely monitoring the fruit's maturity throughout its development is critical to identify the ideal time for harvesting. Automated machine vision techniques have become prevalent for this purpose, aiding in detecting and categorizing growth stages. For instance, Wendel et al. (2018) introduced a novel approach that combined light detection and ranging (LIDAR) sensing, hyperspectral imaging, and a navigation system on a ground vehicle to predict the dry matter content of individual fruits in mango orchards. They achieved substantial accuracy using CNNs and PLS methods. Another study employed a CNN to assess citrus fruit maturity by analyzing the fluorescent spectrum signal (Itakura et al., 2018). They utilized fluorescence spectroscopy to estimate the Brix/acid ratio, demonstrating significant improvements in accuracy compared to earlier methodologies.

These investigations highlighted that various physical and chemical attributes of fruits and vegetables, including nutrient content, hardness, disease, and maturity levels, can be assessed using RGB images, spectral data, and acoustic information (Pu et al., 2015; Pathmanaban et al., 2019; Hernández-Sánchez et al., 2016). Leveraging deep-learning models for training enhances the prediction and classification accuracy, as demonstrated in the studies mentioned in this context (Figure 11.4).

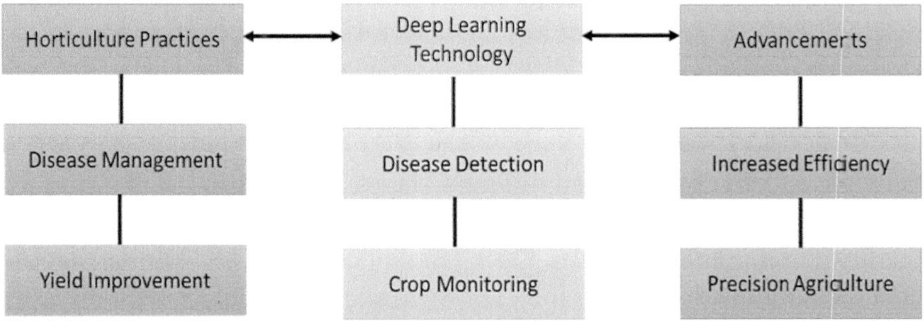

FIGURE 11.4 Impact of deep learning on horticulture advancements.

11.5.4 DETECTING AND MITIGATING BIOTIC AND ABIOTIC STRESSES IN HORTICULTURAL CROPS THROUGH ADVANCED MONITORING AND ANALYSIS

Researchers have explored deep learning's potential in identifying stress in horticultural plants. This involves detecting issues like pests, diseases, and other stressors without relying on subjective human assessments. By utilizing CNNs and related architectures, studies have achieved impressive accuracies in recognizing stress symptoms in various crops. For instance, tomatoes have been a focus for stress detection. Using datasets like PlantVillage, models employing CNNs could accurately pinpoint diseases like leaf mold, early blight, and late blight, achieving high accuracy rates (Umar et al., 2023).

Similar approaches have been applied to cucumbers, where sophisticated CNN-based segmentation methods could precisely identify issues like powdery mildew on leaves, showcasing substantial accuracy (Joseph et al., 2023; Ngugi et al., 2021). Eggplants, onions, potatoes, olives, and apples have also been subjects of study. Using different CNN architectures, researchers reached impressive accuracies in detecting diseases in these crops, highlighting the versatility of deep learning in identifying various stressors across different horticultural species.

These studies reflect the growing application of deep learning models in automating stress detection in crops, offering a more reliable and efficient means of identifying and managing stress-related issues in horticulture. Deep learning has emerged as a powerful tool for detecting and categorizing diseases in plants across various crops. Different deep learning models like multilayer CNNs, LeNet, Faster R-CNN, and deep belief networks have been employed to identify diseases in mangoes, bananas, tulips, peaches, and other plants with remarkable accuracy. For example, a multilayer CNN demonstrated an accuracy of 97.13% in identifying anthracnose fungal disease in mango leaves (Singh et al., 2019). LeNet architecture was effective in recognizing different diseases in banana leaves. The Faster R-CNN model efficiently detected the Tulip Breaking Virus, processing images swiftly at 0.13 s per image (Yang & Xu, 2021). Additionally, a deep belief network effectively classified different stages of decay in peach samples. Early and precise detection of plant diseases is crucial for maintaining crop quality and minimizing losses. Deep learning methods are gaining significant recognition globally for their ability to accurately detect plant diseases in the field, presenting a promising solution for better crop management.

11.5.5 MONITORING THE GROWTH TRAJECTORY OF HORTICULTURAL CROPS THROUGH ADVANCED TRACKING AND ANALYSIS

Deep-learning methods play a crucial role in understanding and managing crop traits, aiding in decision-making for breeders and producers. For instance, researchers have devised intelligent systems to monitor mushroom growth, accurately localizing and tracking their development stages through specialized algorithms. Similarly, in orchards, innovative models have been proposed to detect and segment apple flowers at different stages, facilitating smart management practices.

Advancements in deep learning have led to the creation of systems capable of identifying fruit maturity levels. For example, a method employing natural outdoor images combined with specific algorithms and classifiers was designed to detect and classify passion fruit maturity with high accuracy (Tu et al., 2018, 2020). Additionally, in the case of blueberries, a deep-learning approach utilizing models like Mask R-CNN has been instrumental in monitoring fruit maturity, demonstrating effective predictive capabilities (Rizzo et al., 2023). These technologies significantly enhance the understanding and monitoring of crop growth, supporting better decision-making in agriculture.

11.5.6 OTHER APPLICATIONS

Deep learning has emerged as a valuable tool in predicting untested genetic attributes using genome-wide marker information, a process known as genomic prediction (GP). While GP has seen widespread use in improving breeding efficiency for plants and animals, its application in horticultural crops is still in its early stages. Zingaretti et al. (2020, 2021) evaluated the performance of deep learning in GP for autotetraploid blueberries and allooctoploid strawberries. They found that deep learning shows potential superiority over standard linear approaches by considering various genetic interactions, particularly relevant in polyploids.

In the realm of horticultural product sales, price forecasting plays a critical role due to significant fluctuations in market prices. Weng et al. (2019) used methods like autoregressive integrated moving average

(ARIMA), backpropagation (BP), and RNN to predict daily, weekly, and monthly average prices of horticultural products. Gathering extensive price data through web crawlers, they discovered that neural network methods, particularly BP and RNN, exhibited higher accuracy in price forecasting compared to the ARIMA model. This suggests that deep-learning methods could become the primary approach for predicting horticultural product prices in the future (Figure 11.5).

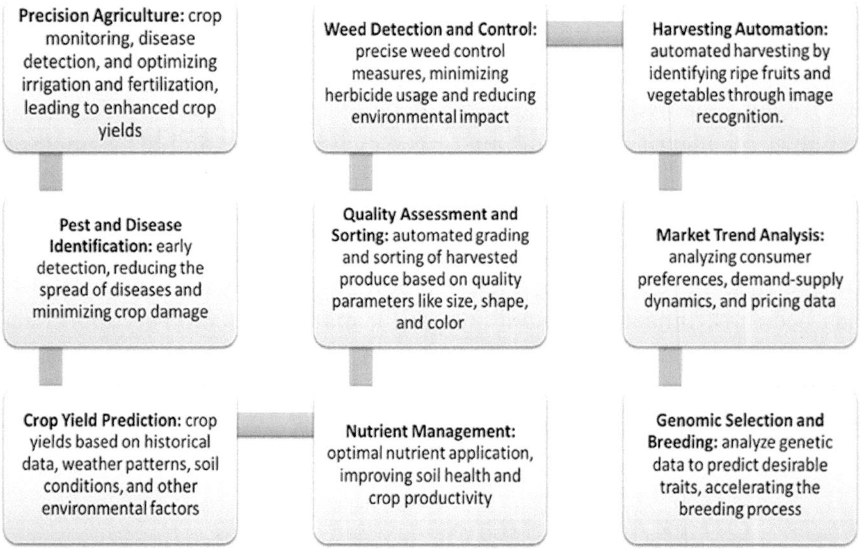

FIGURE 11.5 Pathway of machine learning and deep learning advancements for horticultural development.

11.6 SUMMARY, DISCUSSIONS, AND FUTURE PERSPECTIVES

The fusion of deep learning with horticulture has sparked a transformative era in agriculture, redefining how we perceive, manage, and optimize crop cultivation. This synergy harnesses the potency of advanced machine learning techniques, particularly deep neural networks, to revolutionize various facets of horticultural practices. At the forefront of this integration is the realm of disease detection and classification in crops. Researchers have deployed CNNs with remarkable success to detect and categorize diseases across multiple crops. These models leverage machine vision to analyze plant images, offering rapid and accurate identification of diseases, empowering timely interventions to prevent crop damage and loss. From tomatoes afflicted by various bacterial, viral, and fungal infections to cucumbers suffering from powdery mildew or anthracnose, the applications of deep learning in disease detection span diverse horticultural domains.

Beyond disease identification, deep learning models have significantly impacted growth monitoring in crops. These models excel in recognizing growth stages, measuring flower panicles, assessing fruit maturity, and tracking crop quantity and distribution. For instance, monitoring the growth stages of apples or mangoes through instance segmentation models has proven effective in differentiating between various growth phases, aiding in precise crop management decisions and optimizing harvest timelines.

Furthermore, the utilization of deep learning techniques extends to GP in horticulture. These models decipher complex genetic interactions, enhancing our understanding of traits like fruit size, firmness, and yield potential. In particular, for polyploid crops like blueberries or strawberries, the ability to consider genetic intricacies, including epistasis and dominance, is a significant advantage that deep learning offers over traditional linear methods.

Market analysis and price forecasting in horticulture have also been revolutionized by deep learning. Neural network models, such as RNNs and backpropagation, leverage large datasets collected through web crawlers to predict price fluctuations of horticultural products. These models exhibit higher accuracy in price forecasting compared to conventional methods, heralding deep learning as the future of market analysis in horticulture.

Looking forward, the future of deep learning in horticulture holds promising avenues for further innovation and advancement. Refinement of disease detection algorithms for increased accuracy, expansion of GP models to encompass a wider array of horticultural crops, exploration of novel applications for growth monitoring, and continued enhancements in market forecasting techniques are among the prospective trajectories. As this synergy between deep learning and horticulture continues to evolve, it is poised to fundamentally transform agricultural practices, fostering more sustainable, efficient, and productive crop cultivation systems.

KEYWORDS

- **precision agriculture**
- **crop monitoring**
- **disease detection**
- **pest identification**
- **yield prediction**
- **weed control**
- **quality assessment**
- **sorting automation**
- **harvesting automation**
- **market analysis**
- **genomic selection**

REFERENCES

Altaheri, H., Alsulaiman, M., & Muhammad, G. (2019). Date fruit classification for robotic harvesting in a natural environment using deep learning. *IEEE Access*, 7, 117115–117133.

Che, W., et al. (2018). Pixel based bruise region extraction of apple using Vis-NIR hyperspectral imaging. *Computers and Electronics in Agriculture*, 146, 12–21.

Chen, F., et al. (2019). Genome sequences of horticultural plants: Past, present, and future. *Horticulture Research*, 6, 112.

Colaço, A.F., et al. (2018). Application of light detection and ranging and ultrasonic sensors to high-throughput phenotyping and precision horticulture: Current status and challenges. *Horticulture Research*, 5, 35.

Colmer, J., et al. (2020). SeedGerm: A cost-effective phenotyping platform for automated seed imaging and machine-learning based phenotypic analysis of crop seed germination. *New Phytologist*, 228(2), 778–793.

de Ridder, D., et al. (2017). Healthy diet: Health impact, prevalence, correlates, and interventions. *Psychology & Health*, 32(8), 907–941.

Edwards, E.J., & Moghadam, P. (2020). Intelligent systems for commercial application in perennial horticulture. *Multidisciplinary Digital Publishing Institute Proceedings*, 36(1), 59.

He, Y., et al. (2022). Recent progress of nondestructive techniques for fruits damage inspection: A review. *Critical Reviews in Food Science and Nutrition*, 62(20), 5476–5494.

Hernández-Sánchez, N., et al. (2016). Assessment of internal and external quality of fruits and vegetables. In: *Imaging Technologies and Data Processing for Food Engineers*, pp. 269–309.

Hewett, E. (2012). Postharvest innovation: Current trends and future challenges in the global market. In: *Southeast Asia Symposium on Quality Management in Postharvest Systems and Asia Pacific Symposium on Postharvest Quality,* 989.

Itakura, K., et al. (2018). Estimation of citrus maturity with fluorescence spectroscopy using deep learning. *Horticulturae*, 5(1), 2.

Joseph, D.S., Pawar, P.M., & Pramanik, R. (2023). Intelligent plant disease diagnosis using convolutional neural network: A review. *Multimedia Tools and Applications*, 82(14), 21415–21481.

Kamilaris, A., & Prenafeta-Boldú, F.X. (2018). A review of the use of convolutional neural networks in agriculture. *The Journal of Agricultural Science*, 156(3), 312–322.

Kamilaris, A., & Prenafeta-Boldú, F.X. (2018). Deep learning in agriculture: A survey. *Computers and Electronics in Agriculture*, 147, 70–90.

Koirala, A., et al. (2019). Deep learning–Method overview and review of use for fruit detection and yield estimation. *Computers and Electronics in Agriculture*, 162, 219–234.

LeCun, Y., Bengio, Y. (1998). The handbook of brain theory and neural networks. In: *Chapter Convolutional Networks for Images, Speech, and Time Series*. Cambridge, MA, USA: MIT Press, Vol. 3, pp. 255–258.

LeCun, Y., Bengio, Y., & Hinton, G. (2015). Deep learning. *Nature*, 521(7553), 436–444.

Leiva-Valenzuela, G.A. & Aguilera, J.M. (2013). Automatic detection of orientation and diseases in blueberries using image analysis to improve their postharvest storage quality. *Food Control*, 33(1), 166–173.

López-Maestresalas, A., et al. (2016). Non-destructive detection of blackspot in potatoes by Vis-NIR and SWIR hyperspectral imaging. *Food Control*, 70, 229–241.

Mai, X., et al. (2020). Faster R-CNN with classifier fusion for automatic detection of small fruits. *IEEE Transactions on Automation Science and Engineering*, 17(3), 1555–1569.

Majeed, Y., et al. (2020). Deep learning based segmentation for automated training of apple trees on trellis wires. *Computers and Electronics in Agriculture*, 170, 105277.

Mei, M. & Li, J. (2023). An overview on optical non-destructive detection of bruises in fruit: Technology, method, application, challenge and trend. *Computers and Electronics in Agriculture*, 213, 108195.

Neupane, B., Horanont, T., & Hung, N.D. (2019). Deep learning based banana plant detection and counting using high-resolution red-green-blue (RGB) images collected from unmanned aerial vehicle (UAV). *PLoS One*, 14(10), e0223906.

Ng, H.-W., et al. Deep learning for emotion recognition on small datasets using transfer learning. In: *Proceedings of the 2015 ACM on International Conference on Multimodal Interaction*. 2015.

Ngugi, L.C., Abelwahab, M., & Abo-Zahhad, M. (2021). Recent advances in image processing techniques for automated leaf pest and disease recognition—A review *Information Processing in Agriculture*, 8(1), 27–51.

Noda, K., et al. (2015). Audio-visual speech recognition using deep learning. *Applied Intelligence*, 42, 722–737.

Nturambirwe, J.F.I. & Opara, U.L. (2020). Machine learning applications to non-destructive defect detection in horticultural products. *Biosystems Engineering*, 189, 60–83.

Nturambirwe, J.F.I., Perold, W.J., & Opara, U.L. (2021). Classification learning of latent bruise damage to apples using shortwave infrared hyperspectral imaging. *Sensors*, 21(15), 4990.

Okere, E., (2023). Hyperspectral/Multispectral imaging technology: Application for pomegranate fruit internal quality evaluation and bruise detection. Thesis submitted, Department of Electrical and Electronic Engineering, Stellenbosch University

Pan, L., et al. (2016). Detection of cold injury in peaches by hyperspectral reflectance imaging and artificial neural network. *Food Chemistry*, 192, 134–141.

Pan, S.J., & Yang, Q. (2009). A survey on transfer learning. *IEEE Transactions on Knowledge and Data Engineering*, 22(10), 1345–1359.

Pandiselvam, R., et al. (2022). Recent advancements in NIR spectroscopy for assessing the quality and safety of horticultural products: A comprehensive review. *Frontiers in Nutrition*, 9, 973457.

Pathmanaban, P., Gnanavel, B., & Anandan, S.S. (2019). Recent application of imaging techniques for fruit quality assessment. *Trends in Food Science & Technology*, 94, 32–42.

Pérez-Pérez, B.D., Garcia Vazquez, J.P., & Salomón-Torres, R. (2021). Evaluation of convolutional neural networks' hyperparameters with transfer learning to determine sorting of ripe medjool dates. *Agriculture*, 11(2), 115.

Prasad, M., et al. (2018). An efficient classification of flower images with convolutional neural networks. International *Journal of Engineering and Technology (UAE)*, 7(1.1), 384–391.

Pu, Y.Y., Feng, Y.Z. & Sun, D.W. (2015). Recent progress of hyperspectral imaging on quality and safety inspection of fruits and vegetables: A review. *Comprehensive Reviews in Food Science and Food Safety*, 14(2), 176–188.

Rebortera, M., & Fajardo, A. (2019). Forecasting banana harvest yields using deep learning. In: *2019 IEEE 9th International Conference on System Engineering and Technology (ICSET)*. IEEE.

Rizzo, M., et al. (2023). Fruit ripeness classification: A survey. *Artificial Intelligence in Agriculture*, 7, pp. 44–57.

Selvaraj, M.G., et al. (2020). Detection of banana plants and their major diseases through aerial images and machine learning methods: A case study in DR Congo and Republic of Benin. *ISPRS Journal of Photogrammetry and Remote Sensing*, 169, 110–124.

Singh, A., et al. (2016). Machine learning for high-throughput stress phenotyping in plants. *Trends in Plant Science*, 21(2), 110–124.

Singh, A.K., et al. (2018). Deep learning for plant stress phenotyping: Trends and future perspectives. *Trends in Plant Science*, 23(10), 883–898.

Singh, U.P., et al. (2019). Multilayer convolution neural network for the classification of mango leaves infected by anthracnose disease. *IEEE Access*, 7, 43721–43729.

Tu, S., et al. (2018). Detection of passion fruits and maturity classification using Red-Green-Blue Depth images. *Biosystems Engineering*, 175, 156–167.

Tu, S., et al. (2020). Passion fruit detection and counting based on multiple scale faster R-CNN using RGB-D images. *Precision Agriculture*, 21, 1072–1091.

Tulbure, A.-A., Tulbure, A.-A., & Dulf, E.-H. (2022). A review on modern defect detection models using DCNNs—Deep convolutional neural networks. *Journal of Advanced Research*, 35, 33–48.

Umar, M., et al. (2023). Multi-disease recognition in tomato plants: Evaluating the performance of CNN and improved YOLOv7 models for accurate detection and classification. https://doi:10.21203/rs.3.rs-3245718/v1

van Dael, M., et al. (2016). A segmentation and classification algorithm for online detection of internal disorders in citrus using X-ray radiographs. *Postharvest Biology and Technology*, 112, 205–214.

Walsh, K.B., et al. (2020). Visible-NIR "point" spectroscopy in postharvest fruit and vegetable assessment: The science behind three decades of commercial use. *Postharvest Biology and Technology*, 168, 111246.

Wendel, A., Underwood, J., & Walsh, K. (2018). Maturity estimation of mangoes using hyperspectral imaging from a ground based mobile platform. *Computers and Electronics in Agriculture*, 155, 298–313.

Weng, Y., et al. (2019). Forecasting horticultural products price using ARIMA model and neural network based on a large-scale data set collected by web crawler. *IEEE Transactions on Computational Social Systems*, 6(3), 547–553.

Wrzesień, M., et al. (2019). Prediction of the apple scab using machine learning and simple weather stations. *Computers and Electronics in Agriculture*, 161, 252–259.

Yang, B., & Xu, Y. (2021). Applications of deep-learning approaches in horticultural research: A review. *Horticulture Research*, 8, 123.

Yang, X., & Sun, M. (2019). A survey on deep learning in crop planting. In: *IOP Conference Series: Materials Science and Engineering*. IOP Publishing.

Yu, X., Lu, H., & Wu, D. (2018). Development of deep learning method for predicting firmness and soluble solid content of postharvest Korla fragrant pear using Vis/NIR hyperspectral reflectance imaging. *Postharvest Biology and Technology*, 141, 39–49.

Zhang, Y., et al. (2022). Real-time strawberry detection using deep neural networks on embedded system (RTSD-Net): An edge AI application. *Computers and Electronics in Agriculture*, 192, 106586.

Zingaretti, L.M., et al. (2020). Exploring deep learning for complex trait genomic prediction in polyploid outcrossing species. *Frontiers in Plant Science*, 11, 25.

Zingaretti, L.M. (2021). Deep and Shallow Learning Solutions for Modern Agriculture, *Doctoral Thesis*.

CHAPTER 12

Biotechnological Integration in Smart Horticulture Production

PRASANN KUMAR* and JAISURIYAN K

Department of Agronomy, School of Agriculture, Lovely Professional University, Phagwara, Punjab, India

*Corresponding author

ABSTRACT

Using biotechnology in horticulture redirects conventional agriculture to sustainable path. This research uses genomics, molecular markers, etc., to improve crop yield, and resilience. This study emphasizes data-driven decision-making, resource optimization, and environmental sustainability through the synergistic integration of intelligent technologies. This exploration relies on genome sequencing, providing unprecedented crop improvement insights. Clustered Regularly Interspaced Short Palindromic Repeats/Cas9 gene editing improves targeted traits while precision breeding develops characteristics that increase yield and quality. These strategies encourage the use of DNA-based and RNA-based molecular markers which help to select and modify stress tolerance, quality, and resilience traits at the molecular level. Smart applications, such as sensors for real-time monitoring, automated irrigation and nutrient delivery, are revolutionizing farming. Data analytics and decision support systems provide models for crop and yield management. The study examines how genomic insights and real-time environmental data coexist to create strategies adapting to changing conditions. Case studies examine successful applications of this integrated approach, highlighting issues and potential challenges of this approach. This research envisions a paradigm shift in

Smart Horticulture Production: Integrating Advanced Technologies.
Rajni Rajan and Duniya Ram Singh (Eds.)

DOI: 10.1201/9781998511679-12

horticulture practices driven by biotechnological innovations and smart technologies driven by environmental consciousness and sustainable agriculture.

12.1 INTRODUCTION

Horticulture, an essential part of agriculture, encounters complex challenges that require creative solutions to achieve sustainable production (Kostas et al., 2021; Sharma et al., 2022). A significant challenge lies in effectively managing the delicate equilibrium between increasing global food needs and reducing environmental repercussions. Conventional horticulture faces challenges such as limited resources, climate change, and changing patterns of pests and diseases. These challenges pose significant risks to crop production, quality, and the overall sustainability of agriculture. The importance of incorporating biotechnology into horticulture lies in its capacity to transform traditional agricultural methods and address enduring obstacles. Biotechnological tools provide accurate techniques to improve crop characteristics, strengthen resistance, and optimize the use of resources (Adetunji et al., 2022; Kaur et al., 2023a; Sandhu et al., 2023). Genomics is a field that explores the genetic composition of plants, offering a fundamental comprehension that is essential for making specific enhancements. Molecular marker-guided precision breeding enables the targeted selection of favorable traits, resulting in the cultivation of crops that exhibit enhanced resistance to diseases, pests, and unfavorable environmental conditions. Molecular markers, encompassing DNA- and RNA-based techniques, play a crucial role in horticulture by improving our comprehension of plant biology. Researchers can use DNA-based markers such as simple sequence repeats (SSRs) and single-nucleotide polymorphisms (SNPs) to identify specific genes linked to desired traits. RNA-based markers, specifically microRNA (miRNA), provide valuable information about gene expression patterns. This information can be used to develop crops that have increased tolerance to stress and improved quality. Incorporating intelligent technologies into horticulture enhances the capacity for implementing sustainable agricultural practices. The applications of the Internet of Things (IoT), which include sensor networks and automated systems, enable the real-time monitoring of environmental conditions. This allows farmers to make data-based decisions, optimizing irrigation, nutrient delivery, and overall resource management. In addition, machine learning algorithms drive data analytics and decision support systems, which aid in predictive modeling for effective crop management. Biotechnological tools and the complex interaction between genetics and intelligent technologies are utilized in horticulture in the laboratory. Adaptive strategies are formulated by integrating genomic insights with real-time data from intelligent systems. These strategies empower horticulturists to address fluctuating environmental conditions adaptively, thereby augmenting crop resilience and productivity (Aggarwal et al., 2024; Boamah et al., 2023; Singh et al., 2023a, 2023b). The utmost importance of this integration is emphasized by its capacity to tackle worldwide challenges in food security, ecological sustainability, and agricultural resilience. Biotechnological advancements establish the basis for customized solutions, while intelligent technologies provide the means for precise agriculture and efficient resource utilization. Given the challenges of climate change and population growth, combining biotechnology and intelligent technologies is crucial in achieving sustainable horticulture. Analyzing case studies sheds light on the tangible uses and achievements of combining biotechnological and smart technologies in horticulture. The combined approach has a transformative impact, as evidenced by the increased crop yields, improved quality, and resource-efficient practices. Nevertheless, some obstacles arise, including ethical considerations and potential environmental consequences. Deploying these practices responsibly and carefully considering their ecological impact is crucial to ensure long-term sustainability. The summary of challenges in horticulture production highlights the need for immediate and transformative solutions. The importance of incorporating biotechnology in tackling these challenges is demonstrated by its capacity to offer accurate genetic remedies. Studying crucial biotechnological instruments, such as genomics and molecular markers, provides insight into the complex realm of plant genetics. The investigation into intelligent technology integration highlights the importance of using data-driven decision-making and resource optimization in sustainable horticulture. The future of horticulture production hinges on the seamless integration of biotechnological and intelligent technologies, resulting in a landscape characterized by robust, efficient, and environmentally mindful agricultural methods (Kaur et al., 2023b; Moreno et al., 2024; Zhang et al., 2023b).

12.2 BIOTECHNOLOGICAL TOOLS IN HORTICULTURE

Biotechnological tools have revolutionized horticulture, offering innovative solutions to enhance crop traits, increase yield, and improve overall agricultural sustainability. Among these tools, genomic approaches stand out as a cornerstone, providing a deep understanding of the genetic makeup of plants. This section explores two pivotal aspects of genomic approaches in horticulture: genome sequencing for crop improvement and gene editing techniques, particularly the Clustered Regularly Interspaced Short Palindromic Repeats (CRISPR)/Cas9 system, for targeted modifications (Gaffey et al., 2023; Haris et al., 2023; Rai et al., 2023).

12.2.1 GENOME SEQUENCING FOR CROP IMPROVEMENT

Genome sequencing is a powerful biotechnological tool that involves determining the complete DNA sequence of an organism. This approach has proven instrumental in unraveling the genetic code of various crops in horticulture, providing a comprehensive map of their genes, regulatory elements, and functional elements. The information obtained from genome sequencing serves as a foundation for crop improvement strategies, allowing scientists and breeders to make informed decisions for enhancing desirable traits (Awasthi et al., 2020; Kerry, 2014; Misra & Ghosh, 2024; Ray et al., 2023; Sodhi et al., 2022). One significant advantage of genome sequencing in horticulture is its ability to identify genes associated with key characteristics such as yield, disease resistance, and stress tolerance. By understanding the genetic basis of these traits, researchers can develop targeted breeding programs to accelerate the development of new crop varieties. This expedites the breeding process and enables the selection of attributes that are challenging to assess through traditional breeding methods. Furthermore, genome sequencing facilitates the identification of genetic variations within a species, known as polymorphisms. These polymorphisms serve as molecular markers, allowing for the selection of specific traits with high precision. Breeders can use this information to develop crops with improved nutritional content, extended shelf life, and resistance to environmental stresses. Genome sequencing has been successfully applied to various horticultural crops, including fruits, vegetables, and ornamental plants. For example, the sequencing of the tomato genome has provided insights into the genetic basis of fruit ripening and flavor development, leading to the development of tomatoes with improved taste and nutritional content (Camere & Karana, 2018; Chen et al., 2022; Gohari et al., 2024; Liu et al., 2024; Zhang et al., 2023) (Figure 12.1).

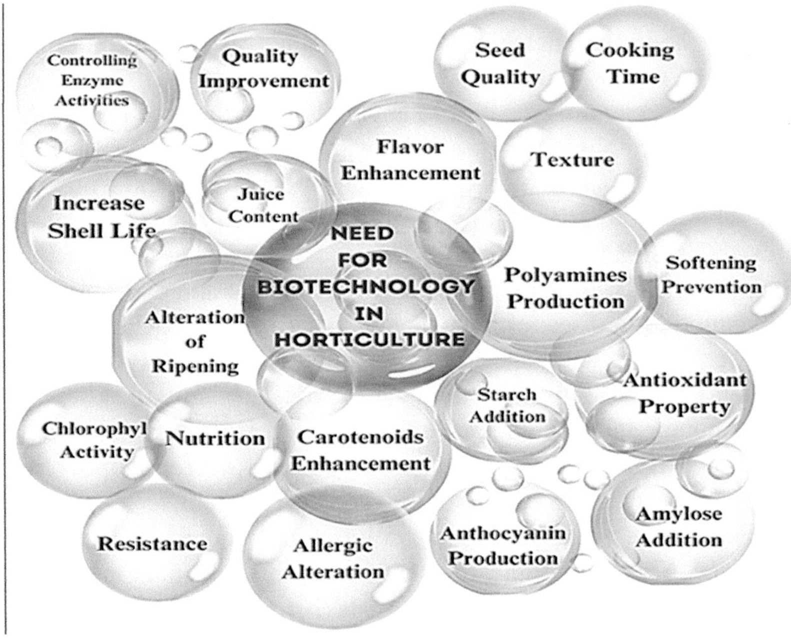

FIGURE 12.1 Biotechnological integration in smart horticulture production.

Source: Created by authors © permission is not required from other sources.

12.2.2 GENE EDITING TECHNIQUES (CRISPR/CAS9) FOR TARGETED MODIFICATIONS

Gene editing techniques, particularly the CRISPR/Cas9 system, have emerged as revolutionary tools in horticulture for making precise modifications to the plant genome. CRISPR is a naturally occurring system in bacteria that defends against viruses. Scientists have harnessed this system to edit the genes of various organisms, including plants, with unparalleled accuracy and efficiency. In horticulture, the CRISPR/Cas9 system allows researchers to target specific genes within a plant's genome and make precise modifications. This could involve introducing new traits, modifying existing ones, or disrupting undesirable characteristics.

The versatility and precision of CRISPR/Cas9 make it a game-changer in crop improvement, enabling breeders to address specific challenges in a targeted manner. One of the critical advantages of CRISPR/Cas9 in horticulture is its speed and efficiency compared to traditional breeding methods (Fasciolo et al., 2023; Mukherjee et al., 2023; Rai et al., 2021a; Uyanga et al., 2024; Yang et al., 2023). Conventional breeding often relies on the slow and unpredictable process of natural genetic recombination. In contrast, CRISPR/Cas9 allows for the direct modification of genes, significantly accelerating the development of new crop varieties. CRISPR/Cas9 has been employed in horticulture for various purposes, including improving crop resistance to diseases and pests, enhancing nutritional content, and extending shelf life. For example, researchers have successfully used CRISPR/Cas9 to develop tomatoes with increased resistance to a common fungal pathogen, providing a sustainable and environmentally friendly solution to disease management. Genomic approaches, including genome sequencing and gene editing techniques like CRISPR/Cas9, have become indispensable biotechnological tools in horticulture. These tools provide a comprehensive understanding of the genetic basis of crop traits and enable precise and targeted modifications for crop improvement. The application of these technologies holds great promise for the future of horticulture, offering sustainable solutions to address the challenges of food security, environmental sustainability, and changing climatic conditions. Precision breeding represents a revolutionary approach in horticulture, harnessing advanced biotechnological tools to expedite the development of crops with enhanced traits. Two key strategies have emerged as significant contributors to trait improvement within precision breeding: marker-assisted selection (MAS) and hybrid breeding strategies. This exploration delves into these methodologies, shedding light on their applications, benefits, and contributions to increasing yield and improving the quality of horticultural crops (Ali et al., 2023; Behera et al., 2020; Day & Sigrimis, 2020; Hemalatha et al., 2023; Karaca & Ince, 2023).

12.2.2.1 MAS FOR TRAIT IMPROVEMENT

MAS is a sophisticated technique that allows breeders to identify and select plants with specific traits at the molecular level. Unlike traditional breeding methods, which rely on observable characteristics, MAS leverages molecular markers associated with desired traits, enabling a more precise and efficient selection process. Molecular markers, such as SSRs and SNPs, serve as genetic signposts within a plant's genome. These markers are linked to particular traits of interest, ranging from disease resistance and stress tolerance to nutritional content and fruit quality. By analyzing the presence or absence of these markers, breeders can identify plants carrying the desired traits without waiting for the characteristics to manifest physically. The application of MAS in horticulture expedites the breeding process, significantly reducing the time required to develop new crop varieties (Faling, 2020; Laxman & Ansari, 2023; Pérez-Mesa et al., 2023; Ravisankar & Mathew, 2022; Tandon et al., 2023). This is particularly advantageous for traits that are challenging to evaluate visually or require multiple breeding cycles. For instance, in growing disease-resistant crops, MAS allows breeders to select plants with the desired resistance genes early in the breeding program, leading to the rapid development of resistant varieties. Moreover, MAS enhances breeding precision, enabling the selection of plants with desirable traits. This is especially valuable in complex traits governed by multiple genes, such as yield and quality. By simultaneously considering various molecular markers, breeders can develop crops with a tailored combination of traits, optimizing the plant's overall performance. The success of MAS is evident in multiple horticultural crops. MAS has been employed to select traits such as flavor, color, and disease resistance in developing fruits like apples and grapes. The technique has also proven effective in treating vegetables with improved nutritional content and resistance to environmental stressors (Forastero, 2023; Kozai et al., 2020; Schröder et al., 2019; Verpoorte et al., 1998; Zielińska et al., 2023).

12.2.2.2 HYBRID BREEDING STRATEGIES FOR INCREASED YIELD AND QUALITY

Hybrid breeding strategies represent another critical facet of precision breeding, aiming to capitalize on the heterosis or hybrid vigor observed in the offspring of genetically diverse parents. Mixed vigor results in superior yield, quality, and overall adaptability compared to nonhybrid varieties. This strategy involves carefully crossing two genetically distinct parent lines to harness the advantageous traits present in each. In horticulture, hybrid breeding is widely employed to enhance crop characteristics, primarily focusing on increasing yield and improving quality. The selection of parent lines is a critical aspect of mixed breeding, with breeders choosing parents that complement each other in desirable traits. These traits may include disease resistance, stress tolerance, and specific quality attributes such as size, color, or flavor. One significant advantage of hybrid breeding is the potential for a substantial increase in yield. The phenomenon of heterosis results in hybrid plants exhibiting greater vigor, leading to higher productivity (Aguilar, et al., 2013; Ghasemnezhad et al., 2019; Ram & Saroj, 2022; Rettore de Araujo et al., 2020; Saleh et al., 2024; Sardoei et al., 2024; Sheikh et al., 2023). This increased yield is particularly crucial in addressing the global challenge of food security, where the crop demand continues to rise. Beyond yield, hybrid breeding contributes to the improvement of crop quality. Breeders can develop hybrids with enhanced taste, nutritional content, and visual appeal by carefully selecting parent lines with desirable quality traits. This is particularly relevant in producing fruits and vegetables, where consumer preferences for high-quality produce drive market demand. Additionally, hybrid breeding contributes to developing crops with improved resilience to environmental stressors. The genetic diversity introduced through hybridization enhances the adaptability of the resulting plants, making them better equipped to withstand challenges such as pests, diseases, and adverse climatic conditions. The application of hybrid breeding has seen notable success in various horticultural crops. For instance, hybrid tomatoes have been developed to exhibit improved fruit size, disease resistance, and shelf life. Similarly, mixed peppers and cucumbers have been bred for enhanced yield and quality attributes, meeting consumer expectations for flavorful and visually appealing produce. Precision breeding, encompassing MAS and hybrid breeding strategies, stands at the forefront of advancements in horticulture. These approaches offer targeted and efficient means of trait improvement, with MAS providing molecular-level precision in trait selection and hybrid breeding, harnessing the power of genetic diversity for increased yield and improved quality. The successful application of these precision breeding strategies underscores their transformative impact on addressing the challenges of modern agriculture, contributing to the development of resilient, high-yielding, and quality crops essential for sustainable food production (Ait-Oubahou, et al., 2019; Farooq et al., 2023; Fu et al., 2022; Full et al., 2019; Gheorghiu et al., 2017; Mahapatra et al., 2022; Zulfiqar et al., 2019).

12.3 MOLECULAR MARKERS FOR CROP IMPROVEMENT

Molecular markers have become essential in crop enhancement, fundamentally transforming the agricultural domain by facilitating accurate selection and augmentation of favorable characteristics. This comprehensive investigation examines the uses and importance of molecular markers, classifying them into two types: DNA-based markers, precisely SSRs and SNPs, and RNA-based markers, such as miRNA markers for stress tolerance and gene expression profiling for improving quality (Bhaskar et al., 2023; Kour et al., 2022; Philp & Winickoff, 2017; Samal & Behera, 2022; Sarmah et al., 2024).

12.3.1 GENETIC MARKERS BASED ON DNA

12.3.1.1 SIMPLE SEQUENCE REPEATS

Microsatellites, also known as SSRs, are a type of DNA marker commonly used in crop improvement programs. These markers are composed of brief, recurring DNA sequences, exhibiting variations in the number of repetitions among individuals. SSRs exhibit significant polymorphism, rendering them valuable for evaluating genetic diversity within a population and facilitating the enhancement of crops with superior characteristics. SSRs function as molecular markers in crop enhancement, enabling scientists to differentiate between various genotypes by analyzing their distinct microsatellite patterns. This capability is essential for breeding programs that choose plants with precise characteristics, such as resistance to diseases, high nutritional value, and ability to adapt to environmental stressors. Moreover, SSRs play a crucial role in creating genetic maps, which assist in identifying genes linked to favorable characteristics and formulate MAS techniques (Aditika et al., 2022; Hackfort, 2023; Keshavarz & Sharafi, 2023; Kumar & Dixit, 2021; Nguyen et al., 2023; Rout et al., 2006; Sagar et al., 2022; Sastry et al., 2010). The versatility of solid state relays (SSRs) is apparent in their application to a wide range of crops. SSR markers have played a crucial role in rice breeding programs by facilitating the development of varieties that exhibit heightened disease resistance and improved yield characteristics. SSRs have also been used in apples and grapes to evaluate genetic variability and aid in identifying parent lines for hybridization. SNPs are DNA-based markers important in crop improvement. SNPs are genetic variations occurring at specific genome positions. They are valuable markers for studying genetic diversity and identifying connections with particular traits. SNPs have multiple benefits, such as their widespread presence in the genome and compatibility with high-throughput genotyping techniques. This enables researchers to simultaneously analyze various SNPs, facilitating efficient genotyping in breeding programs. SNP analysis provides valuable insights into identifying genomic regions linked to traits such as drought tolerance, pest resistance, and nutritional content. Advancements in genomic technologies have facilitated the creation of SNP arrays and high-density SNP maps for different crops in recent years. These tools offer researchers extensive genomic data, facilitating accurate MAS and the creation of crops customized to address specific agricultural obstacles.

12.3.1.2 RNA MARKERS

12.3.1.2.1 miRNA Indicators for Stress Resilience

miRNAs are diminutive RNA molecules that play a pivotal role in governing gene expression. Within crop improvement, miRNAs function as RNA-based markers that have substantial implications for augmenting the ability of plants to withstand stress. Crop productivity is significantly affected by abiotic stresses, such as drought, salinity, and extreme temperatures. Utilizing miRNA markers allows for a focused strategy in developing stress-tolerant crops. miRNA markers modulate the expression of genes implicated in stress response pathways. miRNAs play a role in the plant's ability to endure unfavorable environmental conditions by regulating the activity of specific target genes. Scientists have discovered and described stress-responsive miRNAs in cereals, legumes, and vegetables. This finding opens up the possibility of incorporating these miRNAs into breeding programs. miRNA markers linked to drought tolerance have been identified and integrated into crop breeding strategies such as wheat and maize. These markers facilitate the identification of plants that have improved water-use efficiency and resilience to water scarcity, which are essential characteristics in areas susceptible to drought conditions. Similarly, researchers have investigated miRNA markers associated with the ability of rice plants to tolerate salinity. This research aims to develop rice varieties that grow well in soils with high salt content (Appelqvist et al.,

2022; Beckles & Roessner, 2012; Gruda et al., 2019; Kumar et al., 2023; Oliveri et al., 2023; Ramesh et al., 2023; Ruiz-Vanoye et al., 2022; Sasi et al., 2023; Trejo-Perea et al., 2018; Wang et al., 2023).

12.3.2 UTILIZING GENE EXPRESSION PROFILING TO IMPROVE QUALITY

Gene expression profiling, which includes methods such as RNA sequencing and microarray analysis, offers valuable information about the activity of genes at the transcriptional level in response to different conditions. Gene expression profiling is a strategy that uses RNA-based markers to improve the quality attributes of agricultural produce in crop improvement. Improving the quality of crops entails maximizing their flavor, nutritional value, and postharvest attributes. Gene expression profiling enables researchers to identify crucial genes implicated in synthesizing flavor compounds, antioxidants, and other molecules related to quality. This knowledge can be utilized to choose or develop crops with enhanced flavor, nutritional content, and durability. Gene expression profiling has played a crucial role in understanding the molecular processes that drive fruit ripening and flavor development in tomatoes and strawberries. Through comprehending the gene networks implicated in these processes, breeders can selectively manipulate specific genes, resulting in the development of fruits with prolonged shelf life and improved taste. Furthermore, gene expression profiling is used to identify genes linked to enhanced nutrient content in vegetables like broccoli and spinach, thereby facilitating the creation of more nourishing varieties. Molecular markers, regardless of whether they are DNA-based or RNA-based, are highly effective tools in the field of crop enhancement. SSRs and SNPs offer valuable information about genetic variation and aid in MAS, thereby facilitating the development of crops with enhanced characteristics (Crisp et al., 2022; Hassoun et al., 2023; He et al., 2023; Ijaz et al., 2023; Muhie, 2022; Peng & Simko, 2023; Rai et al., 2021a, 2021b, Tao et al., 2022; Vishnoi & Goel, 2024). RNA-based markers, such as miRNAs for stress tolerance and gene expression profiling for quality enhancement, provide focused methods to tackle specific challenges in agriculture. The incorporation of these molecular markers into breeding programs has the potential to expedite the cultivation of crops that possess resilience, high yield, and superior quality. This, in turn, will contribute to global food security and sustainable agriculture as technology progresses.

12.4 SMART TECHNOLOGIES IN HORTICULTURE

The field of horticulture has been transformed by intelligent technologies, leading to a new era characterized by enhanced efficiency, accuracy, and sustainability. IoT applications, advanced data analytics, and decision support systems drive this transformation. Sensor networks have become essential tools for real-time monitoring in horticulture within the realm of IoT applications. These networks comprise numerous sensors strategically positioned across farms and greenhouses, gathering data on vital environmental variables such as temperature, humidity, soil moisture, and light intensity. These data are transmitted in real time to a central system, allowing farmers to obtain a comprehensive overview of their crop conditions. Real-time monitoring of these variables allows for prompt responses to any deviations from optimal conditions, thereby averting potential problems and ensuring the welfare of the crops. An influential use of IoT in horticulture is the adoption of automated irrigation and nutrient delivery systems. Conventional irrigation techniques frequently lead to excessive or insufficient utilization of water and fertilizers, resulting in the squandering of resources and possible environmental damage (Sharma et al., 2022; Awana et al., 2020; Dagys et al., 2023; Gouthami et al., 2023; Kiran et al., 2023; Kumar et al., 2022; Suganya et al., 2022). Intelligent irrigation systems, equipped with sensors and actuators, allow for accurately regulating the quantity and timing of water and nutrients provided to the crops. By examining up-to-date information gathered from soil moisture sensors and weather forecasts, these systems can autonomously modify irrigation schedules to align with the unique requirements of each crop. This process enhances the utilization of resources and encourages the efficient use of water and nutrients. Regarding data analytics and decision support systems, predictive modeling is a potent tool for crop management in intelligent horticulture. Predictive modelling utilizes historical data, current environmental conditions, and machine learning algorithms to anticipate future trends and outcomes. Within the field of horticulture, this encompasses activities such as determining the ideal time for planting, projecting potential crop production, and anticipating occurrences of pests and diseases. Farmers can make well-informed decisions, strategically plan, and minimize risks through predictive models, resulting in enhanced crop yields and optimized resource utilization. Machine learning algorithms improve data analytics in horticulture, specifically in optimizing crop yield. These algorithms can analyze extensive datasets, encompassing data on soil composition, weather patterns, and crop characteristics, to identify patterns and correlations that may be difficult for humans to perceive. Machine learning algorithms can use these data to suggest the most favorable planting densities, nutrient concentrations, and harvesting schedules to optimize crop yield. Moreover, these algorithms can adjust and enhance their performance over time by assimilating new data, offering a flexible and reactive approach to crop management. Incorporating intelligent technologies in horticulture is not merely a fad but an imperative for ensuring the industry's long-term viability and expansion. The numerous advantages include resource efficiency, increased crop productivity, and decreased environmental impact. Utilizing sensor networks and automated systems for real-time monitoring guarantees that crops are supplied with the precise amount of water and nutrients at the exact moment they require it, thereby reducing wastage and encouraging sustainable agricultural practices (Figure 12.2).

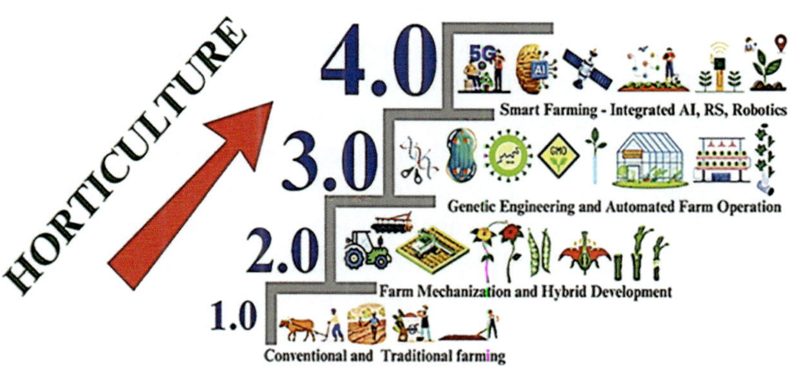

FIGURE 12.2 Technology advancement in horticulture.

Source: Created by authors © permission is not required from other sources.

The deployment of diverse sensors in sensor networks is crucial for obtaining comprehensive insights into the environmental conditions that impact crops. The temperature sensors aid in monitoring the surrounding temperatures, ensuring that crops are not exposed to extreme conditions that could harm their growth. Humidity sensors ensure that crops with specific humidity needs are kept at optimal humidity levels. Soil moisture sensors, which are crucial components, allow for accurate irrigation management by measuring the moisture level in the soil. Light intensity sensors are used to evaluate the presence and strength of sunlight, which is essential for the process of photosynthesis and the overall well-being of plants (Sharma et al., 2022; Awana et al., 2020; He et al., 2023; Muhie, 2022; Priya et al., 2023; Rai et al., 2021b; Saraswathi et al., 2018; Vishnoi & Goel, 2024). The importance of automated irrigation and nutrient delivery systems in innovative horticulture cannot be emphasized enough. Conventional irrigation techniques frequently lack efficiency, resulting in the squandering of water and the runoff of nutrients. Intelligent systems enable farmers to customize irrigation and nutrient provision using real-time data, avoiding excessive watering or nutrient surplus. This practice preserves valuable resources and reduces environmental pollution from the undue entry of fertilizers into water bodies. The automated nature of these systems minimizes the need for manual labor in irrigation, allowing farmers to allocate more time to other crucial aspects of crop management. Predictive modelling plays a central role in the data-driven transformation of horticulture within data analytics and decision support systems. Predictive models utilize past data, present environmental conditions, and sophisticated algorithms to predict crop growth and development. These models can forecast the ideal timing for planting by analyzing past weather patterns and soil conditions. By providing estimations of potential yields, they assist farmers in making well-informed choices regarding the allocation of resources and market strategies. Machine learning algorithms enhance the capabilities of data analytics in horticulture. These algorithms are highly proficient in handling extensive datasets and detecting complex patterns that may be difficult for humans to notice (Aditika et al., 2022; Gruda et al., 2019; Kour et al., 2022; Kumar et al., 2023; Kumar & Dixit, 2021; Mahapatra et al., 2022; Philp & Winickoff, 2017; Samal & Behera, 2022; Sasi et al., 2023; Sastry et al., 2010). Machine learning algorithms in yield optimization analyze various factors, including soil composition, weather fluctuations, and crop characteristics, to suggest the most favorable planting arrangements. The algorithms are dynamic, enabling them to adjust and enhance their recommendations as time progresses, consistently refining strategies to optimize yield. The advantages of intelligent technologies in horticulture go beyond immediate improvements in efficiency and productivity. Incorporating IoT applications and data analytics facilitates the implementation of sustainable agricultural practices in line with international endeavors to tackle climate change and environmental issues. Innovative horticulture is crucial in establishing a sustainable and resilient food production system by minimizing resource wastage, optimizing crop management, and reducing environmental impact. An essential benefit of implementing IoT applications in horticulture is the capacity to enable precision agriculture. Precision agriculture employs advanced technology to customize farming techniques according to the unique requirements of individual plants or crops. By utilizing sensor networks, farmers can exert precise control and optimize environmental conditions, thanks to the networks' ability to monitor in real time. The precision achieved surpasses conventional farming techniques, enabling specific interventions and allocation of resources. Consequently, farmers can attain increased crop production with reduced resources, thereby diminishing the overall environmental impact of agriculture. Data analytics in horticulture encompasses more than just predictive modeling and yield optimization. Data analytics-enabled decision support systems aid farmers in making well-informed decisions throughout all aspects of crop management. To generate practical and implementable insights, these systems employ advanced algorithms to examine diverse datasets, encompassing market trends, climate forecasts, and historical performance. For example, a decision support system can suggest particular crop varieties by considering market demand and climate conditions. This assists farmers in making strategic planting choices that are in line with both economic and environmental objectives (Adetunji et al., 2022; Boamah et al., 2023; Gaffey et al., 2023; Haris et al., 2023; Kaur et al., 2023b; Kerry, 2014; Misra & Ghosh, 2024; Rai et al., 2023; Singh et al., 2023a; Zhang, Han, et al., 2023). The profound influence of intelligent technologies in horticulture extends beyond the scope of extensive commercial farms. These innovations are equally applicable and advantageous for small-scale and local farmers. Indeed, the ease of access and ability to expand numerous intelligent technologies make them especially suitable for smaller enterprises. IoT applications and data analytics can provide local farmers with the tools and valuable insights to compete effectively in the contemporary agricultural environment. Inclusivity enhances the horticulture sector's ability to withstand and maintain itself by promoting a varied and flexible agrarian community. To achieve widespread adoption and success, addressing the challenges and considerations associated with innovative horticulture technologies despite their numerous advantages is crucial. A vital obstacle lies in the upfront expenses linked to adopting these technologies. The initial capital required for acquiring sensors, automated systems, and data analytics infrastructure can be significant, particularly for small-scale farmers with limited financial means. Governments, research institutions, and agricultural organizations are essential in supporting farmers by offering subsidies, grants, and educational programs. These initiatives aim to alleviate the financial challenges associated with adopting smart technologies. Furthermore, it is essential to prioritize the implementation of cybersecurity protocols to safeguard the confidential information produced by IoT applications and data analytics within the field of horticulture. With the growing integration of digital technologies in farms, the vulnerability to cyber threats like data breaches and system manipulations is heightened. Establishing strong cybersecurity measures, such as implementing secure data transmission and storage practices, is crucial to protect the integrity of intelligent horticulture systems and ensure the privacy of farmers' data. Moreover, adopting and managing intelligent technologies can challenge farmers due to the learning curve. Training programs and educational resources are crucial for providing farmers with the necessary knowledge and skills to utilize these technologies effectively. We can effectively bridge the knowledge gap by fostering collaboration among technology providers, agricultural extension services, and educational institutions. This will enable farmers, regardless of their scale or location, to fully utilize the capabilities of intelligent technologies. Ultimately, incorporating intelligent technologies in horticulture signifies a fundamental change in our approach to crop management and agricultural methods. IoT applications, utilizing sensor networks and automated systems, enable instantaneous monitoring and precise control, enhancing resource utilization and fostering sustainable farming practices (Day & Sigrimis, 2020; Karaca & Ince, 2023; Pérez-Mesa et al., 2023; Ravisankar & Mathew, 2022; Tandon et al., 2023; Yang et al., 2023; Zielińska et al., 2023). Data analytics, specifically predictive modelling and machine learning, provides farmers valuable insights to make well-informed decisions and optimize crop yield. The advantages of intelligent horticulture go beyond just improving efficiency and encompassing sustainability, resilience, and inclusivity in various farming operations. Despite the need to overcome obstacles like upfront expenses and cybersecurity concerns, the potential benefits of intelligent technologies in horticulture for increasing global food production and promoting environmental responsibility make their ongoing development and adoption crucial for the future of agriculture.

12.5 INTEGRATION OF BIOTECHNOLOGY AND SMART TECHNOLOGIES: SYNERGIES AND INTERACTIONS

Incorporating genomic data into decision-making processes signifies a revolutionary advancement in diverse domains, such as agriculture, healthcare, and environmental management. When making intelligent decisions, using genomic data provides an exceptional understanding of the molecular basis of living organisms, allowing for more accurate and customized approaches. The change in perspective has resulted in new methods, such as flexible strategies that rely on up-to-date environmental information, leading to a decision-making framework that is constantly changing and responsive.

12.5.1 UTILIZING GENOMIC DATA FOR INFORMED DECISION-MAKING

Genomic data, which includes the entirety of an organism's genes, offers a thorough plan for comprehending the fundamental processes of life. Integrating genomic data in intelligent decision-making enables a more comprehensive and refined analysis of biological processes. This is especially clear in agriculture, where genomics enhances crops, develops disease resistance, and promotes sustainable farming methods. Genomic data are utilized in crop science to pinpoint genes linked to favorable characteristics such as drought tolerance, pest resistance, and enhanced nutritional value. By analyzing the genetic composition of various crop varieties, breeders can make well-informed choices regarding which plants to crossbreed, thereby expediting the development of novel and enhanced cultivars (Hackfort, 2023; Keshavarz & Sharafi, 2023; Kumar & Dixit, 2021; Nguyen et al., 2023; Sagar et al., 2022; Sarmah et al., 2024; Sastry et al., 2010). Employing this focused methodology reduces the customary trial-and-error process in crop breeding, resulting in expedited and more effective results. Genomic data play a crucial role in personalized medicine, where customized treatment strategies are developed based on a patient's genetic makeup. Comprehending the genetic foundation of diseases enables healthcare professionals to prescribe medications with greater effectiveness and reduce adverse side effects. Specifically, particular genetic variations can impact the way individuals process drugs, and genomic data aid in the identification of these variations, thereby assisting in the selection of suitable medications and dosages. In addition to agriculture and healthcare, utilizing genomic data also encompasses biodiversity conservation, wherein scientists employ DNA sequencing to ascertain and safeguard endangered species. This method facilitates a more accurate comprehension of the genetic variation among populations, assisting in conservation endeavors to protect the long-term sustainability of ecosystems. The emergence of CRISPR-Cas9 technology in the era of precision biology has enabled precise gene editing with unparalleled accuracy. This innovative tool allows the correction of genetic mutations, the cultivation of disease-resistant crops, and the treatment of human genetic disorders. Using genomic data with gene-editing technologies can fundamentally transform entire ecosystems, resolving previously insurmountable obstacles.

12.5.2 DYNAMIC STRATEGIES INFORMED BY UP-TO-DATE ENVIRONMENTAL DATA

Concurrently utilizing genomic data, adaptive strategies that rely on up-to-date environmental data have emerged as supplementary methods for making intelligent decisions. This entails the constant surveillance of ecological factors and the adaptive modification of tactics in reaction to shifting circumstances. Adaptability in real time is paramount in agriculture, climate science, and resource management. Integrating sensors and data analytics enables farmers to continuously monitor real-time soil moisture, temperature, and weather conditions. Subsequently, this data are utilized to enhance irrigation schedules, modify planting times, and implement precision farming methodologies. Through the process of adjusting to the existing environmental circumstances, farmers can optimize the production of crops while minimizing the usage of resources, thereby promoting sustainable and resilient agriculture. The field of climate science is enhanced by employing adaptive strategies that leverage up-to-date data to model and forecast climate patterns (Crisp et al., 2022; Hassoun et al., 2023; He et al., 2023; Ijaz et al., 2023; Kumar et al., 2022; Muhie, 2022; Oliveri et al., 2023; Rai et al., 2021a; Tao et al., 2022). This entails surveillant sea surface temperatures, atmospheric carbon levels, and other pivotal indicators. Modifying climate models using the most recent data improves the precision of forecasts, facilitating more informed choices regarding the mitigation and adaptation to climate change. Real-time environmental data are also advantageous for resource management, encompassing water and energy. Intelligent grids can modify power allocation according to immediate demand, enhancing energy efficiency and minimizing waste. Likewise, water management systems can adjust to varying circumstances, guaranteeing this valuable asset's effective utilization and preservation. The combination of genomic and environmental data enhances the effectiveness of adaptive strategies. For instance, integrating crop genetics data with real-time agricultural ecological information enables precise and focused interventions. Agricultural practitioners can modify their planting techniques according to projected weather patterns, select crop varieties that flourish in particular conditions, and employ accurate pest control methods informed by genomic knowledge of pest resilience.

12.5.3 EFFECTIVE UTILIZATION OF INTEGRATED BIOTECHNOLOGICAL AND IMAGINATIVE APPROACHES

Combining biotechnological advancements and creative approaches has led to many successful applications in different fields. Combining genomic data and adaptive strategies showcases the effectiveness of interdisciplinary collaboration and inventive thinking. Effective use of integrated biotechnological and creative methodologies in agriculture is apparent in cultivating genetically modified (GM) crops. GM crops frequently integrate genes from different organisms to bestow characteristics such as pest resistance, herbicide tolerance, or improved nutritional value. The imaginative element involves conceiving innovative gene combinations to tackle specific challenges, while the biotechnological aspect entails precise genetic manipulation to attain the desired results. Using CRISPR-Cas9 technology is a prime example of effectively implementing integrated methodologies in the biotechnology field (Aditika et al., 2022; Bhaskar et al., 2023; Fu et al., 2022; Full et al., 2019; Mahapatra et al.,

2022; Philp & Winickoff, 2017; Sastry et al., 2010; Zulfiqar et al., 2019). Scientists can accurately modify genes to improve crops' characteristics or rectify animal genetic mutations. The integration of this technology, along with a profound comprehension of genomics, presents opportunities for developing crops that exhibit greater adaptability to shifting climates, necessitate reduced resources, and possess enhanced nutritional characteristics. Integrating genomic data and innovative approaches in healthcare has led to the successful development of personalized medicine. Healthcare professionals can customize interventions based on an individual's genetic composition rather than using a standardized treatment approach. This paradigm shift has resulted in the development of more efficient and focused treatments, which have the potential to reduce adverse effects and enhance patient results. Integrated approaches are also advantageous for environmental monitoring and conservation efforts. Researchers utilize genomic data to monitor endangered species' migration patterns and genetic variability. Innovative approaches entail formulating conservation strategies considering genetic diversity, guaranteeing populations' sustained survival. This comprehensive approach takes into account both genetic and environmental factors, leading to the implementation of conservation practices that are more efficient and enduring.

12.5.4 KEY TAKEAWAYS AND POSSIBLE OBSTACLES

Utilizing genomic data and adaptive strategies in decision-making has yielded valuable lessons and insights. The acquisition of these lessons and comprehension of potential obstacles directs the continuous development of these strategies. An essential takeaway is the significance of ethical considerations when utilizing genomic data. With the progress of technology, the capacity to arrange and scrutinize genetic information becomes increasingly attainable, prompting concerns regarding privacy, consent, and the conscientious utilization of genetic data. Maintaining a delicate equilibrium between utilizing the potential of genomic information and protecting individual rights is crucial to ensure these methods' ongoing effectiveness. Furthermore, it is imperative to emphasize the necessity of continuous collaboration and interdisciplinary research. The utilization of genomic data is significantly enhanced when integrated with expertise from various disciplines, including data science, environmental science, and ethics. Collaborative endeavors promote a comprehensive comprehension of intricate problems and propel the creation of more efficient and enduring solutions (Awasthi et al., 2020; Gaffey et al., 2023; Kaur et al., 2023a, 2023b; Kerry, 2014; Misra & Ghosh, 2024; Ray et al., 2023; Sandhu et al., 2023; Sharma et al., 2022; Singh et al., 2023b). The challenges related to integrating genomic data and adaptive strategies stem from the complicated nature of biological systems and the unpredictable dynamics of the environment. Despite the progress made in genomics, comprehending the complex interactions within living organisms remains difficult. Various factors influence ecological systems, and accurately predicting their behavior with complete certainty is inherently challenging. Recognizing these difficulties highlights the significance of humility when dealing with intricate situations and the necessity of ongoing education and adjustment. In addition, the possible misapplication of biotechnological instruments, such as gene-editing technologies, raises ethical and safety apprehension. Establishing guidelines and regulations to promote responsible usage and mitigate unintended repercussions is crucial. It is essential to maintain a harmonious equilibrium between innovation and preventive measures to navigate the ever-changing field of genomic technologies successfully. The future is promising for integrating genomic data and adaptive strategies in decision-making as lessons are learned and challenges are tackled. The intersection of biotechnology, data science, and creative thinking provides a means to tackle urgent global issues, such as ensuring food security, improving healthcare, and promoting environmental sustainability. Incorporating genomic data into intelligent decision-making processes signifies a groundbreaking transformation in our capacity to comprehend and manipulate living systems. Integrating genomic knowledge and adaptive approaches informed by up-to-date environmental data is crucial for tackling intricate agriculture, healthcare, and environmental conservation challenges. The success of GM crops and personalized medicine demonstrates the transformative power of combining biotechnological advancements with creative strategies (Ali et al., 2023; Chen et al., 2022; Faling, 2020; Fasciolo et al., 2023; Gohari et al., 2024; Karaca & Ince, 2023; Laxman & Ansari, 2023; Pérez-Mesa et al., 2023; Ravisankar & Mathew, 2022; Uyanga et al., 2024). Nonetheless, it is imperative that we rely on valuable insights regarding ethics, collaboration, and the inherent intricacy of biological and environmental systems to steer our ongoing investigation of these frontiers. To ensure that genomics and decision-making positively impact the well-being of individuals and the sustainability of our planet, it is crucial to prioritize responsible innovation, interdisciplinary collaboration, and continuous learning.

12.6 ENVIRONMENTAL SUSTAINABILITY AND ETHICAL CONSIDERATIONS

Contemporary discussions prioritize environmental sustainability and ethical considerations when evaluating the effects of technological interventions on ecosystems. As society needs creative solutions to ecological problems, it is crucial to assess the impact of biotechnological interventions on the environment and address the ethical concerns related to using intelligent technologies.

12.6.1 ECOSYSTEM IMPACTS

Despite their potential to effectively tackle environmental problems, biotechnological interventions can also significantly impact ecosystems. Understanding and reducing the adverse effects of these impacts is crucial for attaining sustainable results. An essential factor to consider is the ecological footprint, which quantifies the extent of human impact on nature and the environment. The introduction of GM crops in agriculture is a biotechnological intervention that has implications for the ecosystem. Although these crops may provide advantages such as enhanced pest and disease resistance, assessing the potential ecological ramifications is crucial. For example, cultivating GM crops resistant to certain herbicides can result in modification in farming methods, which may affect nontarget plants and insects. The dissemination of genetic alterations to indigenous counterparts of cultivated crops is also a matter of concern, giving rise to inquiries regarding the enduring ecological ramifications on biodiversity. Evaluating the ecological footprint of biotechnological interventions necessitates thoroughly examining their environmental ramifications, utilization of resources, and possible unforeseen outcomes. Life cycle assessments (LCAs), which comprehensively analyze a product's or process's complete life cycle, are valuable tools for evaluating the environmental impact (Full et al., 2019; Kour et al., 2022; Philp & Winickoff, 2017; Sagar et al., et al., 2022; Sarmah et al., 2024; Sastry et al., 2010). This methodology allows scientists to measure the environmental inputs and outputs linked to biotechnological interventions, facilitating well-informed decision-making and formulating strategies to mitigate any adverse effects on ecosystems. Furthermore, biotechnological instruments, such as gene-editing techniques like CRISPR-Cas9, present difficulties with unforeseen ecological repercussions. The deliberate modification of genes in organisms, although providing possibilities for improving desirable characteristics, raises concerns regarding the potential environmental consequences. Introducing GM organisms into the environment, intentionally or accidentally, may have uncertain impacts on ecosystems, thus requiring thorough risk evaluations and ethical deliberations. Marine ecosystems are investigating biotechnological interventions to tackle problems like overfishing and habitat degradation. One example of technological advancement in aquaculture is the creation of GM fish that exhibit improved growth rates and resistance to diseases. Nevertheless, there are concerns regarding the possible release of these GM organisms into natural ecosystems, which could result in ecological disturbances and alterations in the genetic variability of wild populations. The ethical considerations in deploying intelligent technologies go beyond the immediate advantages and encompass the broader implications for biodiversity, ecological resilience, and ecosystem interconnections. Achieving a harmonious blend of technological advancement and environmental conservation necessitates a sophisticated comprehension of the complex interconnections within ecosystems and a steadfast dedication to ethical methodologies.

12.6.2 EVALUATING THE ENVIRONMENTAL IMPACT OF BIOTECHNOLOGICAL INTERVENTIONS

The assessment of the ecological impact of biotechnological interventions requires a comprehensive analysis that takes into account various dimensions, including environmental, social, and economic factors. LCAs are valuable tools that offer a comprehensive perspective on the ecological consequences of a particular product or technology. LCAs encompass multiple stages, such as the extraction of raw materials, the manufacturing process, transportation, product utilization, and disposal. LCAs are useful in agriculture, particularly in biotechnological interventions, as they accurately measure the environmental consequences of GM crops. This entails evaluating the energy and resource inputs necessary for cultivating them, the ecological consequences of related agricultural practices, and potential repercussions on soil health and water quality (Sharma et al., 2022; Crisp et al., 2022; Dagys et al., 2023; Gouthami et al., 2023; Hassoun et al., 2023; He et al., 2023; Ijaz et al., 2023; Muhie, 2022). LCAs allow researchers and policymakers to evaluate and compare the environmental impact of various farming systems. This information helps them make well-informed decisions regarding the adoption of biotechnological interventions. Furthermore, LCAs play a crucial role in evaluating the environmental consequences of biofuels, particularly those that rely on biotechnological methods like fermenting plant-derived feedstocks. Researchers can analyze the complete life cycle of biofuel production, including the growth of feedstocks, the conversion process, and the final use, to identify ways to enhance the sustainability of bioenergy systems and reduce negative ecological impacts. Bioremediation, a process that utilizes microorganisms to cleanse contaminated environments, necessitates an evaluation of the environmental effect by comprehending the interplay between introduced organisms and Indigenous ecosystems. LCAs for bioremediation projects consider various factors, including the energy demands for cultivating microorganisms, the possible discharge of GM microorganisms, and the lasting ecological impacts on soil and water ecosystems. In addition, as synthetic biology progresses, involving the development of artificial organisms and biological systems, environmental impact evaluation becomes more intricate. To comprehend the possible interactions between synthetic organisms and natural ecosystems, it is necessary to conduct a thorough assessment of the ecological implications. This assessment should encompass the potential risks of unintended consequences and the ethical considerations arising from modifying fundamental aspects of biological systems.

12.6.3 ETHICAL CONSIDERATIONS IN IMPLEMENTING INTELLIGENT TECHNOLOGIES

Analyzing ethical considerations is essential to ensure responsible and sustainable practices when deploying smart technologies, including biotechnological interventions. Ethical considerations encompass a range of principles, such as transparency, accountability, equity, and respect for autonomy and biodiversity. An essential ethical consideration in biotechnology is the principle of informed consent. Ensuring comprehensive awareness of potential risks and benefits of interventions is crucial in human-related applications, such as genetic testing and personalized medicine. Transparent communication with individuals, while respecting their autonomy and privacy rights, is essential for the responsible utilization of genomic data. Within agriculture, ethical considerations encompass the effects of biotechnological interventions on farmers, specifically those residing in developing nations. Concerns regarding equity and social justice arise from seed sovereignty, access to technology, and the potential reliance on patented GM seeds (Ait-Oubahou et al., 2019; Bhaskar et al., 2023; Farooq et al., 2023; Gheorghiu et al., 2017; Karaca & Ince, 2023; Rai et al., 2021b; Sardoei et al., 2024; Zielińska et al., 2023). To reconcile the interests of multinational biotechnology companies and local farmers, it is necessary to establish ethical frameworks that prioritize fair access to advantageous technologies and safeguard traditional agricultural practices. The ethical aspects of biotechnological interventions become more pronounced when considering their potential impact on biodiversity. Introducing GM organisms into the environment prompts concerns regarding unforeseen outcomes, ecological robustness, and the conservation of natural genetic variability. When developing ethical frameworks, it is crucial to carefully evaluate the long-term effects of biotechnological interventions on ecosystems and acknowledge the possibility of irreversible alterations to biodiversity. Furthermore, the ethical ramifications of gene-editing technologies, specifically human germline editing, incite profound, honest discussions. The capacity to modify the genetic composition of succeeding generations raises concerns regarding the possibility of unforeseen repercussions, the advent of customized infants, and the societal ramifications of manipulating the human germline. Ethical frameworks must adapt to address these intricate and multifaceted concerns as technology progresses. Within environmental conservation, using intelligent technologies raises ethical concerns regarding the choice between intervention and nonintervention approaches. Using biotechnological interventions, such as genetic modification or assisted reproduction, to address declining populations of endangered species raises ethical considerations regarding preserving natural biodiversity compared to the potential risks and uncertainties associated with human-mediated interventions (Boamah et al., 2023; Gaffey et al., 2023; Haris et al., 2023; Kaur et al., 2023a; Rai et al., 2023; Sandhu et al., 2023; Sodhi et al., 2022).

12.6.4 KEY TAKEAWAYS AND POSSIBLE OBSTACLES

The convergence of environmental sustainability and ethical considerations in biotechnological interventions provides valuable insights that inform our ethical and responsible involvement with these technologies. The lessons learned to emphasize the importance of comprehending the complex interconnections within ecosystems, recognizing the uncertainty linked to interventions, and prioritizing transparency and inclusivity. An essential lesson is understanding the interdependence of ecosystems and the possible far-reaching consequences of biotechnological interventions. Ecosystems' intricate and ever-changing nature frequently results in unanticipated outcomes, underscoring the significance of modesty and prudence when implementing intelligent technologies. Consequently, adopting a precautionary approach, which entails thoroughly assessing potential hazards before implementing interventions, becomes an essential guiding principle. Transparency and public engagement are essential elements in dealing with ethical considerations. The significance of including a wide range of stakeholders in decision-making is emphasized by the lessons learned from previous instances where a lack of transparency affected public perception and trust. Engaging in well-informed discussions that include a wide range of viewpoints helps to develop stronger ethical guidelines that accurately represent the values and concerns of communities impacted by biotechnological interventions. The lessons learned also emphasize the necessity of interdisciplinary collaboration to tackle the intricate challenges of environmental sustainability and ethical considerations (Forastero, 2023; Ghasemnezhad et al., 2019; Karaca & Ince, 2023; Kozai et al., 2020; Pérez-Mesa et al., 2023; Ram & Saroj, 2022; Sardoei et al., 2024; Schröder et al., 2019; Verpoorte et al., 1998). By assembling specialists from various disciplines, such as biology, ecology, ethics, and social sciences, we can improve our comprehension of intricate matters and expedite the creation of thorough and situation-specific ethical frameworks. Difficulties remain in understanding and navigating the ethical aspects of biotechnological interventions, especially as technological capabilities progress. Achieving a harmonious equilibrium between innovation and responsible practices necessitates continuous introspection, adjustment, and the formulation of ethical principles that adapt to the dynamic nature of biotechnology. The potential challenges encompass the swift rate of technological advancement, the worldwide scope of biotechnological effects, and the necessity for regulatory frameworks that align with scientific progress. Another obstacle involves resolving the potential conflicts of interest that may arise between commercial interests and ethical considerations. Given that biotechnological interventions are frequently propelled by research and development endeavors undertaken by private enterprises, prioritizing adherence to ethical principles in decision-making processes becomes crucial. Ethical frameworks should encompass issues about the patenting of living organisms, the exertion of corporate authority over genetic resources, and the fair allocation of benefits resulting from biotechnological progress. Moreover, the global nature of biotechnological interventions presents difficulties concerning governance, the alignment of regulations, and the establishment of ethical standards on a worldwide scale. Biotechnological interventions have wide-ranging effects that go beyond the boundaries of individual countries. This requires international cooperation to address ethical concerns consistently. Coordinating the efforts of governments, international organizations, and the scientific community is necessary to establish global standards for responsible biotechnological practices. The challenge in environmental sustainability is to balance the pressing need to address environmental issues and the requirement for thoughtful and ethical consideration of interventions. Swift reactions to ecological challenges, such as climate change or biodiversity loss, may unintentionally result in unforeseen outcomes if ethical considerations are not sufficiently considered (Adetunji et al., 2022; Aggarwal et al., 2024; Boamah et al., 2023; Haris et al., 2023; Kerry, 2014; Moreno et al., 2024; Ray et al., 2023; Sandhu et al., 2023; Zhang, Han, et al., 2023). Achieving lasting and sustainable outcomes requires a delicate equilibrium between a sense of urgency and sincere introspection. Humanity will rely on lessons from previous experiences and a dedication to continuous ethical contemplation to shape our approach to biotechnological interventions and intelligent technologies. To effectively address environmental sustainability challenges and ensure biotechnological interventions' positive impact on human societies and ecosystems, it is crucial to incorporate ethical considerations into decision-making processes and promote interdisciplinary collaboration. The convergence of environmental sustainability and ethical considerations within biotechnological interventions poses an intricate and diverse challenge. Evaluating the ecological impact of these interventions, comprehending their potential effects on ecosystems, and addressing the ethical considerations linked to implementing intelligent technologies necessitate a nuanced and comprehensive strategy. Our journey toward responsible and sustainable engagement with biotechnological advancements is guided by the lessons we have learned from past experiences and our ongoing reflections on potential challenges. Transparency, inclusivity, and global collaboration are essential for guiding the direction of biotechnological interventions to ensure ethical considerations and environmental sustainability. This commitment will benefit both current and future generations.

12.7 FUTURE PERSPECTIVES

The future of biotechnology is filled with promising and groundbreaking opportunities as emerging technologies are positioned to transform multiple domains profoundly. At the forefront of these advancements is the ongoing development of CRISPR and its applications, which are pushing the limits of genetic engineering. The future holds numerous possibilities for revolutionary advancements in cutting-edge horticulture technology, marking the beginning of a new age of environmentally friendly and productive crop management techniques.

12.7.1 ADVANCEMENTS IN BIOTECHNOLOGY

The swift advancements in biotechnological research and development have resulted in the emergence of numerous cutting-edge technologies that hold the potential to revolutionize our comprehension of life and living systems. CRISPR-Cas9 is an innovative and revolutionary tool for accurate and specific genetic modification. Derived from the bacterial immune system, CRISPR-Cas9 has evolved into a versatile and extensively utilized technology in genetic engineering. The potential applications of CRISPR span various fields, including agriculture, medicine, and environmental conservation. CRISPR technology in agriculture allows for accurate manipulation of plant genomes, resulting in improved crop characteristics such as increased disease resistance, drought tolerance, and nutritional value. The precise nature of CRISPR-Cas9 enables unparalleled accuracy in genetic alterations, minimizing unintended effects and expediting the cultivation of crops with desired characteristics. Furthermore, the medical implications of CRISPR are equally significant. Scientists are currently investigating the potential of CRISPR for gene therapy, aiming to address or rectify genetic disorders by precisely altering or modifying defective genes. While presenting ethical complexities, the ability to change the human germline can potentially prevent hereditary diseases and promote the progress of personalized medicine. The utilization of CRISPR in cancer research significantly impacts the medical field, as it allows for the development of targeted therapies by genetically modifying cancer cells. CRISPR extends its influence beyond the domains of agriculture and medicine, encompassing implications for environmental conservation. Scientists are investigating the potential of CRISPR technology to tackle the decline in biodiversity and the degradation of ecosystems. Genetically modifying organisms to increase their ability to withstand changing environmental conditions or to combat invasive species presents novel solutions to urgent ecological problems. In the future, the development of CRISPR technology is progressing with innovations such as base editing and prime editing. Base editing facilitates the direct transformation of a specific DNA base pair into another without causing double-strand breaks, resulting in more accurate modifications with decreased mistakes. Prime editing, in contrast, broadens the range of available techniques by allowing the incorporation of novel genetic information without the need for donor DNA templates. These advancements augment the capacities of CRISPR, rendering it an increasingly versatile and potent instrument for genetic manipulation. Aside from CRISPR, other nascent biotechnologies are positioned to influence the future profoundly. Synthetic biology, a field encompassing the design and construction of novel biological entities, is promising to develop artificial organisms with customized functionalities. Synthesizing principles from biology, engineering, and computer science, this interdisciplinary field aims to manipulate biological systems for targeted objectives, such as generating biofuels or creating innovative materials. Moreover, progress in gene therapy, encompassing RNA-based treatments and gene editing techniques surpassing CRISPR, offers the potential for treating various genetic disorders. Precisely, targeting and editing genes linked to genetic disorders offers opportunities for developing curative therapies and enhancing patient outcomes. As these advancing biotechnologies continue to build, ethical considerations become increasingly important. Prudent utilization of these formidable instruments necessitates thoughtful deliberation regarding the possible hazards, ethical principles, and societal ramifications. Maintaining a balance between innovation and ethical considerations is crucial to ensure these technologies improve humanity and the environment.

12.7.2 ADVANCEMENTS IN CRISPR AND BEYOND

The progressive development of CRISPR technology signifies a dynamic boundary in genetic engineering, with ongoing progressions broadening its applications and enhancing its accuracy. In addition to the fundamental CRISPR-Cas9 system, the field has observed substantial advancements in various CRISPR-based tools, each possessing distinct capabilities and tackling specific obstacles. Significant progress has been made in the emergence of base editing, a technique that can directly convert a particular DNA base pair into another without causing double-strand breaks. Conventional CRISPR-Cas9 editing is dependent on the inherent repair mechanisms of the cell, which may lead to the occurrence of inaccuracies. Base editing addresses this problem by precisely converting one nucleotide to another, thereby minimizing the occurrence of unintended mutations. This technological progress improves the precision and safety of genetic editing, creating new opportunities for therapeutic medical applications and precise modifications in agriculture. Furthermore, prime editing signifies a notable advancement in CRISPR technology. Prime editing enables researchers to make specific modifications to the DNA sequence without the need for donor DNA templates, a requirement in conventional CRISPR-Cas9 editing. This innovation broadens the range of genetic alterations by allowing the precise introduction of new genetic information. Prime editing has significant potential for rectifying genetic mutations linked to diseases and propelling the field of gene therapy forward. CRISPR-based technologies are fueling advancements in crop enhancement within the agricultural industry. In addition to traditional genetic modification methods, CRISPR enables the accurate modification of particular genes responsible for crops' favorable characteristics. This focused strategy expedites the advancement of crops with augmented nutritional composition, heightened resistance to pests and diseases, and enhanced tolerance to environmental pressures. The transformative impact of CRISPR-based crop improvements is evident in their potential to tackle global food security challenges effectively (Camere & Karana, 2018; Fasciolo et al., 2023; Haris et al., 2023; Kaur et al., 2023b; Liu et al., 2024; Rai et al., 2021b; Ray et al., 2023; Sodhi et al., 2022; Uyanga et al., 2024). In addition to CRISPR, the investigation into alternative gene-editing technologies persists. Zinc finger nucleases (ZFNs) and transcription activator-like effector nucleases (TALENs) are earlier gene-editing techniques developed before CRISPR-Cas9. CRISPR has become widely adopted because of its simplicity and efficiency. However, ZFNs and TALENs still have relevance in specific applications, demonstrating the range of available genetic engineering tools. In the future, the combination of CRISPR technology and other advancements in biotechnology, such as synthetic biology, shows potential for developing new biological systems and tackling intricate problems. Artificial biology techniques like DNA synthesis and assembly enable researchers to meticulously design and fabricate novel physical entities. The integration of CRISPR with synthetic biology provides opportunities to develop synthetic organisms with customized functionalities, spanning from the production of bioenergy to the remediation of the environment. As the field progresses, addressing the ethical considerations related to utilizing CRISPR and other gene-editing technologies is crucial. Continued contemplation and adherence to responsible methods are necessary due to the possibility of unforeseen outcomes, unintended impacts, and the ethical considerations associated with altering the human germline. Establishing ethical frameworks that govern the reliable deployment of these powerful tools relies heavily on the collaborative efforts of scientists, ethicists, policymakers, and the public.

12.7.3 POSSIBLE ADVANCES IN CUTTING-EDGE HORTICULTURE TECHNOLOGY

The future of horticulture is characterized by the potential of cutting-edge technologies that seek to transform conventional methods, improve sustainability, and optimize the use of resources. From implementing precision farming techniques to adopting vertical agriculture methods, these advancements can revolutionize how we grow, oversee, and utilize crops (Adetunji et al., 2022; Aggarwal et al., 2024; Boamah et al., 2023; Haris et al., 2023; Kerry, 2014; Moreno et al., 2024; Ray et al., 2023; Sandhu et al., 2023; Zhang, Han, et al., 2023). A notable advancement involves incorporating cutting-edge technologies and data analytics into horticulture methods. Precision farming is possible using sensor networks, automated systems, and real-time data analysis. This technology allows farmers to monitor and accurately control environmental conditions closely. Such a high degree of accuracy improves resource efficiency by customizing irrigation, nutrient distribution, and pest management to suit the unique requirements of each crop. The outcome is enhanced crop production and minimized ecological footprint by efficiently utilizing resources. Vertical farming is an innovative method of horticulture that is especially useful in urban areas with limited space. Vertical farms employ multitiered cultivation systems within vertically inclined structures, typically in controlled indoor settings. This method reduces the requirement for cultivable land and enables cultivation throughout the year, regardless of external weather conditions. Vertical farming achieves sustainable urban agriculture through advanced lighting systems, climate control, and hydroponic or aeroponic cultivation methods. Using robotics in horticulture is positioned to enhance efficiency and tackle labor-related difficulties. Autonomous vehicles, equipped with precise tools, can carry out tasks such as planting, harvesting, and monitoring crops. This automation process decreases expenses related to human labor and allows for uninterrupted, 24-hour operations, thereby enhancing overall efficiency. In addition, robotic systems can traverse crop fields while causing minimal soil compaction, enhancing soil health and promoting sustainability. Genomic technologies, such as advancements based on CRISPR, are essential in influencing the future of horticulture. The capacity to accurately modify the genetic makeup of plants presents possibilities for cultivating crops with augmented characteristics, such as enhanced nutritional composition, heightened disease resistance, and increased resilience to environmental pressures. GM crops possess the capacity to effectively tackle worldwide food security issues by enhancing crop yields and diminishing reliance on chemical inputs (Haris et al., 2023; Kaur et al., 2023b; Rai et al., 2023; Ray et al., 2023; Sharma et al., 2022; Singh et al., 2023a, 2023b; Sodhi et al., 2022). The integration of hydroponics and aeroponics signifies a transition toward cultivation methods that are more efficient in their use of resources and promote sustainability. These hydroponic cultivation methods entail growing plants in nutrient-dense solutions or suspensions, reducing water consumption and minimizing nutrient runoff. Hydroponic and aeroponic systems can be utilized in various settings, from conventional farms to urban environments, to promote water conservation and sustainable agricultural practices. Moreover, progress in seed technology, such as the creation of GM seeds with targeted characteristics, is propelling innovation in horticulture. These seeds are genetically engineered to possess resistance against pests, diseases, and environmental stressors, resulting in increased agricultural productivity and decreased reliance on chemical interventions. Seed technologies are essential for maintaining crop diversity and resilience, contributing to the overall sustainability of horticulture practices. Using artificial intelligence (AI) and machine learning in horticulture signifies a fundamental change in decisions. These technologies utilize advanced data analysis to examine large quantities of data, encompassing environmental conditions, crop performance, and market trends, to offer practical and implementable insights for farmers. AI systems can forecast the most favorable periods for planting crops, suggest accurate schedules for irrigation, and detect possible pest infestations. This equips farmers with up-to-date information to make well-informed decisions. As horticulture progresses, incorporating these advanced technologies necessitates a thoughtful examination of ethical and environmental consequences (Appelqvist et al., 2022; Awana et al., 2020; Gouthami et al., 2023; Kaur et al., 2023a; Keshavarz & Sharafi, 2023; Kumar et al., 2022, 2023; Kumar & Dixit, 2021; Muhie, 2022; Nguyen et al., 2023; Peng & Simko, 2023; Trejo-Perea et al., 2018; Vishnoi & Goel, 2024). Ensuring the responsible implementation of genetic technologies, upholding ethical standards in treating farm labor amidst automation, and promoting the sustainable utilization of resources is of utmost importance. Effective collaboration among scientists, policymakers, farmers, and consumers is crucial for successfully navigating the intricate terrain of cutting-edge horticulture technology and guaranteeing its beneficial influence on worldwide food systems. The prospects of biotechnology are characterized by the ongoing advancement of CRISPR and other emerging technologies, providing unparalleled opportunities in genetic engineering. The integration of CRISPR advancements with synthetic biology and other gene-editing tools drives the field forward, opening up new possibilities for groundbreaking applications in agriculture, medicine, and environmental conservation. Concurrently, advancements in cutting-edge horticulture technology hold the potential to completely transform methods of crop management, improve sustainability, and tackle the challenges of global food security. As these technologies develop, it is important to consider ethics to responsibly implement them and ensure their benefits positively impact humanity's and the environment's well-being.

12.8 CONCLUSION

To summarize, incorporating biotechnological innovations into brilliant horticulture production signifies significant and revolutionary progress in sustainable and efficient agricultural methods. Combining biotechnology and intelligent technologies can tackle crucial issues confronting worldwide food systems, such as rising demand, limited resources, and climate fluctuations. As we explore this new territory, it is clear that the accuracy and abilities provided by biotechnological tools, such as CRISPR-based gene editing and advanced seed technologies, improve the ability of crops to withstand challenges and increase their productivity in intelligent horticulture systems. Using sensor networks and intelligent automation in irrigation and nutrient delivery processes allows for real-time monitoring, optimizing resources and promoting environmental sustainability. Integrating genetic knowledge with data analytics and decision support systems gives farmers exceptional authority over crop management, resulting in enhanced productivity, diminished ecological consequences, and heightened overall effectiveness. Moreover, the ethical considerations associated with biotechnological integration in innovative horticulture cannot be overstated. Achieving a harmonious equilibrium between innovation and responsible practices is crucial to guarantee equitable distribution of the advantages offered by biotechnological tools while effectively managing environmental consequences. Researchers, policymakers, and the agricultural community must collaborate to tackle biodiversity, genetic diversity, and the possible unforeseen repercussions of genetic modifications. Looking ahead, the ongoing advancement of biotechnological integration in innovative horticulture shows potential for tackling emerging challenges in a swiftly transforming world. The advancement of crops with augmented nutritional composition enhanced resilience against pests and diseases and heightened adaptability to fluctuating environmental conditions exemplify the potentially beneficial influence on worldwide food security. In addition, using intelligent technologies to optimize resource allocation promotes the adoption of sustainable agricultural methods, thereby enhancing the ability to withstand changing climate conditions. Combining biotechnology and brilliant horticulture production forms a strong partnership beyond conventional agrarian limits. It not only provides solutions to current challenges but also creates opportunities that were previously considered impossible to achieve. Incorporating biotechnological advancements in brilliant horticulture production, achieved through responsible innovation, ethical consideration, and collaborative efforts, is a guiding light toward a more sustainable, resilient, and productive agricultural future.

KEYWORDS

- **biotechnology**
- **crop**
- **environment**
- **innovation**
- **sustainable technology**
- **horticulture**
- **ecological**
- **ethical**

REFERENCES

Adetunji, C.O., Anani, O.A., Olaniyan, O.T., Bodunrinde, R.E., Osemwegie, O.O., & Ubi, B.E. (2022). Chapter 14—Sustainability of biofertilizers and other allied products from genetically modified microorganisms. In: S. Varjani, A. Pandey, T. Bhaskar, S. V. Mohan, & D C.W. Tsang. *Biomass, Biofuels, Biochemicals*, Eds., Elsevier, pp. 363–393.

Aditika, B.K., Singh, S., & Kumar, P. (2022). Taro (Colocasia Esculenta): Zero wastage orphan food crop for food and nutritional security. *South African Journal of Botany*, 145, 157–169. https://doi.org/10.1016/j.sajb.2021.08.014.

Aggarwal, B., Rajora, N., Raturi, G., Dhar, H., Kadam, S.B., Mundada, P.S., Shivaraj, S.M., Varshney, V., Deshmukh, R., Barvkar, V.T., Salvi, P., & Sonah, H. (2024). Biotechnology and urban agriculture: A partnership for the future sustainability. *Plant Science*, 338, 111903. https://doi.org/10.1016/j.plantsci.2023.111903.

Aguilar, A., Magnien, E., & Thomas, D. (2013). Thirty years of European biotechnology programmes: From biomolecular engineering to the bioeconomy. *New Biotechnology*, 30(5), 410–425. https://doi.org/10.1016/j.nbt.2012.11.014.

Ait-Oubahou, A., Nur Hanani, Z.A., & Jamilah, B. (2019). Chapter 11—Packaging. In: E.M. Yahia, Ed., *Postharvest Technology of Perishable Horticultural Commodities*. Woodhead Publishing, pp. 375–399.

Ali, S., Tyagi, A., Rajarammohan, S., Mir, Z.A., & Bae, H. (2023). Revisiting alternaria-host interactions: New insights on its pathogenesis, defense mechanisms and control strategies. *Scientia Horticulturae*, 322, 112424. https://doi.org/10.1016/j.scienta.2023.112424.

Appelqvist, I., De Wet, H., Marusic, S., & Janakievski, F. (2022). 20—Industry 4.0 and the Impact on the Agrifood Industry. In: P. Juliano, R. Buckow, M. H. Nguyen, K. Knoerzer, & J.B. Sellahewa, Eds., *Food Engineering Innovations Across the Food Supply Chain*. Academic Press, pp. 343–356.

Sharma, A., Sharma, S., Bashir, S., Kasi, R., & Subramaniam, R.T. (2022). The significance of graphene based composite hydrogels as smart materials: A review on the fabrication, properties, and its applications. *FlatChem*, 33, 100352. https://doi.org/10.1016/j.flatc.2022.100352.

Awana, M., Jain, N., Samota, M.K., Rani, K., Kumar, A., Ray, M., Gaikwad, K., Praveen, S., Kumar Singh, N., & Singh, A. (2020). Protein and gene integration analysis through proteome and transcriptome brings new insight into salt stress tolerance in pigeonpea (*Cajanus cajan* L.). *International Journal of Biological Macromolecules*, 164, 3589–3602. https://doi.org/10.1016/j.ijbiomac.2020.08.223.

Awasthi, G., Singh, T., Tiwari, Y., Awasthi, A., Tripathi, R.D., Shrivastava, S., Vajpayee, P., Kumar, A., & Awasthi, K.K. (2020). A review on nanotechnological interventions for plant growth and production. *Materials Today: Proceedings*, 31, 685–693. https://doi.org/10.1016/j.matpr.2020.07.255.

Beckles, D.M., & Roessner, U. (2012). 5—Plant metabolomics: Applications and opportunities for agricultural biotechnology. In: A. Altman and P.M. Hasegawa, Eds., *Plant Biotechnology and Agriculture*. San Diego: Academic Press, pp. 67–81.

Behera, S., Rana, G., Satapathy, S., Mohanty, M., Pradhan, S., Kumar Panda, M., Ningthoujam, R., Hazarika, B.N., & Singh, Y.D. (2020). Biosensors in diagnosing COVID-19 and recent development. *Sensors International*, 1, 100054. https://doi.org/10.1016/j.sintl.2020.100054.

Bhaskar, R., Pandey, S.P., Narayanan, K.B., Gupta, M.K., & Han, S.S. (2023). Chapter 4—Nanobionics in plant production: A novel approach to regulate plant functions. *Nanomaterial–Plant Interactions*, In: S. Seena, A. Rai, S. Kumar, Eds., Academic Press, pp. 111–133.

Boamah, P.O., Onumah, J., Aduguba, W.O., & Santo, K.G. (2023). Application of depolymerized chitosan in crop production: A review. *International Journal of Biological Macromolecules*, 235, 123858. https://doi.org/10.1016/j.ijbiomac.2023.123858.

Camere, S., & Karana, E. (2018). Fabricating materials from living organisms: An emerging design practice. *Journal of Cleaner Production*, 186, 570–584. https://doi.org/10.1016/j.jclepro.2018.03.081.

Chen, C., Wang, C., Liu, Y., Shan, T., Shi, X., & Gao, X. (2022). Integration analysis of pacbio SMRT- and Illumina RNA-Seq Reveals P450 genes involved in thiamethoxam detoxification in *Bradysia odoriphaga*. *Pesticide Biochemistry and Physiology*, 186, 105176. https://doi.org/10.1016/j.pestbp.2022.105176.

Crisp, P.A., Bhatnagar-Mathur, P., Hundleby, P., Godwin, I.D., Waterhouse, P.M., & Hickey, L.T. (2022). Beyond the gene: Epigenetic and cis-regulatory targets offer new breeding potential for the future. *Current Opinion in Biotechnology*, 73, 88–94. https://doi.org/10.1016/j.copbio.2021.07.008.

Dagys, K., Agipar, B., Tsolmon, S., Ringler, C., Bellisario, K., & Fanzo, J. (2023). Maximizing nutrition in key food value chains of Mongolia under climate change. *Food Policy*, 117, 102468. https://doi.org/10.1016/j.foodpol.2023.102468.

Day, B., & Sigrimis, N. (2020). An invited editorial interview with Professor Nick Sigrimis, Agricultural University of Athens, on smart agriculture and the digital revolution. *Biosystems Engineering*, 198, 350–354. https://doi.org/10.1016/j.biosystemseng.2020.09.001.

Faling, M. (2020). Framing agriculture and climate in Kenyan policies: A longitudinal perspective. *Environmental Science & Policy*, 106, 228–239. https://doi.org/10.1016/j.envsci.2020.01.014.

Farooq, M., Gogoi, N., & Michele, B.T. (2023). Sustainable agriculture and the environment. In: Pisante, Eds. *Index*. Academic Press, pp. 557–566.

Fasciolo, B., Awouda, A., Bruno, G., & Lombardi, F. (2023). A smart aeroponic system for sustainable indoor farming. *Procedia CIRP,* 116, 636–641. https://doi.org/10.1016/j.procir.2023.02.107.

Forastero, Á.G. (2023). Resources, conservation & recycling advances circular economy in Andalusia: A review of public and non-governmental initiatives. *Resources, Conservation & Recycling Advances*, 17, 200133. https://doi.org/10.1016/j.rcradv.2023.200133.

Fu, X., Zheng, Ze, Sha, Z., Cao, H., Yuan, Q., Yu, H., & Li, Q. (2022). Biorefining waste into nanobiotechnologies can revolutionize sustainable agriculture. *Trends in Biotechnology*, 40(12), 1503–1518. https://doi.org/10.1016/j.tibtech.2022.09.013.

Full, J., Miehe, R., Kiemel, S., Bauernhansl, T., & Sauer, A. (2019). The biological transformation of energy supply and storage—Technologies and scenarios for biointelligent value creation. *Procedia Manufacturing*, 39, 1204–1214. https://doi.org/10.1016/j.promfg.2020.01.349.

Gaffey, J., Rajauria, G., McMahon, H., Ravindran, R., Dominguez, C., Ambye-Jensen, M., Souza, M.F., Meers, E., Macias Aragonés, M., Skunca, D., & Sanders, JP.M. (2023). Green biorefinery systems for the production of climate-smart sustainable products from grasses, legumes and green crop residues. *Biotechnology Advances*, 66, 108168. https://doi.org/10.1016/j.biotechadv.2023.108168.

Ghasemnezhad, A., Ghorbanpour, M., Sohrabi, O., & Ashnavar, M. (2019). Chapter three—A general overview on application of nanoparticles in agriculture and plant science , In: S. K. Verma and A.K. Das, Eds., *Engineered Nanomaterials and Phytonanotechnology: Challenges for Plant Sustainability*. Elsevier, Vol. 87, pp. 85–110.

Gheorghiu, R., Andreescu, L., Zulean, M., & Curaj, A. (2017). Chapter 14—Entrepreneurial discovery as a foresight for smart specialization: Trade-offs of inclusive and evidence-based consensus. In: S. Radosevic, A. Curaj, R. Gheorghiu, L. Andreescu, & I. Wade,

Eds., *Advances in the Theory and Practice of Smart Specialization*. Academic Press, pp. 319–344.

Gohari, G., Jiang, M., Manganaris, G.A., Zhou, J., & Fotopoulos, V. (2024). Next generation chemical priming: With a little help from our nanocarrier friends. *Trends in Plant Science*, 2024;29:150–166. https://doi.org/10.1016/j.tplants.2023.11.024.

Gouthami, K., Lakshminarayana, L., Faniband, B., Veeraraghavan, V., Bilal, M., Bhargava, R.N., Ferreira, L.F.R., Rahdar, A., Kakkameli, S., & Mulla, S.I. (2023). 1—Introduction to polymeric nanomaterials. In: N. Ali, M. Bilal, A. Khan, T. A. Nguyen, & R.K. Gupta, Eds., *Micro and Nano Technologies*. Elsevier, pp. 3–25.

Gruda, N., Bisbis, M., & Tanny, J. (2019). Influence of climate change on protected cultivation: Impacts and sustainable adaptation strategies—A review. *Journal of Cleaner Production*, 225, 481–495. https://doi.org/10.1016/j.jclepro.2019.03.210.

Hackfort, S. (2023). Unlocking sustainability? The power of corporate lock-ins and how they shape digital agriculture in Germany. *Journal of Rural Studies*, 101, 103065. https://doi.org/10.1016/j.jrurstud.2023.103065.

Haris, M., Hussain, T., Mohamed, H.I., Khan, A., Ansari, M.S., Tauseef, A., Khan, A.A., & Akhtar, N. (2023). Nanotechnology—A new frontier of nano-farming in agricultural and food production and its development. *Science of the Total Environment*, 857, 159639. https://doi.org/10.1016/j.scitotenv.2022.159639.

Hassoun, A., Kamiloglu, S., Garcia-Garcia, G., Parra-López, C., Trollman, H., Jagtap, S., Aadil, R.M., & Esatbeyoglu, T. (2023). Implementation of relevant fourth industrial revolution innovations across the supply chain of fruits and vegetables: A short update on traceability 4.0. *Food Chemistry*, 409, 135303. https://doi.org/10.1016/j.foodchem.2022.135303.

He, Z., Lan, Y., Zhou, X., Yu, B., Zhu, T., Yang, Fa, Fu, L-Yu, Chao, H., Wang, J., Feng, R-Xu, Zuo, S., Lan, W., Chen, C., Chen, M., Zhao, X., Hu, K., & Chen, D. (2023). Single-cell transcriptome analysis dissects LncRNA-associated gene networks in arabidopsis. *Plant Communications*, 5(2), 100717. https://doi.org/10.1016/j.xplc.2023.100717.

Hemalatha, P., Abda, E.M., Shah, S., Venkatesa Prabhu, S., Jayakumar, M., Karmegam, N., Kim, W., & Govarthanan, M. (2023). Multi-Faceted CRISPR-Cas9 strategy to reduce plant based food loss and waste for sustainable bio-economy—A review. *Journal of Environmental Management*, 332, 117382. https://doi.org/10.1016/j.jenvman.2023.117382.

Ijaz, M., Khan, F., Ahmed, T., Noman, M., Zulfiqar, F., Rizwan, M., Chen, J., Siddique, KH.M., & Li, B. (2023). Nanobiotechnology to advance stress resilience in plants: Current opportunities and challenges. *Materials Today Bio*, 22, 100759. https://doi.org/10.1016/j.mtbio.2023.100759.

Karaca, M., & Ince, A.G. (2023). Chapter 8—Revisiting sustainable systems and methods in agriculture. In: M. Farooq, N. Gogoi, & M. Pisante, Eds., *Sustainable Agriculture and the Environment*. Academic Press, pp. 195–246.

Kaur, B., Mansi, .Dimri, S., Singh, J., Mishra, S., Chauhan, N., Kukreti, T., Sharma, B., Singh, SP., Arora, S., Uniyal, D., Agrawal, Y., Akhtar, S., Rather, M.A., Naik, B., Kumar, V., Gupta, A.K., Rustagi, S., & Preet, M.S. (2023a). Insights into the harvesting tools and equipment's for horticultural crops: From then to now. *Journal of Agriculture and Food Research*, 14, 100814. https://doi.org/10.1016/j.jafr.2023.100814.

Kaur, H., Hussain, S.J., Mir, R.A., Verma, V.C., Naik, B., Kumar, P., & Dubey, R.C. (2023b). Nanofertilizers—Emerging smart fertilizers for modern and sustainable agriculture. *Biocatalysis and Agricultural Biotechnology*, 54, 102921. https://doi.org/10.1016/j.bcab.2023.102921.

Kerry, J.P. (2014). Chapter 23—New packaging technologies, materials and formats for fast-moving consumer products. In: J.H. Han, Ed., *Innovations in Food Packaging (Second Edition)*. San Diego: Academic Press, pp. 549–584.

Keshavarz, M., & Sharafi, H. (2023). Scaling up climate-smart regenerative agriculture for the restoration of degraded agroecosystems in developing countries. *Sustainable Production and Consumption*, 38, 159–173. https://doi.org/10.1016/j.spc.2023.04.003.

Kiran, B.R., Prasad, M.N.V., & Venkata Mohan, S. (2023). Chapter 2—Farm to fork: Sustainable agrifood systems. In: M.N. Vara Prasad, & M. Smol, Eds., *Sustainable and Circular Management of Resources and Waste Towards a Green Deal*. Elsevier, pp. 25–38.

Kostas, E.T., Adams, JM.M., Ruiz, H.A., Durán-Jiménez, G., & Lye, G.J. (2021). Macroalgal biorefinery concepts for the circular bioeconomy: A review on biotechnological developments and future perspectives. *Renewable and Sustainable Energy Reviews*, 151, 111553. https://doi.org/10.1016/j.rser.2021.111553.

Kour, H., Khan, S.S., Kour, D., Singh, S., Kumari, S., Kaur, M., Khan, R.T., & Yadav, A.N. (2022). Chapter 9—Nanotechnologies for microbial inoculants as biofertilizers in the horticulture. In: M. Seymen, E. S. Kurtar, C. Erdinc, & A. Kumar, Eds., *Developments in Applied Microbiology and Biotechnology*. Academic Press, pp. 201–261.

Kozai, T., Niu, G., & Michiko B. T. Plant Factory (Second Edition) Takagaki. eds. (2020). *Chapter 3—PFAL Business and R&D in Asia and North America: Status and Perspectives*. Academic Press, pp. 35–76.

Kumar, M., Barbhai, M.D., Puranik, S., Radha, S.N., Senapathy, M., Dhumal, S., Singh, S., Kumar, S., Deshmukh, V.P., Anitha, T., Pandiselvam, R., & Lorenzo, J.M. (2023). Combination of green extraction techniques and smart solvents for bioactives recovery. *TrAC Trends in Analytical Chemistry*, 169, 117286. https://doi.org/10.1016/j.trac.2023.117286.

Kumar, N., & Dixit, A. (2021). Chapter 2—Nanotechnology in agriculture., In *Micro and Nano Technologies*, Elsevier, pp. 33–68.

Kumar, P., Singh, A., Rajput, V.D., Yadav, A.K.S., Kumar, P., Singh, A.K., & Minkina, T. (2022). Chapter 36—*Role of artificial intelligence, sensor technology, big data in agriculture: Next-generation farming*. In: P. Sharma, D. Yadav, & R.K. Gaur. Eds., Academic Press, pp. 625–639.

Laxman, L., & Ansari, A.H. (2023). Chapter 4—*Negotiating the Safe Use of Biotechnology to Conserve Agrobiodiversity: Issues and Challenges Impacting Universal Sustainability*. In: L. Laxman, & A.H. Ansari, Eds., Woodhead Publishing, pp. 107–141.

Liu, X., Wang, S., Zeng, K., Li, W., Wang, S., Huang, S., Wang, H., & Yang, X. (2024). N-Myristoyltransferase1 regulates biomass accumulation in cucumber (*Cucumis sativus* L.). *Journal of Integrative Agriculture*. https://doi.org/10.1016/j.jia.2024.01.013.

Mahapatra, D.M., Satapathy, K.C., & Panda, B. (2022). Biofertilizers and nanofertilizers for sustainable agriculture: Phycoprospects and challenges. *Science of The Total Environment*, 803, 149990. https://doi.org/10.1016/j.scitotenv.2021.149990.

Misra, S., & Ghosh, A. (2024). Chapter 6—Agriculture paradigm shift: A journey from traditional to modern agriculture. In: K. Singh, M.C. Ribeiro, Ö. Calicioglu, Eds., Elsevier, pp. 113–141.

Moreno, J.C., Berenguel, M., Donaire, J.G., Rodríguez, F., Sánchez-Molina, J.A., Luis Guzmán, J., & Giagnocavo, C.L. (2024). A pending task for the digitalisation of agriculture: A general framework for technologies classification in agriculture. *Agricultural Systems*, 213, 103794. https://doi.org/10.1016/j.agsy.2023.103794.

Muhie, S.H. (2022). Novel approaches and practices to sustainable agriculture. *Journal of Agriculture and Food Research*, 10, 100446. https://doi.org/10.1016/j.jafr.2022.100446.

Mukherjee, P.K., Das, B., Bhardwaj, P.K., Tampha, S., Singh, H.K., Chanu, L.D., Sharma, N., & Devi, S.I. (2023). Socio-economic sustainability with circular economy—An alternative approach. *Science of the Total Environment*, 904, 166630. https://doi.org/10.1016/j.scitotenv.2023.166630.

Nguyen, M.T.P., Knowling, M., Tran, N.N., Burgess, A., Fisk, I., Watt, M., Escribà-Gelonch, M., This, H., Culton, J., & Hessel, V. (2023). Space farming: Horticulture systems on spacecraft and outlook to planetary space exploration. *Plant Physiology and Biochemistry*, 194, 708–721. https://doi.org/10.1016/j.plaphy.2022.12.017.

Olıverı, L.M., Arfò, S., Matarazzo, A., D'Urso, D., & Chıacchıo, F. (2023). Improving the composting process of a treatment facility via an industry 4.0 monitoring and control solution: Performance and economic feasibility assessment. *Journal of Environmental Management*, 345, 118776. https://doi.org/10.1016/j.jenvman.2023.118776.

Peng, H., & Simko, I. (2023). Extending lettuce shelf life through integrated technologies. *Current Opinion in Biotechnology*, 81, 102951. https://doi.org/10.1016/j.copbio.2023.102951.

Pérez-Mesa, J.C., García Barranco, Ma.C., Giagnocavo, C., & Rubio, J.H. (2023). Seeking new strategic options for promotion of intermodal transport in perishables: The use of short sea shipping. *Research in Transportation Business & Management*, 50, 101034. https://doi.org/10.1016/j.rtbm.2023.101034.

Philp, J., & Winickoff, D.E. (2017). Clusters in industrial biotechnology and bioeconomy: The roles of the public sector. *Trends in Biotechnology*, 35(8), 682–686. https://doi.org/10.1016/j.tibtech.2017.04.004.

Priya, S.A., Afzal, O., Khalid, M., Ahmad, MdF, Upadhyay, A., Kumar, S., Garg, A., Ramzan, M., Hussain, A., Altamimi, M.A., Altamimi, A.S.A., Webster, T.J., & Khanam, A. (2023). Biogenic nanoparticles from waste fruit peels: Synthesis, applications, challenges and future perspectives. *International Journal of Pharmaceutics*, 643, 123223. https://doi.org/10.1016/j.ijpharm.2023.123223.

Rai, S.K., Apoorva, & Pandey-Rai, S. (2021a). Chapter 5—Marker-assisted breeding for abiotic stress tolerance in horticultural crops. In: A. Chandra Rai, A. Rai, K. Kumar Rai, V. P. Rai, & A. Kumar, Eds., Woodhead Publishing, pp. 63–74.

Rai, K.K., Kumar, A., Rai, A., Rai, V.P., & Rai, A.C. (2021b). Chapter 2—Conventional breeding approaches for abiotic stress management in horticultural crops. In: A. Chandra Rai, A. Rai, K. Kumar Rai, V. P. Rai, & A. Kumar, Eds., Woodhead Publishing, pp. 21–32.

Rai, P.K., Sonne, C., & Kim, Ki-H. (2023). Heavy metals and arsenic stress in food crops: Elucidating antioxidative defense mechanisms in hyperaccumulators for food security, agricultural sustainability, and human health. *Science of the Total Environment*, 874, 162327. https://doi.org/10.1016/j.scitotenv.2023.162327.

Ram, C., & Saroj, P.L. (2022). Chapter 4—Application of omics in arid fruit crops: Present status, challenges, and future perspectives. In: G.R. Rout and K.V. Peter, Eds., Academic Press, pp. 55–74.

Ramesh, S.V., Rajesh, S., & Radhamani, T. (2023). Chapter 6—Plant small RNAs: Biogenesis, mechanistic functions and applications. In: P. Guleria, V. Kumar, & B. Mo, Eds., Academic Press, pp. 129–181.

Ravisankar, V., & Mathew, D. (2022). Chapter 3—Multiomics bioinformatics approaches in horticultural crops. G.R. Rout, & K.V. Peter, Eds., Academic Press, pp. 27–54.

Ray, S., Sneha, K., & Jangid, C. (2023). CRISPR-Cas9 for sustainable food production: Impacts, recent advancements and future perspectives. *Food and Humanity*, 1, 1458–1471. https://doi.org/10.1016/j.foohum.2023.10.014.

Rettore de Araujo, Z., Eduardo da Silva, A., & Albini, L.C.P. (2020). Security challenges to smart agriculture: Current state, key issues, and future directions. *Array*, 8, 100048. https://doi.org/10.1016/j.array.2020.100048.

Rout, G.R., Mohapatra, A., & Jain, S.M. (2006). Tissue culture of ornamental pot plant: A critical review on present scenario and future prospects. *Biotechnology Advances*, 24(6), 531–560. https://doi.org/10.1016/j.biotechadv.2006.05.001.

Ruiz-Vanoye, J.A., Barrera-Cámara, R.A., Díaz-Parra, O., Ramos-Fernández, J.C., Fuentes-Penna, A., Ochoa-Zezzatti, A., Alberto Hernández-Aguilar, J., & Campero-Jurado, I. (2022). Chapter 14—Surveying smart farming for smart cities. In: A. Abraham, S. Dash, J.J.P.C. Rodrigues, B. Acharya, & S.K. Pani, Eds., *Intelligent Data-Centric Systems, AI, Edge and IoT-based Smart Agriculture*. Academic Press, pp. 241–262.

Sagar, N.A., Kumar, N., Choudhary, R., Bajpai, V.K., Cao, H., Shukla, S., & Pareek, S. (2022). Prospecting the role of nanotechnology in extending the shelf-life of fresh produce and in developing advanced packaging. *Food Packaging and Shelf Life*, 34, 100955. https://doi.org/10.1016/j.fpsl.2022.100955.

Saleh, M., Ashqar, H.I., Alary, R., Bouchareb, E.M., Bouchareb, R., Dizge, N., & Balakrishnan, D. (2024). Chapter 5—Biodiversity for ecosystem services and sustainable development goals. In: K. Singh, M.C. Ribeiro, & Ö. Calicioglu, Eds., Elsevier, pp. 81–110.

Samal, K.C., & Behera, L. (2022). Chapter 28—Omics databases in horticultural crops. In: G.R. Rout, & K.V. Peter, Eds., Academic Press, pp. 603–618.

Sandhu, R., Chaudhary, N., Bindia, R.S., Singh, K., & Pandey, V.K. (2023). A critical review on integrating bio fortification in crops for sustainable agricultural development and nutritional security. *Journal of Agriculture and Food Research*, 14, 100830. https://doi.org/10.1016/j.jafr.2023.100830.

Saraswathi, M.S., Kalaiponmani, K., Uma, S., & Backiyarani, S. (2018). Chapter 14—Critical evaluation of the benefits and risks of genetically modified horticultural crops. In: G.R. Rout, & K.V. Peter, Eds., Academic Press, pp. 315–351.

Sardoei, A.S., Sharifani, M., Sarmast, M.K., Ghasemnejhad, M., & Ghorbanpour, M. (2024). In silico characterization of the limonene synthase gene in citrus species under freezing tolerance. *Industrial Crops and Products*, 208, 117915. https://doi.org/10.1016/j.indcrop.2023.117915.

Sarmah, B., Baruah, V.J., Dutta, M., Begum, M., & Deka, B. (2024). Chapter 16—Precision nutrient management for field and horticultural crops. In: S. Lamine, P. K. Srivastava, A. Kayad, F. Muñoz-Arriola, & P.C. Eds., *Earth Observation*. Academic Press, pp. 317–344.

Sasi, S., Prasad, K., Weerasinghe, J., Bazaka, O., Ivanova, E.P., Levchenko, I., & Bazaka, K. (2023). Plasma for aquaponics. *Trends in Biotechnology*, 41(1), 46–62. https://doi.org/10.1016/j.tibtech.2022.08.001.

Sastry, R.K., Rashmi, H.B., Rao, N.H., & Ilyas, S.M. (2010). Integrating nanotechnology into agri-food systems research in India: A conceptual framework. *Technological Forecasting and Social Change*, 77(4), 639–648. https://doi.org/10.1016/j.techfore.2009.11.008.

Schröder, P., Sauvêtre, A., Gnädinger, F., Pesaresi, P., Chmeliková, L., Doğan, N., Gerl, G., Gökçe, A., Hamel, C., Millan, R., Persson, T., Ravnskov, S., Rutkowska, B., Schmid, T., Szulc, W., Teodosiu, C., & Terzi, V. (2019). Discussion paper: Sustainable increase of crop production through improved technical strategies, breeding and adapted management—A European perspective. *Science of the Total Environment*, 678, 146–161. https://doi.org/10.1016/j.scitotenv.2019.04.212.

Sharma, V., Kumar Tripathi, A., & Mittal, H. (2022). Technological revolutions in smart farming: Current trends, challenges & future directions. *Computers and Electronics in Agriculture*, 201, 107217. https://doi.org/10.1016/j.compag.2022.107217.

Sheikh, M., Iqra, F., Ambreen, H., Kumar, A.P., Ikra, M., & Chung, Y.S. (2023). Integrating artificial intelligence and high-throughput phenotyping for crop improvement. *Journal of Integrative Agriculture*. https://doi.org/10.1016/j.jia.2023.10.019.

Singh, A., Rajput, V.D., Varshney, A., Ghazaryan, K., & Minkina, T. (2023a). Small tech, big impact: Agri-nanotechnology journey to optimize crop protection and production for sustainable agriculture. *Plant Stress*, 10, 100253. https://doi.org/10.1016/j.stress.2023.100253.

Singh, A., Pandey, H., Pal, A., Chauhan, D., Pandey, S., Gaikwad, D.J., Sahu, C., & Atta, K. (2023b). Linking the role of melatonin in plant stress acclimatization. *South African Journal of Botany*, 159, 179–190. https://doi.org/10.1016/j.sajb.2023.05.034.

Sodhi, A.S., Sharma, N., Bhatia, S., Verma, A., Soni, S., & Batra, N. (2022). Insights on sustainable approaches for production and applications of value added products. *Chemosphere*, 286, 131623. https://doi.org/10.1016/j.chemosphere.2021.131623.

Suganya, T., Devi, N.R., Vignesh, S., Rajendran, S., Dorothy, R., & Nguyen, T.A. (2022). 3—Microbiology in agriculture: An introduction. In: A. Denizli, T. A. Nguyen, S. Rajendran, G. Yasin, & A. K. B. T.-N. for S. A. Nadda, Eds., *Micro and Nano Technologies*. Elsevier, pp. 41–51.

Tandon, A., Singh, A., Thakur, A., & Sharma, V. (2023). Nanomaterial mediated genome engineering for sustainable food production: Current status and future prospects. *Biocatalysis and Agricultural Biotechnology*, 54, 102891. https://doi.org/10.1016/j.bcab.2023.102891.

Tao, H., Xu, S., Tian, Y., Li, Z., Ge, Y., Zhang, J., Wang, Yu, Zhou, G., Deng, X., Zhang, Ze, Ding, Y., Jiang, D., Guo, Q., & Jin, S. (2022). Proximal and remote sensing in plant phenomics: 20 years of progress, challenges, and perspectives. *Plant Communications*, 3(6), 100344. https://doi.org/10.1016/j.xplc.2022.100344.

Trejo-Perea, M., Rios-Moreno, J.G., Herrera-Ruiz, G., Feregrino-Pérez, A.A., Torres-Pacheco, I., & Guevara-Gonzalez, R.G. (2018). Chapter 1—Role of biotechnology in the agrofood industry. In: A.M. Holban, & A.M. Grumezescu, Eds., *Handbook of Food Bioengineering*. Academic Press, pp. 1–26.

Ulfiqar, F., Navarro, M., Ashraf, M., Akram, N.A., & Munné-Bosch, S. (2019). Nanofertilizer use for sustainable agriculture: Advantages and limitations. *Plant Science*, 289, 110270. https://doi.org/10.1016/j.plantsci.2019.110270.

Uyanga, V.A., Bello, S.F., Bosco, N.J., Jimoh, S.O., Mbadianya, I.J., Kanu, U.C., Okoye, C.O., Afriyie, E., Mak-Mensah, E., Agyenim-Boateng, K.G., Ogunyemi, S.O., Nkoh, J.N., Olasupo, I.O., Karikari, B., & Ahiakpa, J.K. (2024). Status of agriculture and food security in post-COVID-19 Africa: Impacts and lessons learned. *Food and Humanity*, 2, 100206. https://doi.org/10.1016/j.foohum.2023.100206.

Verpoorte, R., van der Heijden, R., & Memelink, J. (1998). Chapter 12 plant biotechnology and the production of alkaloids: Prospects of metabolic engineering. In: G.A. Cordell, Eds., Academic Press, Vol. 50, pp. 453–508.

Vishnoi, S., & Goel, R.K. (2024). Climate smart agriculture for sustainable productivity and healthy landscapes. *Environmental Science & Policy*, 151, 103600. https://doi.org/10.1016/j.envsci.2023.103600.

Wang, Ya-H., Zhang, R.-R., Yin, Y., Tan, G.-F., Wang, G.-L., Liu, H., Zhuang, J., Zhang, J., Zhuang, F.-Y., & Xiong, Ai-S. (2023). Advances in engineering the production of the natural red pigment lycopene: A systematic review from a biotechnology perspective. *Journal of Advanced Research*, 46, 31–47. https://doi.org/10.1016/j.jare.2022.06.010.

Yang, B., Pan, F., Yasmeen, F., Shan, L., Pan, J., Zhang, M., Weng, X., Wang, M., Li, M., Wang, Q., & Cheng, K. (2023). Integrated multi-omic analysis reveals the cytokinin and sucrose metabolism-mediated regulation of flavone glycoside biosynthesis by MeJA exposure in ficus pandurata hance. *Food Research International*, 174, 113680. https://doi.org/10.1016/j.foodres.2023.113680.

Zhang, M., Xu, S., Han, Y., Li, D., Yang, S., & Huang, Y. (2023a). High-throughput horticultural phenomics: The history, recent advances and new prospects. *Computers and Electronics in Agriculture*, 213, 108265. https://doi.org/10.1016/j.compag.2023.108265.

Zhang, M., Han, Y., Li, D., Xu, S., & Huang, Y. (2023b). Smart horticulture as an emerging interdisciplinary field combining novel solutions: Past development, current challenges, and future perspectives. *Horticultural Plant Journal*. https://doi.org/10.1016/j.hpj.2023.03.015.

Zielińska, A., Szalata, M., Wielgus, K., Szalata, M., Gorczyński, A., Alves, T.F.R., Chaud, M.V., Souto, E.B., & Słomski, R. (2023). 9—Nanostructured polymeric tools for the treatment and diagnosis of plant diseases and applications in field crops. In: A.F. Eds., *Micro and Nano Technologies*. Elsevier, pp. 189–237.

CHAPTER 13

Dehumanizing Extension Advisory Services in Horticulture through the Application of Internet of Things

SAPNA JARIAL[1*], PANKAJ KUMAR[2], RUPESH KAUSHIK[2], JATINDER SINGH[2], and HARPINDER SINGH[3]

[1]*Department of Agricultural Economics and Extension, School of Agriculture, Lovely Professional University, Phagwara, Punjab, India*

[2]*Department of Horticulture, School of Agriculture, Lovely Professional University, Phagwara, Punjab, India*

[3]*School of Mechanical Engineering, Lovely Professional University, Phagwara, Punjab, India*

*Corresponding author

ABSTRACT

This review examines the potential impact of the Internet of Things (IoT) on extension advisory services in horticulture and explores the implications for dehumanization. The utilization of IoT technology in horticulture has the potential to revolutionize extension advising services by offering real-time monitoring, data collection, and analysis. Concerns have been raised, however, about the possibility that human contact and specialized knowledge would be lost in the advising process. This review delves into the advantages and obstacles linked to the utilization of IoT within the realm of horticultural extension advisory services. Additionally, the importance of maintaining a balance between technological advancements and human involvement to ensure effective and sustainable advisory practices is emphasized.

Smart Horticulture Production: Integrating Advanced Technologies.
Rajni Rajan and Duniya Ram Singh (Eds.)

DOI: 10.1201/9781998511679-13

13.1 INTRODUCTION

Horticulture, an expansive scientific field spanning from decorative plants to edible crops, is in a perpetual state of development. A central aim of horticultural advancement is to enhance the uniformity and foreseeability of products. Among the main conduits bridging novel innovations and technologies with cultivators is extension services (Lamm et al., 2023). Extension advisory services in horticulture face several challenges that hinder their effectiveness and impact. One of the main challenges is the decline in investment in extension services. Donor organizations have displayed growing apprehension regarding the declining returns on their investments within this sector, leading to reduced funding (Olorunfemi & Oladele, 2018). This lack of financial support limits the resources available for extension services to reach and assist farmers effectively. Although Extension experts are commonly thought to have a close affiliation with and understanding of the needs of their clientele, there exist information gaps. The farmer and extension agent ratio is very wide in India (1:5000) as against China (1:5813). Another challenge is the gender gap in access to extension services. Discrimination against women, particularly in terms of inheritance and property rights, exacerbates their limited access to productive resources (Agholor, 2019). Even though there have been endeavors to integrate gender into agricultural extension services, women continue to encounter limitations when it comes to accessing the learning opportunities offered by rural advisory services. This gender disparity hinders the equitable distribution of extension services and limits the potential impact on women farmers. Limited academic and technical capacity in horticulture is another challenge. In countries like Myanmar, there is a shortage of well-qualified graduates in the agricultural sector, including horticulture (Davis & Bost, 2017). The lack of qualified professionals in the field hampers the delivery of quality extension services and the dissemination of up-to-date knowledge and practices to farmers. Furthermore, wars and pandemics pose challenges. COVID-19 lockdown policies and restrictions on public gatherings and close contact activities have disrupted the traditional modes of extension delivery (Baffoe-Bonnie et al., 2021). To overcome these challenges, extension personnel have had to adapt to unconventional means of communication, such as mobile phones, radio, television, and Internet of Things (IoT) to continue providing support to farmers. It means dehumanize extension advisory services.

13.2 IOT TECHNOLOGIES IN HORTICULTURE

The concept of the IoT pertains to a network comprising interconnected physical devices, vehicles, structures, and various items that are equipped with sensors, software, and network connectivity. This configuration empowers these entities to amass and share data (Ben-Daya et al., 2017). These devices can interact with one another and with humans, facilitating a smooth amalgamation and automation of diverse processes. IoT serves as the foundational technology facilitating the delivery of real-time information, which is essential for the optimal functioning of other technologies. Functioning as an open network of intelligent devices, it possesses the capability for self-organization, exchanging resources, information, and data, and adeptly reacting and adjusting to a range of situations and fluctuations in the surroundings (Kaburuan & Jayadi, 2020). The IoT designates the interconnected network of tangible devices, automobiles, household gadgets, and other articles that incorporate electronics, software, sensors, and connectivity, allowing these entities to establish connections and share data.

Machine learning (ML) and deep learning (DL) constitute subdivisions within the realm of artificial intelligence (AI), focusing on crafting algorithms and models that empower computers to acquire knowledge from data and generate forecasts or choices without the need for explicit programming. ML algorithms are formulated to scrutinize and interpret data patterns, recognize correlations, and formulate predictions or classifications rooted in the patterns they uncover. These algorithms learn from labeled or unlabeled data through a process of training, where they adjust their internal parameters to optimize their performance on a specific task. ML algorithms find utility across a broad spectrum of challenges, encompassing tasks such as image recognition, natural language processing, recommendation systems, and predictive analytics. Conversely, DL constitutes a segment within the realm of ML centered on creating and refining artificial neural networks with multiple layers. These intricate neural networks possess the ability to acquire layered understandings of data, allowing them to encapsulate intricate structures and associations. DL has attained noteworthy triumphs across diverse fields, such as computer vision, speech comprehension, and natural language processing. Notably, models within DL, such as convolutional neural networks (CNNs) and recurrent neural networks, have showcased cutting-edge effectiveness in undertakings like image categorization, object identification, and language translation.

The techniques of ML and DL share a strong connection with IoT owing to their capacity to handle and scrutinize the extensive volumes of data produced by IoT devices. These devices continuously generate streams of sensor data, which can be subjected to ML and DL algorithms to derive valuable insights. As an illustration, ML and DL techniques find application within IoT setups for purposes such as anomaly identification, anticipatory maintenance, intelligent energy administration, and tailoring individual user interactions. Within the domain of horticulture, the integration of IoT holds the potential for monitoring and overseeing various facets of horticultural practices. For instance, IoT devices offer the capacity to amass data on factors like soil moisture levels, temperature, humidity, and other environmental conditions. This real-time information proves invaluable for efficient irrigation, crop management, and assessments of soil nutrients. A notable illustration is a study conducted by Postolache et al. (2022), which proposed an IoT-driven wireless smart sensor capable of measuring soil attributes such as soil texture, fertilizer distribution, and soil fertility status. Additionally, IoT presents opportunities for monitoring plant health, detecting pests, and preventing diseases through the utilization of sensors and data analysis, as illustrated by Singh et al. (2022). Another notable study by Selvaraj et al. (2019) showcased the effectiveness of employing a deep convolutional neural network (DCNN) for identifying banana diseases and pests in digital images. By employing a pretrained model to recognize diseases, the researchers adeptly employed deep transfer learning to construct a network capable of accurate predictions. The integration of IoT into horticulture empowers farmers to enhance their agricultural efficiency while also curbing adverse environmental impacts, as highlighted by Kumar et al. (2020). With IoT, farmers now gain access to comprehensive insights concerning variables like rainfall patterns, water cycles, and fertilizer requirements. Companies can capitalize on this valuable knowledge to make informed decisions regarding crop selection and optimal harvesting periods, thereby maximizing their profits, as demonstrated by Coble et al. (2018). According to Kondratieva et al. (2021), the introduction and utilization of IoT in horticulture can potentially lead to substantial increases in crop yields, ultimately resulting in elevated food production and decreased reliance on food imports. This transformation stands to significantly contribute to enhancing a country's food security.

In the context of IoT security, ML and DL techniques can be employed for intrusion detection, anomaly detection, and threat identification.

By analyzing network traffic patterns and device behavior, ML and DL algorithms can identify abnormal activities and potential security breaches (Luo, 2022; Popoola et al., 2021). Employing these methods can bolster the security of IoT systems by facilitating continuous monitoring and preemptive actions against security vulnerabilities. Furthermore, ML and DL techniques have the potential to play a role in optimizing resource allocation within IoT networks. With the heterogeneous and dynamic nature of IoT, efficient allocation and utilization of network resources are crucial. ML and DL algorithms can optimize resource allocation, improve network performance, and enable adaptive decision-making based on changing network conditions (Hussain et al., 2020). Moreover, ML and DL techniques find application in data analysis and decision support within precision horticulture (Figure 13.1). By scrutinizing data amassed from IoT devices and sensors, ML and DL algorithms offer valuable insights into the wellbeing of crops, soil conditions, and environmental variables. This knowledge equips farmers with the ability to make well-informed choices regarding irrigation, fertilization, pest management, and other facets of crop administration (Singh et al., 2021). In the field of horticulture, IoT can be harnessed to oversee and regulate diverse aspects of plant growth and progression, encompassing factors like temperature, humidity, soil moisture, and nutrient levels. By collecting and analyzing this real-time data, cultivators can make judicious determinations about the timing and quantity of watering, fertilization, and harvesting, thereby leading to heightened efficiency and enhanced yields.

IoT in horticultural extension refers to the application of IoT in the field of horticultural extension services (Figure 13.2). Horticultural extension services aim to disseminate knowledge, information, and technologies to farmers, helping them improve their farming practices, productivity, and profitability. Leveraging IoT within horticultural extension can amplify the efficacy of extension services through the provision of immediate data and insights to farmers. Deploying IoT devices like sensors and monitoring systems in the field facilitates data collection across diverse metrics, including soil moisture, temperature, humidity, and crop development. These data can be wirelessly relayed to extension personnel or farmers, empowering them to make well-grounded choices concerning irrigation, fertilization, pest management, and other elements of crop oversight (Kitinoja et al., 2011). By leveraging IoT technologies, extension workers can remotely monitor and assess the conditions of crops, identify

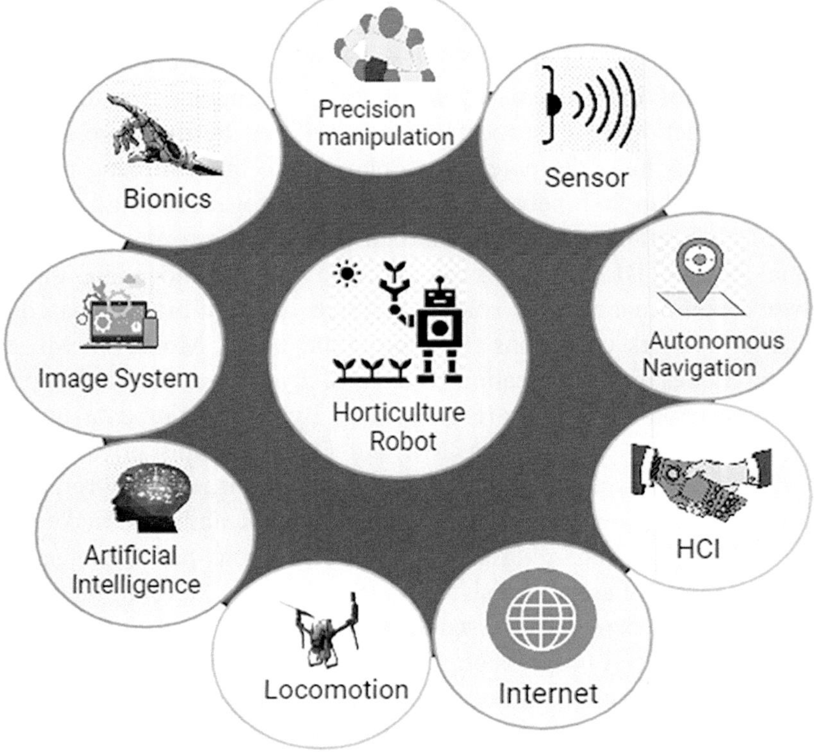

FIGURE 13.1 Fundamental technologies implicated in horticultural robotic applications.
Source: Authors.

potential issues or risks, and provide timely advice and recommendations to farmers. This can help farmers optimize their farming practices, reduce losses, and improve overall productivity. For example, IoT-based systems can provide alerts about pest infestations, disease outbreaks, or adverse weather conditions, allowing farmers to take preventive measures or implement appropriate interventions (Eyssartier et al., 2008).

Furthermore, IoT can facilitate the analysis and collection of data on postharvest practices as well as technologies. Extension workers can use IoT devices to monitor storage conditions, track product quality, and provide guidance on postharvest handling, storage, and transportation. This can help reduce postharvest losses and improve the marketability of horticultural produce (Fatty et al., 2021). However, the adoption of IoT in horticultural extension may face challenges. These include the

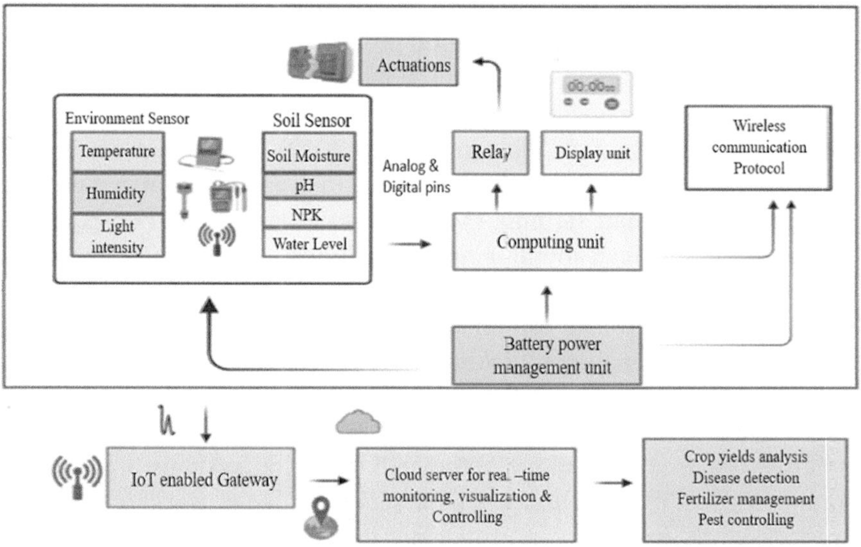

FIGURE 13.2 Application of IoT in horticulture.

Source: Authors.

cost of IoT devices and infrastructure, the need for technical expertise to deploy and manage IoT systems, and ensuring reliable connectivity in rural areas. Additionally, there may be a need for capacity building and training programs to enable extension workers and farmers to effectively utilize IoT technologies (Lamm et al., 2023). The IoT emerges as a promising technology with the potential to provide effective and dependable enhancements across various sectors, including horticultural extension services (Farooq et al., 2019). Within this realm, IoT-driven solutions are under development to streamline and supervise farms while minimizing human intervention (Farooq et al., 2019). These solutions employ network technologies such as low-power wide area networks, including Sigfox, narrowband IoT, and LoRa, recognized for their low energy consumption, extensive coverage, and cost-efficiency (Nagaraju & Chawla, 2019). The implementation of IoT in agriculture and allied has been shown to enhance access to agricultural advisory and extension services, improving productivity and reducing risk and uncertainties among farmers (Tumbo et al., 2018; Paudel et al., 2018). The adoption of IoT in horticulture can enhance supply chain management by facilitating product traceability and improving transparency along the supply chain (Tarmizi et al., 2020).

IoT technologies can enable the tracking of products from farm to table, ensuring the quality and safety of food products (Tarmizi et al., 2020).

IoT technologies in horticulture involve the utilization of sensors and devices to record parameters like environmental conditions, soil moisture, and crop health (Raju & Vijayaraghavan, 2022). These sensors collect data that can be used for decision-making in areas like management of water, crop, and smart farming (Madushanki et al., 2019). For example, IoT-based systems can monitor field conditions and automate precision agriculture practices to optimize crop production (Raju & Vijayaraghavan, 2022). The collected data can undergo analysis through cloud computing and extensive data analytics, yielding valuable insights tailored for farmers (Farooq et al., 2019). Through the automation and monitoring of agricultural processes, IoT can enable farmers to make data-driven decisions and optimize resource utilization. Automated irrigation systems are a prime example of IoT in horticulture. These setups utilize sensors to track soil moisture levels and autonomously adapt irrigation schedules in response. This not only saves water but also ensures that crops receive the optimal amount of moisture for healthy growth. Smart sensors are another important IoT application in horticulture. These sensors have the capability to oversee temperature, humidity, light, and other environmental variables that exert an influence on plant growth. Through the analysis of this data, farmers can make well-founded choices concerning the optimal timing for planting, harvesting, and applying fertilizers to their crops. Precision farming techniques also rely heavily on IoT technology. By using drones, satellite imagery, and other high-tech tools, farmers can gather detailed information about their fields and make precise adjustments to maximize crop yields. This can include everything from adjusting fertilizer levels to identifying areas that need water. The swift demands of the horticultural products market exceed the capacity of human labor. As an alternative, intelligent robotics emerges as an effective solution to address this demand. Within this section, we will introduce five key categories of fruit and vegetable robots, encompassing transplanting robots, patrolling robots, pesticide spraying robots, gardening robots, and picking robots.

13.2.1 ROBOTS FOR TRANSPLANTING

The field of horticulture has been undergoing a transformation with the increasing adoption of robotic technologies (Table 13.1). Transplanting robots have gained significant attention due to their potential to automate the seedling transplantation process, leading to improved efficiency, reduced labor demands and significantly accelerate the speed of transplantation. The quality of seedling transplantation directly impacts crop yield, making it a crucial stage in the automation of protected horticulture.

TABLE 13.1 Robots in Horticulture

Company	Services	Benefits	Limitations
Ecorobotix	Autonomous movements and orchard/vineyard Market leader	Precise mowing and weeding, reduced fuel consumption, and labor costs. Ideal market niche: Vineyards and orchards	Limited effectiveness on steep terrain, high initial investment
Iron Ox Solution	Crop management and indoor Market leader	Maximized space usage, reduced manual labor, increased production efficiency. Ideal market niche: Indoor vertical farms, urban agriculture	High initial investment, scalability for large-scale operations
Root AI (AppHarvest) Solution	Harvest and indoor Market leader	Reduced labor costs, increased efficiency, and minimized crop damage during harvest. Ideal market niche: Indoor farms, high-value produce	Adapting to various crop types, scalability
Vision Robotics Corporation	Crop management and orchard/ vineyard Market leader	Pruning, thinning, monitoring crop growth, reduced labor costs, improved crop quality. Ideal market niche: Orchard and vineyard owners	Adapting to different crop varieties, high initial investment
Crux Agribotics	Packaging and grading	Helping farmers improve efficiency, reduce labor costs, and ensure consistent product quality	High energy consumption and cost of fabrication. GNSS accuracy for precise localization Maintenance of robots or robotic systems

Recent studies have showcased the feasibility and benefits of using robotic technology in transplanting. Khadatkar et al. (2023) presented a compact robotic transplanter for plug seedlings, while Liu et al. (2022) developed a two-degree-of-freedom sweet potato transplanting robot, both aiming to enhance transplanting success rates and increase yield. Moreover, indoor agriculture robots have emerged as indispensable tools in controlled environments like greenhouses, vertical farms, and hydroponic facilities. These robots can execute a range of tasks, including seeding, watering, monitoring, and harvesting, contributing to optimal space utilization, precise environmental control, and the production of high-quality yields (Bagagiolo et al., 2022). Moreover, Yang et al. (2020) created a transplantation robot featuring a triad of degrees of freedom. A subsequent experiment was carried out, demonstrating that the transplantation robot accomplished an impressive success rate of 95.13%. Jin et al. (2020) introduced a sophisticated control method employing manipulators for hydraulic transplanting robots. An innovative control approach for the transplanting manipulator is introduced by Jin et al. (2020), utilizing a fuzzy PID control strategy. The outcomes demonstrated that the application of fuzzy PID control effectively addressed the inherent challenges of nonlinearity and limited control precision within the hydraulic seedling picking-up system. Moreover, this implementation bolstered the system's stability and dynamic capabilities. These innovations led to enhanced precision and stability in the transplantation process. The integration of robotic technologies in horticulture holds promise for further advancements in the industry, laying the way for more efficient and sustainable agricultural practices in the future.

13.2.2 ROBOTS FOR DETECTION OF DISEASE AND PEST

Significant advancements have been made in leveraging IoT and ML within the realm of plant pathology. Diverse technologies such as sensors, drones, robots, and intelligent monitoring systems have found extensive utility in this field. Notably, the use of computer vision-based phenotyping has emerged as a promising method for identifying and gauging plant diseases, as well as quantifying the intensity of plant stress. These breakthroughs have gained notable traction, particularly in horticultural crops, as highlighted by the work of Prabha et al. (2022). The swift and efficient identification of plant diseases or pests is crucial for minimizing their negative effects. Presently, utilizing AI algorithms to identify plant diseases and pests in real-world scenarios with varying conditions is a particularly complex challenge. Traditionally, researchers create algorithms trained on meticulously controlled exposure images, which often fall short when applied in actual field conditions. Moreover, processing real-time video demands specialized computing resources located near the central point of irrigation management (Debauche et al., 2020).

There are numerous numbers of studies that highlight the potential of IoT and AI in the diagnosis of pest and diseases. For instance, Khan & Srivastava (2023) invented a PlantDoc for the detection of disease in horticultural gardens. It employs an AI model built upon the CNN algorithm from the realm of DL. This model is educated using diverse images of plant leaves, enabling it to discern plants that exhibit defects. The primary function of PlantDoc revolves around detecting diseases in plants. Moreover, a study (Ramya & Mohana, 2021) underscores the technique employed to differentiate plant diseases from leaves, which involves a process known as plant disease recognition. The key motivation behind identifying these plant ailments is to enhance both the quality and quantity of crop yields, thereby reducing the necessity for pesticides and cutting down crop-related expenses in the agricultural field. Furthermore, in a research carried out by Mahmud et al. (2018), the identification of strawberry powdery mildew was executed through the utilization of image texture analysis. The ailment was categorized by employing support vector machines (SVM) and k-nearest neighbor algorithms. The outcomes illustrated that SVM achieved the highest overall precision of 91.86% when categorizing the powdery mildew disease. In essence, the convergence of IoT, ML, and innovative methodologies is propelling the field of plant pathology forward, offering transformative solutions for disease detection, and stress quantification, and ultimately, bolstering the agricultural sector's productivity and sustainability.

13.2.3 PESTICIDE SPRAYING ROBOT

Spraying pesticides on horticultural crops has raised environmental concerns due to excessive use. To tackle this issue, considerable efforts have been invested in designing pesticide spraying robots with advanced technologies, servo-operated sprayers, flow management mechanisms, and ultrasonic detectors. The research community has shown great interest in developing these robots to enhance efficiency and reduce environmental impact in pesticide application. To overcome the challenges posed by large-scale spraying tasks, multirobot systems have been explored. Kim & Son (2020) formulated a task distribution framework utilizing a Voronoi diagram for an orchard-based multirobot spraying system. Their study demonstrated the feasibility of employing multiple robots to improve spraying performance in vast agricultural settings. Cantelli et al. (2019) addressed the issue of spraying accuracy and safety by developing an autonomous spraying robot with a vehicle and a spraying control system. The successful collaboration between these two components was demonstrated through experimental tests, highlighting the robot's enhanced precision and safety during spraying operations. Seol et al. (2020) introduced a smart spraying robot, incorporating a flow control system based on semantic segmentation. In field experiments, this system outperformed existing control methods, proving its efficiency and superiority in pesticide spraying tasks.

Furthermore, Kim et al. (2020) introduced a robotic spraying system for orchards that utilized the SegNet model for semantic segmentation, combined with depth data fused with trained RGB data. The system's hardware configuration and smart spraying capabilities were evaluated in field experiments, showing its environmentally friendly and satisfactory performance. Additionally, Hejazipoor et al. (2021) presented a four-degree-of-freedom robot for spraying, using plant bulk volume calculations. The robot moved along plant rows, stopping at plant stems, and gradually opening its manipulator to capture deep RGB1 colored images. After calculating the plant's bulk volume, precise spraying operations were initiated, ensuring optimal quality throughout the plant's height. The creation of an economical agricultural robot designed for spraying fertilizers and pesticides in horticultural fields and conducting general crop monitoring. The prototype includes a mobile-based two-wheeled robot featuring a spraying mechanism, wireless movement control, and a camera for monitoring crop health, growth, and detecting pests within orchards (Ghafar et al., 2023). These studies collectively contribute to the ongoing efforts to develop advanced robotic spraying systems that address environmental concerns, improve efficiency, and ensure accurate pesticide application in horticultural practices.

13.2.4 GARDENING ROBOTS

Gardening robots offer more than just basic tasks like planting, weeding, and watering. They can play an important role in monitoring and maintaining plant health by integrating sensors to measure essential parameters like humidity, temperature, and light levels. This continuous monitoring enables the robots to provide real-time data and alerts to gardeners, empowering them to take timely actions to prevent plant diseases or nutrient deficiencies (Zhang et al., 2022). Employing appropriate control units like Raspberry Pi controllers or Arduino boards, gardening robots have the potential to alleviate the arduous tasks and responsibilities of farming and gardening. This technology proves invaluable to farmers and plant enthusiasts who are dedicated to nurturing their plants and staying informed about their growth and well-being.

Several studies have demonstrated the potential of robotics in gardening. Like, Tangtisanon et al. (2019) showcased the capabilities of compact gardening robots designed to assist users in small garden tasks, complemented by an automated watering system, which can be controlled remotely. Gravalos et al. (2019) presented the development of a versatile robotic irrigation system featuring a mobile bridge manipulator and a sensor-based platform, offering adaptability and expandability through open-source components and software. Indoor gardening received special attention as Lamsen et al. (2020) designed a robotic irrigation system tailored specifically for indoor gardens. The system's intelligent design utilizes an Arduino microcontroller to regulate water flow based on moisture sensor data. Additionally, the system addresses the need for fertilization with the development of an automatic fertilizer sprayer. Furthermore, Zhang et al. (2022) focused on jujube cultivation and designed a pruning manipulator with five degrees of freedom. The subsequent performance test demonstrated exceptional automated capabilities, showing minimal positioning error and an average success rate of over 85.16%. Rhaman et al. (2022) presented the Line Follower Gardener Robot, a cost-effective solution based on Arduino. This robot navigates along the designated path encircling the garden until it reaches the desired destination. Once the water supply is depleted, the robot autonomously returns to its initial position, ready for the next set of instructions. These studies collectively highlight the significant potential of robotic technology in modern gardening practices, revolutionizing the way we care for and maintain our plants with greater precision and efficiency.

13.2.5 ROBOTIC SYSTEMS FOR MATURITY DETECTION

Robotic systems designed for the assessment of fruit maturity have found diverse applications across industries, notably in horticulture and food processing. These systems play a pivotal role in determining the readiness of products like fruits, vegetables, and other perishable items. By harnessing advanced technologies, these systems accurately pinpoint the ideal time for harvest, thereby enhancing product quality, minimizing wastage, and optimizing overall production efficiency. Numerous researchers are exploring the potential of developing AI-driven tools based on the IoT to detect fruit maturity. For example, one study delved into the realm of computer vision and image processing techniques to differentiate mature and transitioning tomato fruits (Al-Mashhadani et al., 2020). In a similar vein, another research effort by Zhou et al. (2021) introduced an innovative real-time approach within a natural greenhouse setting, enabling the identification and assessment of tomato fruit maturity at a rate of five to six frames per second using a scouting robot. In a different line of investigation, Sahu et al. (2017) conducted a study focusing on mango fruit maturity and defect identification through digital image analysis. This method involves capturing individual images of mangoes from a single angle to detect defects and evaluate the fruit's maturity. The mangoes are categorized into two groups based on a quality ratio, wherein if the ratio surpasses a predetermined threshold, the fruit is classified as spoiled, while a lower quality ratio identifies the fruit as sound. This algorithmic approach allows for the systematic classification of mango fruits based on their quality, thus enhancing their value in the production process. Additionally, researchers have integrated cutting-edge techniques to address the challenges of fruit detection and localization. In an illustrative study by He et al. (2021), the fusion of an object detection framework, YOLOv4, with a classification model, AlexNet, led to a robust method for identifying and precisely pinpointing strawberries. This technique shows substantial potential for accurate strawberry detection, particularly for robotic harvesters equipped with a cooperative dual-manipulator system. Furthermore, fruit maturity assessment has been facilitated through the deployment of the mask region-based convolutional neural network object detection algorithm, as evidenced by the work of Hsieh et al. (2021). This algorithmic innovation further contributes to the arsenal of tools available for effective and efficient fruit maturity evaluation. These developments and integrations of robotic systems and AI-driven tools, such as computer vision algorithms and DL models, have opened new avenues for accurately assessing fruit maturity. These advancements hold significant promise for industries seeking to enhance quality control, reduce waste, and optimize production processes within the realm of horticulture and food processing.

13.2.6 PICKING ROBOTS/ORCHARD AND VINEYARD ROBOTS

In the realm of fruit trees and grapevine cultivation, specialized robots have been designed to overcome unique challenges. These robots are proficient in tasks such as pruning, canopy management, and fruit thinning. Utilizing advanced vision systems and delicate manipulators, they can precisely identify and harvest ripe fruits without causing harm to the plant, leading to increased yield and reduced labor costs. The implementation of machine vision and its corresponding algorithms enhances the efficiency, functionality, intelligence, and remote interaction capabilities of harvesting robots in intricate horticultural settings (Tang et al., 2020). Numerous studies have been dedicated to developing and evaluating specific robotic harvesting systems. For instance, Xiong et al. (2019) introduced an autonomous strawberry-harvesting robot, equipped with an obstacle-separation algorithm and an improved vision system to efficiently pick strawberries in cluttered environments. Likewise, Arad et al. (2020) successfully field-tested a sweet pepper harvesting robot in a commercial greenhouse, achieving an average cycle time of 24 s per fruit harvest. Perception and detection algorithms play a pivotal role in the effectiveness of robotic harvesting. Lehnert et al. (2017) highlighted the challenges posed by changing lighting conditions, crop variability, and occlusions encountered in robotic harvesting tasks. Kuznetsova et al. (2020) implemented the YOLOv13 algorithm for apple detection in a fruit-harvesting robot, significantly improving the robot's ability to perceive and identify apples. Efforts are also being made to enhance the autonomy and efficiency of robotic harvesting systems. Yang et al. (2022) discussed the utilization of reinforcement learning for intelligent path planning in agricultural robots, while Kiktev et al. (2022) developed a control system to increase the level of autonomy in a robotic platform designed for apple picking. Furthermore, Kim et al. (2022) made significant contributions to achieving autonomous tomato harvesting in controlled horticulture environments. Their work introduced crucial technologies, integrating AI and machine vision to create Deep-ToMaToS, a DL network explicitly engineered to accurately identify and classify ripe tomatoes at the optimal harvesting time. Moreover, a study by Tang et al. (2020) explores the integration of stereo vision technology with harvesting robots. The research primarily focuses on crop identification and localization, as well as the cooperative behavior control of the robot manipulator, and error processing using stereo vision systems. The system employs a set of images to identify the crops and extract relevant information about the target. In study by Park et al. (2021) introduced a holistic system encompassing IoT, edge computing, AI, and cloud technologies to oversee the environmental parameters in hydroponic strawberry cultivation. This unified system not only acquires and retains data pertaining to the strawberry's growth conditions but also offers graphical representations of the collected information. Furthermore, they propose a harvest decision mechanism that utilizes a DL algorithm for categorizing the maturity stages of strawberries via image analysis. These advancements in robotic harvesting technologies hold great promise for revolutionizing the agricultural industry by increasing efficiency, reducing labor requirements, and ensuring higher quality harvests.

13.2.7 PACKAGING AND GRADING

Optimizing packaging and grading in horticulture through the integration of robotics and AI is a crucial aspect of enhancing efficiency and quality in the industry. Several studies have explored the potential of these technologies in this context, showcasing their significance in modern agricultural practices. For instance, a study by Mahawar et al. (2019) centered around developing a mass model for kinnow mandarin fruits utilizing physical characteristics, enabling the development of mathematical models to predict fruit mass. This information can be used to optimize packaging formations and enhance marketability. Dewi et al. (2020) demonstrated the use of an arm robot as a sorting and grading system, relying on image processing techniques to recognize the color and size of fruits like tomatoes and grapes. This approach offers a more automated and efficient sorting process for commercial applications. Birania et al. (2022) explored the use of AI and robotics in grading strawberries by developing mass models based on physical attributes. These models aid in optimizing packaging and reducing wastage, contributing to improved efficiency in strawberry handling.

Wang et al. (2023) developed a prototype for a peach packaging robot utilizing DL, which significantly improved positioning accuracy and grasping robustness. The successful implementation of this robot suggests potential applicability in postharvest commercialization. Kumari et al. (2022) introduced an intelligent automation system specifically tailored for grading Kesar mangoes. By integrating advanced methods such as discrete Fourier transform, local binary pattern, gray-level co-occurrence matrix, and shape feature extraction, the system achieves precise and accurate classification, streamlining the grading process. Additionally, Rizwan & Hakim (2022) utilized DL to revolutionize mango variety classification and quality assessment. The implementation of CNNs enabled automatic learning of crucial image features, leading to a more efficient and accurate grading process for mangoes. Another research presents a novel approach wherein the researchers introduced a robot system seamlessly integrating AI capabilities for the purpose of grading and sorting garlic post-root trimming. This developed robotic system employs an advanced DL model, specifically a DCNN, for the automatic grading and sorting of root-trimmed garlic by analyzing images. Together, these studies underscore the importance of AI and robotics in the horticulture industry, offering innovative solutions to enhance efficiency, improve product quality, and optimize postharvest processes. The biggest advantage of using IoT in horticulture is enhanced efficiency. By automating postharvest tasks, farmers can save time and resources. This not only reduces labor costs but also helps to optimize the quality of the produce.

13.3 CHALLENGES AND CONCERNS

The execution of IoT in horticultural extension offers numerous advantages, but it also introduces several challenges and concerns that must be thoroughly addressed the deployment of IoT in extension advisory services transforms the society. It covers the possible loss of human connection and knowledge, as well as the need for efficient data administration and the safeguarding of personal information. We are required to consider ethical, social, and philosophical dimensions. In addition to this, it investigates the significance of farmer acceptance and implementation of IoT technologies. However, the adoption of IoT in agriculture also presents challenges. Security issues in IoT agriculture systems need to be addressed to protect against cyber threats (Farooq et al., 2019). Additionally, there are open research issues and challenges in the field of IoT agriculture, such as the development of standardized regulations and policies, as well as the integration of AI and blockchain technologies for improved control, management, and security in supply-chain networks (Farooq et al., 2019; Ahmed et al., 2022). However, the adoption of IoT in horticulture and other industries is not without challenges. Some of the challenges include the complexity of integrating different IoT devices and platforms, ensuring data security and privacy, and addressing the digital divide and access to technology (Ben-Daya et al., 2017). Additionally, the adoption of IoT requires investment in infrastructure, training, and technical expertise (Tarmizi et al., 2020).

Situations of conflicting interests can also emerge when agricultural advisory services are linked to agri-input businesses (Wan et al., 2019). This can potentially compromise the impartiality and objectivity of the advice provided to farmers, as dehumanized bots advisors may be programmed by commercial interests. In India, the ratio of extension professionals and farmers gap is very wide. The impact of closed access to research publications extends beyond the research community and can limit the availability of information and knowledge to practitioners, policymakers, and stakeholders in the horticulture industry, hindering evidence-based decision-making and the adoption of innovative practices (Hughes & Foulkes, 2022). The diverse array of IoT devices available in the market often use different communication protocols and standards. Integrating and ensuring smooth communication among these devices can be complex and may require additional development efforts (Rana et al., 2022). Balancing the need for business linkages while ensuring unbiased advice is a challenge that extension services must navigate. Dehumanized horticultural advisory services can render socio-political factors invisible or insignificant. By ignoring the power dynamics, local contexts, and social realities, extension programs will fail to address the diverse needs and aspirations of farmers. This rendering of technical knowledge neglects the complexities of horticulture as a socio-economic system, limiting the potential for holistic development. Furthermore, the perception of smallholder farmers regarding the support provided by extension and advisory officers can be a challenge. In some cases, farmers may perceive that the support provided is inadequate or does not address their specific needs (Nyawo & Mubangizi, 2021). This highlights the importance of understanding and addressing the expectations and concerns of farmers to ensure the relevance and effectiveness of extension services. The substantial data output from IoT devices can become overwhelming systems not equipped to handle it. Effective data processing, storage, and analysis are necessary to extract meaningful insights without compromising system performance

(Sharma et al., 2022). "Widespread concerns regarding automation contributing to rising unemployment, although comprehensible, raise doubts and generally lack support from historical trends. Overall, automation serves to mitigate labor scarcities, enhancing agricultural production's resilience and efficacy. It aids in elevating product quality, optimizing the utilization of resources, fostering decent employment opportunities, and advancing environmental sustainability. Additionally, automation is poised to decrease positions tied to routine tasks like planting and harvesting while bolstering demand for skilled roles necessitating higher education, for instance. In nations characterized by substantial rural labor forces, this employment shift could potentially deepen existing inequalities. Tackling these challenges necessitates the reduction of barriers to the adoption of automated solutions, particularly among small-scale producers, women, and youth. This ensures that automated innovations are neutral in terms of scale, rendering them accessible to agricultural producers of varying sizes—from small to large. Achieving these objectives mandates technological breakthroughs that customize automation to suit the circumstances of small-scale producers."

One of the major challenges of implementing IoT in horticulture is the issue of data privacy. With the vast amounts of data being collected by sensors and other devices, there is a risk that sensitive information about crops, yields, and other factors could be exposed. This could lead to serious consequences for growers, including loss of competitive advantage and damage to their reputation. To counteract this potential hazard, it is crucial to establish resilient security protocols, including encryption and access controls, to safeguard sensitive data against unauthorized entry. Dehumanizing horticulture through robotics technology offers numerous benefits that cannot be ignored. One of the most significant advantages is the increased efficiency of the process. Robots can work tirelessly and without rest, which means they can cover more ground and complete tasks faster than human workers. This results in reduced labor costs and higher productivity levels, ultimately leading to improved profitability for businesses. Another benefit of dehumanizing horticulture is the potential for improved crop yields. Robots are designed to perform specific tasks with precision, which means they can identify and address issues with crops much more quickly than humans. This leads to healthier plants and better yields, which is good news for farmers and consumers alike.

One of the main concerns associated with dehumanizing horticulture via robotics is job displacement. As more and more tasks are automated, there is a risk that human workers may be replaced by machines. This could lead to unemployment and economic instability in some regions. However, proponents argue that this approach could also create new jobs in fields such as robotics engineering and maintenance. Another concern is the potential negative environmental impact of using robots in agriculture. Critics argue that increased use of machinery could lead to greater carbon emissions and other forms of pollution. However, supporters point out that the use of precision robotics can reduce the number of resources needed for farming, leading to a more sustainable and eco-friendly approach. Another significant risk associated with IoT in horticulture is the threat of cyber-attacks. With an increasing number of devices being linked to the internet, they become vulnerable to hacking and other forms of cybercrime. This could result in physical damage to crops or equipment, as well as theft of valuable data. To address this risk, it is essential to have a comprehensive cybersecurity strategy in place. This should encompass routine security evaluations, employee training initiatives, and the utilization of sophisticated tools for detecting and responding to threats. Tackling these challenges necessitates a comprehensive approach involving technological innovation, policy development, and stakeholder engagement. By effectively managing these concerns, the integration of IoT in horticulture can unlock substantial benefits while minimizing risks.

13.4 STRIKING A BALANCE

This section places an emphasis on the significance of maintaining a healthy equilibrium between the progression of technology and the engagement of real people in extension advisory services. The exclusion of human, socio-political, and economic factors from horticultural extension advisory services discussions calls for a broader understanding of the extension discourse that considers the complex realities of rural communities. Researchers should reflect on the balance between technological advancements and preserving the essence of what it means to be human. IoT-based world could further marginalize farmers, stakeholders who do not have access to technology, are wealthy and educated. For example, open source in horticulture plays a crucial role in promoting transparency, collaboration, and innovation. Mawikere (2023) discussed the environmentally friendly preservation of horticultural plants in the Bintuni Bay Regency, emphasizing the need for open access to information and collaboration among stakeholders to ensure the preservation of plant diversity. Lee et al. (2022) emphasized the significance of open-source platforms and data sharing in the evaluation and selection of tomato cultivars for organic cultivation in greenhouses. Petrova et al. (2018) mentioned the importance of open-source data and information sharing in understanding the spread of plant diseases, such as *Alternanthera mosaic potexvirus*. Open access to data facilitates research and helps in developing effective management strategies, enabling researchers to access and analyze data, leading to informed decision-making. Open access to information, data sharing, and the use of open-source technologies contribute to the advancement of knowledge, research, and sustainable practices in horticulture. Researchers should explore the intersection of technology and human augmentation, contemplating the possibilities of enhanced cognition, longevity, and human–machine symbiosis.

13.5 FUTURE PROSPECTS

The prospects of using AI in horticulture are promising and offer numerous opportunities for improving various aspects of the industry. One potential application of AI in horticulture is the use of DL approaches for recognizing images and assessing the quality of horticultural products (Yang & Xu, 2021). This can help farmers and orchard owners in identifying and assessing the quality of their produce more efficiently. Moreover, AI can find application in early detection systems for identifying plant stress and predicting yields, empowering farmers to implement proactive measures to avert crop damage and enhance their harvest optimization. (Yang & Xu, 2021).

The widespread adoption and commercialization of AI technologies, such as the Qualcomm Neural Processing SDK, are expected to make portable intelligent devices and mobile applications for DL more accessible to ordinary farms and orchards soon (Yang & Xu, 2021). This would enable farmers to leverage AI capabilities directly on their mobile devices, facilitating real-time decision-making and enhancing productivity. However, there are challenges when implementing AI algorithms in real-time field applications, notably linked to the comparatively limited image processing capabilities of edge devices (Zhang et al., 2021). To address this, researchers have proposed lightweight fruit-detection algorithms specifically designed for edge devices, ensuring real-time detection while maintaining accuracy (Zhang et al., 2021). This development is crucial for the practical application of AI in horticulture, as it allows for on-site and immediate analysis of plant-related data.

The use of AI in horticulture has mainly favored extensive industrial farming operations, exhibiting a focus on commodity crops and high-value horticulture crops (Williamson et al., 2021). However, advancements in AI methods and the increasing quantity and nature of collected data offer opportunities to address new questions and challenges in plant science and agriculture (Williamson et al., 2021). In terms of data management, the integration of AI in horticulture research and practice requires effective strategies for handling and analyzing large volumes of data (Williamson et al., 2021). The utilization of AI methods, like ML and DL, relies heavily on high-quality and well-curated datasets. Therefore, data management challenges require attention for comprehensive resolution harness the potential of AI in horticulture (Williamson et al., 2021). This opens possibilities for AI to contribute to the sustainable production and consumption of fruits and vegetables, aligning with the United Nations' Sustainable Development Goals (Singh et al., 2022). The future of horticulture is here, and it is all thanks to IoT. Through the assistance of interconnected devices and intelligent sensors, we can now monitor and manage crops in ways that were once impossible. This technology holds the capacity to transform the sector, enhancing its efficiency and sustainability, and profitable than ever before. However, the benefits of IoT go beyond just improving crop yields and reducing costs. It also allows us to collect large quantities of data pertaining to factors like soil moisture levels to weather patterns, giving us insights that were previously impossible to obtain. These data can be utilized to optimize growing conditions, predict production, and even detect and prevent diseases before they spread.

13.6 CONCLUSION

The IoT is a revolutionary notion that encompasses an extensive network of interconnected devices engineered to collect and share data. In the realm of horticulture, the integration of IoT technologies offers an exciting opportunity to revolutionize various facets of agricultural practices. From irrigation management to crop supervision and even supply chain logistics, the application of IoT has the potential to yield a multitude of advantages. One of the standout benefits is the significant enhancement in operational efficiency. By harnessing real-time data from IoT sensors placed strategically throughout farms, growers can make informed decisions about irrigation scheduling, nutrient application, and pest control, resulting in optimized resource utilization and healthier production. Furthermore, IoT implementation in horticulture contributes to heightened product traceability and accountability. Through the continuous monitoring of environmental conditions and cultivation practices, growers can establish a comprehensive digital record of each crop's journey from seed to harvest. This not only enhances consumer trust by offering insights into the origin and handling of produce but also assists in adhering to rigorous industry regulations and quality standards. Despite the immense potential, the widespread adoption of IoT in horticulture is not without its challenges. Ensuring seamless integration of diverse hardware and software components is a complex task that demands careful planning and execution. Additionally, safeguarding the enormous volume of data produced by IoT devices against cyber threats and unauthorized access is of paramount importance. To fully embrace the benefits of IoT, the agricultural community must address these security concerns through robust encryption protocols, multilayer authentication, and continuous monitoring. Turning our attention to the future, the prospects of AI in horticulture hold great promise. AI-powered algorithms can play a pivotal role in image recognition, facilitating the automated identification of pests, diseases, and anomalies in crops. This technology has the potential to revolutionize quality evaluation processes, enabling growers to assess the condition of their produce and make informed decisions about postharvest handling and distribution swiftly and accurately. Moreover, AI-driven early detection systems stand as a powerful tool in the fight against crop diseases. By analyzing data from IoT sensors and satellite imagery, these systems can detect subtle changes in plant health, enabling prompt intervention and mitigation measures. Additionally, AI algorithms can leverage historical and real-time data to predict crop yields, empowering farmers to better anticipate market demand and manage their resources effectively. The commercialization of AI technologies and the development of lightweight algorithms tailored for edge devices represent significant steps toward democratizing AI usage in horticulture. As these advancements progress, farmers of varying technological backgrounds can harness the benefits of AI without requiring extensive technical expertise. However, the integration of AI is not devoid of challenges. Data management, including collection, storage, and analysis, requires robust infrastructure and protocols to ensure data accuracy and security. Furthermore, the successful incorporation of AI methods demands a certain level of technical knowledge and training among farmers and agricultural professionals. The convergence of IoT and AI technologies offers a transformative pathway for the horticulture industry. With careful consideration of challenges and a proactive approach to addressing them, growers can unlock unprecedented efficiency, precision, and sustainability in their practices, contributing to a more resilient and productive agriculture.

KEYWORDS

- **Internet of Things**
- **machine learning**
- **deep learning**
- **extension advisory services**
- **horticulture**

REFERENCES

Agholor, A. (2019). Gender gap in sub-Saharan Africa, Reminiscence of rural extension and advisory services: Delineation, challenges and strategies. *South African Journal of Agricultural Extension*, 47(3). https://doi.org/10.17159/24113-13221/2019/v47n13a514

Ahmed, R., Hemdan, E., El-Shafai, W., El-Rabaie, E., & El-Samie, F. (2022). Climate-smart agriculture using intelligent techniques, blockchain and internet of things: Concepts, challenges, and opportunities. *Transactions on Emerging Telecommunications Technologies*, 33(11) e4607. https://doi.org/10.1002/ett.4607

Al-Mashhadani, Z., & Chandrasekaran, B. (2020). Autonomous ripeness detection using image processing for an agricultural robotic system. In: *2020 11th IEEE Annual Ubiquitous Computing, Electronics & Mobile Communication Conference (UEMCON)*. IEEE, pp. 07413–0748.

Arad, B., Balendonck, J., Barth, R., Ben-Shahar, O., Edan, Y., Hellström, T., & Tuijl, B. (2020). Development of a sweet pepper harvesting robot. *Journal of Field Robotics*, 37(6), 1027–10139. https://doi.org/10.1002/rob.219137

Baffoe-Bonnie, A., Martin, D., & Mrema, F. (2021). Agricultural extension and advisory services strategies during Covid-19 lockdown. *Agricultural & Environmental Letters*, 6(4) e20056. https://doi.org/10.1002/ael2.20056

Bagagiolo, G., Matranga, G., Cavallo, E., & Pampuro, N. (2022). Greenhouse robots: Ultimate solutions to improve automation in protected cropping systems—A review. *Sustainability,* 14(11), 64136.

Ben-Daya, M., Hassini, E., & Bahroun, Z. (2017). Internet of things and supply chain management: A literature review. *International Journal of Production Research,* 57(15–16), 4719–4742. https://doi.org/10.1080/002075413.2017.1402140

Birania, S., Attkan, A.K., Kumar, S., Shukla, A., & Singh, V.P. (2022). Mass modeling of strawberry (Fragaria × Ananasa) based on selected physical attributes. *Journal of Food Process Engineering*, 45, e14023. https://doi.org/10.1111/jfpe.140213

Cantelli, L., Bonaccorso, F., Longo, D., Melita, C.D., Schillaci, G., & Muscato, G. (2019). A small versatile electrical robot for autonomous spraying in agriculture. *AgriEngineering* , 1, 1391–402.

Coble, K.H., Mishra, A.K., Ferrell, S., & Griffin, T. (2018). Big data in agriculture: A challenge for the future. *Applied Economic Perspectives and Policy*, 40, 79–96.

Davis, T., & Bost, E. (2017). Emerging horticultural research and education opportunities in Myanmar. *HortScience*, 52(9), 1141–1144. https://doi.org/10.212713/hortsci11904-17

Debauche, O., Mahmoudi, S., Elmoulat, M., Mahmoudi, S.A., Manneback, P., & Lebeau, F. (2020). Edge AI-IoT pivot irrigation, plant diseases, and pests identification. *Procedia Computer Science*, 177, 40–48.

Dewi, T., Risma, P., & Oktarina, Y. (2020). Fruit sorting robot based on color and size for an agricultural product packaging system. *Bulletin of Electrical Engineering and Informatics*, 9(4), 14138–1445.

Eyssartier, C., Ladio, A.H., & Lozada, M. (2008). Cultural transmission of traditional knowledge in two populations of North-western Patagonia. *Journal of Ethnobiology and Ethnomedicine*, 4, 25. https://doi.org/10.1186/1746-4269-4-25

Farooq, M., Riaz, S., Abid, A., Abid, K., & Naeem, M. (2019). A survey on the role of Iot in agriculture for the implementation of smart farming. *IEEE Access*, 7, 1562137–156271. https://doi.org/10.1109/access.2019.29497013

Fatty, L.K.M., Ode, I.O., & Ahule, B.G. (2021). Horticultural farmers' access to agricultural extension information on post-harvest technology in Kombo Central and North, West Coast region, the Gambia. *International Journal of Advanced Economics*, 2(13), 26–139. https://doi.org/10.51594/ijae.v13i2.229

Ghafar, A.S.A., Hajjaj, S.S.H., Gsangaya, K.R., Sultan, M.T.H., Mail, M.F., & Hua, L.S. (2023). Design and development of a robot for spraying fertilizers and pesticides for agriculture. *Materials Today: Proceedings*, 81, 242–248.

Gravalos, I., Avgousti, A., Gialamas, T., Alfieris, N., & Paschalidis, G. (2019). A robotic irrigation system for urban gardening and agriculture. *Journal of Agricultural Engineering*, 50(4), 198–207.

He, Z., Karkee, M., & Upadhayay, P. (2021). Detection of strawberries with varying maturity levels for robotic harvesting using YOLOv4. In: *2021 ASABE Annual International Virtual Meeting*. American Society of Agricultural and Biological Engineers, p. 1.

Hejazipoor, H., Massah, J., Soryani, M., Vakilian, K.A., & Chegini, G. (2021). An intelligent spraying robot based on plant bulk volume. *Computers and Electronics in Agriculture*, 180, 105859.

Hsieh, K.W., Huang, B.Y., Hsiao, K.Z., Tuan, Y.H., Shih, F.P., Hsieh, L.C., Chen, S., & Yang, I.C. (2021). Fruit maturity and location identification of beef tomato using R-CNN and binocular imaging technology. *Journal of Food Measurement and Characterization*, 15(6), 5170–5180. https://doi.org/10.12688/f1000research.52204.1

Hughes, K., & Foulkes, J. (2022). Reducing environmental impacts at the royal botanic garden Edinburgh. *Sustainability,* 14(14), 87913. https://doi.org/10.131390/su141487913

Hussain, F., Hassan, S.A., Hussain, R., & Hossain, E. (2020). Machine learning for resource management in cellular and IoT networks: Potentials, current solutions, and open challenges. *IEEE Communications Surveys & Tutorials*, 2(22), 1251–1275. https://doi.org/10.1109/comst.2020.29645134

Jin, X., Chen, K., Zhao, Y., Ji, J., & Jing, P. (2020). Simulation of hydraulic transplanting robot control system based on fuzzy PID controller. *Measurement,* 164, 1080213.

Kaburuan, E.R., & Jayadi, R. (2020). A design of IoT-based monitoring system for intelligence indoor micro-climate horticulture farming in Indonesia. *Procedia Computer Science*, 157, 459–464.

Khadatkar, A., Pandirwar, A., & Paradkar, V. (2023). Design, development and application of a compact robotic transplanter with automatic seedling picking mechanism for plug-type seedlings. *Scientific Reports*, 113(1). https://doi.org/10.10138/s41598-0213-28760-4

Khan, A., & Srivastava, A. (2023). PlantDoc-plant disease detection using AI. *Journal of Informatics Electrical and Electronics Engineering (JIEEE)*, 4(1), 1–10.

Kiktev, N., Jewiarz, M., Khort, D., Smirnov, I., Zubina, V., Hutsol, T., Tomasik, M., & Biliuk, M. (2022). Robotic platform for horticulture: Assessment methodology and increasing the level of autonomy. *Sensors*, 22(22), 8901. https://doi.org/10.131390/s22228901

Kim, J., Pyo, H., Jang, I., Kang, J., Ju, B., & Ko, K. (2022). Tomato harvesting robotic system based on Deep-ToMaToS: Deep learning network using transformation loss for 6D pose estimation of maturity classified tomatoes with side-stem. *Computers and Electronics in Agriculture*, 201, 1071300.

Kim, J., & Son, H. (2020). A Voronoi diagram-based workspace partition for weak cooperation of multi-robot system in Orchard. *IEEE Access*, 8, 20676–20686. https://doi.org/10.1109/access.2020.2969449

Kim, J., Seol, J., Lee, S., Hong, S.W., & Son, H.I. (2020). An intelligent spraying system with deep learning-based semantic segmentation of fruit trees in orchards. In: *Proceedings of the 2020 IEEE International Conference on Robotics and Automation (ICRA)*, pp. 139213–13929.

Kitinoja, L., Saran, S., Roy, S.S., & Kader, A.A. (2011). Postharvest technology for developing countries: Challenges and opportunities in research, outreach and advocacy. *Journal of the Science of Food and Agriculture*, 4(91), 597–6013. https://doi.org/10.1002/jsfa.4295

Kondratieva, O.V., Fedorov, A.D., Fedorenko, V.F., & Slinko, O.V. (2021, March). Using digital technologies in horticulture. In: *IOP Conference Series: Earth and Environmental Science*. IOP Publishing, p. 7213, 013201313.

Kumar, P.C., & Bhatnagar, R. (2020). Social internet of things in agriculture: An overview and future scope. In: *Toward Social Internet of Things (SIoT): Enabling Technologies, Architectures and Applications*. Berlin/Heidelberg, Germany: Springer, pp. 1317–13134.

Kumari, N., Dwivedi, R.K., Bhatt, A.K., & Belwal, R. (2022). Automated fruit grading using optimal feature selection and hybrid classification by self-adaptive chicken swarm optimization: Grading of mango. *Neural Computing and Applications*, 34, 1285–1306.

Kuznetsova, A., Maleva, T., & Soloviev, V. (2020). Using Yolov13 algorithm with pre- and post-processing for apple detection in fruit-harvesting robot. *Agronomy*, 7(10), 1016. https://doi.org/10.131390/agronomy10071016

Lamm, K.W., Lamm, A.J., Dobbins, C.E., & Park, J. (2023). Extension networks and dissemination of horticultural advancements: Development and Validation of a professionalization instrument. *Horticulturae*, 9(2), 245. https://doi.org/10.131390/horticulturae9020245.

Lamsen, F.C., Favi, J.C., & Castillo, B.H.F. (2020). Indoor gardening with automatic irrigation system using arduino microcontroller. *ASEAN International Multidisciplinary Research Journal*, 10, 1131–148.

Lee, G., Jang, Y., Choi, K., H.K., Ji, S., Kim, B., Kwon, Y., & Shin, H. (2022). Occurrence, distribution and risk assessment of PAHs in road dust sediment in Daejeon Metropolitan City, South Korea. *Environmental Engineering Research*, 28(2), 2200413. https://doi.org/10.4491/eer.2022.0413

Lehnert, C., English, A., McCool, C., Tow, A., & Perez, T. (2017). Autonomous sweet pepper harvesting for protected cropping systems. *IEEE Robotics and Automation Letters*, 2(2), 872–879. https://doi.org/10.1109/lra.2017.2655622

Liu, Z., Lv, Z., Zheng, W., & Wang, X. (2022). Trajectory control of two-degree-of-freedom sweet potato transplanting robot arm. *IEEE Access*, 10, 26294–261306.

Luo, A. (2022). Intrusion detection system for internet of vehicles based on ensemble learning and CNN. *Journal of Physics: Conference Series*, 2414(1), 012014. https://doi.org/10.1088/1742-6596/2414/1/012014

Madushanki, A., Halgamuge, M., Wirasagoda, W., & Syed, A. (2019). Adoption of the internet of things (IoT) in agriculture and smart farming towards Urban Greening: A review. *International Journal of Advanced Computer Science and Applications*, 10(4), 11–28. https://doi.org/10.14569/ijacsa.2019.0100402

Mahawar, M.K., Bibwe, B., Jalgaonkar, K., & Ghodki, B.M. (2019). Mass modeling of kinnow mandarin based on some physical attributes. *Journal of Food Process Engineering*, 42(5), e13079. https://doi.org/10.1111/jfpe.113079

Mahmud, M.S., Chang, Y.K., Zaman, Q.U., Esau, T.J., Price, G.W., & Prithiviraj, B. (2019). Development of an artificial cloud lighting condition system using machine vision for strawberry powdery mildew disease detection. *Computers and Electronics in Agriculture*, 158, 219–225.

Mawikere, N.L., Irawati, O.P., Taberima, S., Fenetiruma, O.A., & Prabawardani, S. (2023). The lowland area for sustainable conservation of horticultural plants in Bintuni Bay Regency. *IOP Conference Series: Earth and Environmental Science*, 1192(1), 012015. https://doi.org/10.1088/1755-11315/1192/1/012015

Nagaraju, M., & Chawla, D. (2019). IoT implementation and management for smart farming. *International Journal of Innovative Technology and Exploring Engineering*, 8(10), 24813–2491. https://doi.org/10.135940/ijitee.j9545.0881019

Nyawo, J., & Mubangizi, B. (2021). Perception of agricultural smallholder farmers on public extension and advisory services in Ethekwini municipality, Kwazulu-natal Province. *Journal of Southwest Jiaotong University*, 56(3), 499–518. https://doi.org/10.135741/issn.0258-2724.56.13.42

Olorunfemi, O., & Oladele, O. (2018). Knowledge of extension agents regarding professionalization of extension services: Evidence from Southwest Nigeria. *Journal of International Agricultural and Extension Education*, 25(3), 1313–1350. https://doi.org/10.5191/jiaee.2018.2513013

Park, S., & Kim, J. (2021). Design and implementation of a hydroponic strawberry monitoring and harvesting timing information supporting system based on nano AI-cloud and IoT-edge. *Electronics*, 10(12), 1400.

Paudel, R., Baral, P., Lamichhane, S., & Marahatta, B. (2018). ICT based agro-advisory services in Nepal. *Journal of the Institute of Agriculture and Animal Science*, 1(135), 21–28. https://doi.org/10.13126/jiaas.v135i1.22510

Petrova, E., Trifonova, E., Nikitin, N., Atabekov, J., & Karpova, O. (2018). Alternanthera mosaic potexvirus: Several features, properties, and application. *Advances in Virology*, 2018, 1973705. https://doi.org/10.1155/2018/19713705

Popoola, S.I., Adebisi, B., Ande, R., Hammoudeh, M., Anoh, K., & Atayero, A.A. (2021). SMOTE-DRNN: A deep learning algorithm for botnet detection in the internet-of-things networks. *Sensors*, 9(21), 2985. https://doi.org/10.131390/s21092985

Postolache, S., Sebastião, P., Viegas, V., Postolache, O., & Cercas, F. (2022). IoT-based systems for soil nutrients assessment in horticulture. *Sensors*, 213(1), 4013.

Prabha, K. (2021). Disease sniffing robots to apps fixing plant diseases: Applications of artificial intelligence in plant pathology—A mini review. *Indian Phytopathology*, 74(1), 113–120.

Raju, K., & Vijayaraghavan, V. (2022). A self-powered, real-time, NRF24l01 IoT-based cloud-enabled service for smart agriculture decision-making system. *Wireless Personal Communication*, 124, 207–2136. https://doi.org/10.1007/s11277-021-09462-4

Ramya, N., & Mohana, S. (2021). A survey-plant leaf disease in horticulture using classification algorithm. *Annals of the Romanian Society for Cell Biology*, 25(4), 101367–101372.

Rana, S., Verma, J., & Gautam, A.K. (2022). A comprehensive study with challenges of internet of things (IoT) based model for smart farming. *International Journal of Education and Management Engineering*, 12, 413–55.

Rizwan Iqbal, H.M., & Hakim, A. (2022). Classification and grading of harvested mangoes using convolutional neural network. *International Journal of Fruit Science*, 22(1), 95–109.

Sahu, D., & Potdar, R.M. (2017). Defect identification and maturity detection of mango fruits using image analysis. *American Journal of Artificial Intelligence*, 1(1), 5–14.

Selvaraj, M.G., Vergara, A., Ruiz, H., Safari, N., Elayabalan, S., Ocimati, W., & Blomme, G. (2019). AI-powered banana diseases and pest detection. *Plant Methods*, 15, 1–11.

Seol, J., Kim, J., & Son, H.I. (2020). Field evaluations of a deep learning-based intelligent spraying robot with flow control for pear orchards. *International Journal on Advances in Precision Agriculture*, 213, 712–7132.

Sharma, D., Chaudhary, A., & Kumar, A. (2022, November). Internet of things (IoT) in agriculture. In: *2022 7th International Conference on Computing, Communication and Security (ICCCS)*. IEEE, pp. 1–4.

Singh, R., Berkvens, R., & Weyn, M. (2021). Agrifusion: An architecture for IoT and emerging technologies based on a precision agriculture survey. *IEEE Access*, 9, 11362513. https://doi.org/10.1109/access.2021.13116814

Singh, R., Singh, R., Gehlot, A., Akram, S.V., Priyadarshi, N., & Twala, B. (2022). Horticulture 4.0: Adoption of industry 4.0 technologies in horticulture for meeting sustainable farming. *Applied Sciences*, 12(24), 12557.

Tang, Y., Chen, M., Wang, C., Luo, L., Li, J., Lian, G., & Zou, X. (2020). Recognition and localization methods for vision-based fruit picking robots: A review. *Frontiers in Plant Science*, 11, 510.

Tangtisanon, P. (2019). Small gardening robot with decision-making watering system. *Sensors & Materials*, 131(6), 1905–1916.

Tarmizi, H.A., Kamarulzaman, N., Rahman, A.A., & Atan, R. (2020). Adoption of internet of things among Malaysian halal agro-food SMEs and its challenges. *Food Research International*, S1(4), 256–265. https://doi.org/10.26656/fr.2017.4(s1).s26

Tumbo, S., Mwalukasa, N., Fue, K., Mlozi, M., Haug, R., & Sanga, C. (2018). Exploring information seeking behavior of farmers' in information related to climate change adaptation through ICT (Chai). *The International Review of Research in Open and Distributed Learning*, 19(3). https://doi.org/10.191713/irrodl.v19i13.13229

Wan, M., Gu, R., Zhang, T., Zhang, Y., Ji, H., Buyun, W., & Toepfer, S. (2019). Conflicts of interests when connecting agricultural advisory services with agri-input businesses. *Agriculture*, 10(9), 218. https://doi.org/10.131390/agriculture9100218

Wang, Q., Wu, D., Sun, Z., Zhou, M., Cui, D., Xie, L., Hu, D., Rao, X., Jiang, H., & Ying, Y. (2023). Design, integration, and evaluation of a robotic peach packaging system based on deep learning. *Computers and Electronics in Agriculture*, 211, 1080113.

Williamson, H., Brettschneider, J., Caccamo, M., Davey, R., Goble, C., Kersey, P., May, S., Morris, R.J., Ostler, R., Pridmore, T., Rawlings, C., Studholme, D., Tsaftaris, S.A., & Leonelli, S. (2021). Data management challenges for artificial intelligence in plant and agricultural research. *F1000Research*, 10, 1324.

Xiong, Y., Ge, Y., Grimstad, L., & From, P. (2019). An autonomous strawberry-harvesting robot: Design, development, integration, and field evaluation. *Journal of Field Robotics*, 2(137), 202–224. https://doi.org/10.1002/rob.21889

Yang, B., & Xu, Y. (2021). Applications of deep-learning approaches in horticultural research: A review. *Horticulture Research*, 8, 123. https://doi.org/10.10138/s414138-021-00560-9

Yang, J., Ni, J., Li, Y., Wen, J., & Chen, D. (2022). The intelligent path planning system of agricultural robot via reinforcement learning. *Sensors*, 12(22), 41316. https://doi.org/10.131390/s221241316

Yang, Q., Jia, C., Sun, M., Zhao, X., He, M., Mao, H., Hu, J., & Addy, M. (2020). Trajectory planning and dynamics analysis of greenhouse parallel transplanting robot. *International Journal of Agricultural Engineering*, 29, 64–76.

Zhang, B., Chen, X., Zhang, H., Shen, C., & Fu, W. (2020). Design and performance test of a jujube pruning manipulator. *Agriculture*, 12, 552.

Zhang, W., Liu, Y., Chen, K., Li, H., Duan, Y., Wu, W., Shi, Y., & Guo, W. (2021). Lightweight fruit-detection algorithm for edge computing applications. *Frontiers in Plant Science*, 12, 740936. https://doi.org/10.131389/fpls.2021.7409136

Zhang, W., Miao, Z., Li, N., He, C., & Sun, T. (2022). Review of current robotic approaches for precision weed management. *Current Robotics Reports*, 3(3), 139–151.

Zhou, X., Wang, P., Dai, G., Yan, J., & Yang, Z. (2021, July). Tomato fruit maturity detection method based on YOLOV4 and statistical colour model. In: *2021 IEEE 11th Annual International Conference on CYBER Technology in Automation, Control, and Intelligent Systems (CYBER)*. IEEE, pp. 904–908.

CHAPTER 14

Leveraging Industry 4.0 Technologies for Sustainable Horticulture for Operational Excellence

PANKAJ KUMAR[1], SAPNA JARIAL[2,*], RUPESH KAUSHIK[1], ALI ABDULLAHI[3], and AMIRHOSSEIN PIRMORADI[4]

[1]*Department of Horticulture, School of Agriculture, Lovely Professional University, Phagwara, Punjab, India*

[2]*Department of Agricultural Economics and Extension, School of Agriculture, Lovely Professional University, Phagwara, Punjab, India*

[3]*Department of Agricultural Economics and Extension, Bayero University, PMB, Kano, Nigeria*

[4]*Department of Agricultural Extension, Faculty of Agriculture, Razi University, Kermanshah, Iran*

*Corresponding author

ABSTRACT

The horticulture sector provides employment and income. However, it is marred by suboptimal productivity, fruit loss and wastage, limited labor availability, and inadequate quality, compounded by unsustainable practices impacting operational efficiency. Technological advancements in Industry 4.0, including artificial intelligence (AI) and the internet of things, hyperspectral imaging, drones, robots, and block chain, are solutions. Therefore, this chapter aims to examine the utilization of Industry 4.0 in horticultural crop management for operational excellence, its challenges, and opportunities. The conceptual foundation is derived from theories of

Smart Horticulture Production: Integrating Advanced Technologies.
Rajni Rajan and Duniya Ram Singh (Eds.)

DOI: 10.1201/9781998511679-14

disruption and operational excellence. Horticulture 4.0 provides disruption in services and innovations. Customers use disruptive high-cost technology, including AI agricultural models, greenhouse functions, and insect pest and disease prognosis. It improves orchard yield estimation, harvesting, fruit maturity, quality detection, on farm sorting and grading of fruits and vegetables. It streamlines on-farm transportation, storage, and retrieval systems. Although Horticulture 4.0 is plausible, yet, it does not reduce carbon emissions. The high investment raises questions about technology equity, access, and cost across regions and production scales. Building trust in new technologies, data security, safety, and privacy requires proactive steps and ethical frameworks. Future horticulturists must reskill in Industry 4.0 technology competence for operational efficiency.

14.1 INTRODUCTION

The four industrial revolutions mark sequential advancements in technology and industrial practices, starting with mechanization in the first revolution, introduction of mass production in the second, the incorporation of computerization and automation in the third, and finally, evolving to smart, interconnected systems in the fourth phase, referred to as Industry 4.0. Each revolution has progressively brought forward more efficient, sustainable, and intelligent industrial practices.

Using money as the tool of change, society has moved quickly to exploding growth. Resources are being rapidly depleted because of rising carbon emissions and environmental damage (Perçin, 2023). The burgeoning population poses significant challenges to meeting the demands for agricultural and allied, that is, horticultural produce through conventional methods, marked by the sector's prevailing issues like low productivity, limited labor availability, and insufficient marketing strategies (Cappelli et al., 2022; Rathore, 2021). Traditional methods fail in excellence and quantity to meet the requirements of rapidly growing global and local markets (Rahutomo et al., 2019). For example, a substantial percentage of harvested fruits and vegetables, ranging between 20 and 40%, is wasted annually due to challenges in sorting, grading, storage, and transportation, necessitating substantial labor (Luo et al., 2022). Low productivity induced by human fatigue and variability in inspection experience lead to inconsistencies in quality standards (Sushma et al., 2022; Rysz & Mehta, 2021). The pervasiveness of damages from environmental conditions, pest infestations, improper harvesting practices, and postharvest infections exacerbates these losses.

Operational excellence in the horticulture industry can be achieved through the implementation of Industry 4.0 technologies (Lamm et al., 2021). Horticulture 4.0 is a good way to accomplish the sustainable development goals (SDGs). In the evolving era of Industry 4.0, where the merger of internet of things (IoT), cloud computing, and cyber physical systems (Lin et al., 2023) is revolutionizing industries such as engineering (Kalla et al., 2020), business management (Jayashree et al., 2020), healthcare (Siddiqui et al., 2022), supply chain (Al Kurdi et al., 2023), textiles among others (Sharma et al., 2023) horticulture domain is no omission (Brini, 2023). The fusion of intelligent technologies with horticultural practices has the potential to usher in a new era of sustainability and operational excellence, addressing constraints related to productivity, resource utilization, and environmental conservation (Howard et al., 2021). However, along with the potential advantages that horticulture 4.0 can bring, issues and dangers might also be observed (da Silveira et al., 2023a).

Therefore, our research question was how Industry 4.0 technologies create operational excellence for sustainable horticulture? The present chapter provides a better understanding on the application of Industry 4.0 application in the horticultural sector followed by the challenges and opportunities.

14.2 CONCEPTUAL FRAMEWORK

The present study is grounded on operational excellence and disruptive theories. Operational excellence is a concept that aims to optimize business operations and achieve superior performance (Found et al., 2018). Operational excellence is a multidimensional concept that encompasses various theories and approaches. It involves the continuous improvement of processes, the elimination of waste, and the pursuit of excellence in all aspects of an organization. Lean thinking, Six Sigma, and excellence models provide frameworks and tools for achieving operational excellence (Tripathi et al., 2023).

The theory of disruptive innovation emphasizes the skills and behaviors of disruptive innovators to achieve operational excellence. In particular, technology companies satisfy the demands of their most valuable and lucrative customers while falling short of those of others when they optimize their products and services (Christensen, 1997). The resilience framework proposed provides (Figure 14.1) a contextualized approach to deal with disruptions in horticulture sector and achieve operational excellence.

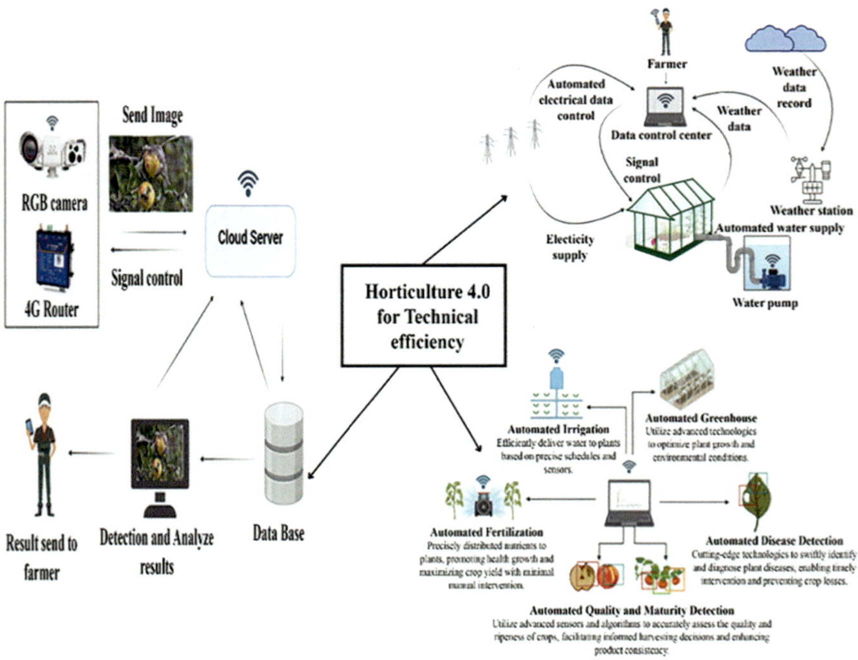

FIGURE 14.1 Industry 4.0 and operational excellence conceptual framework.

14.3 HORTICULTURE 4.0: DRIVING OPERATIONAL EXCELLENCE THROUGH SMART INNOVATION

14.3.1 AUTOMATION FOR PREHARVEST PRACTICES IN HORTICULTURE

Automation in preharvest practices in horticulture is an emerging trend that offers numerous benefits in enhancing efficiency, improving productivity, and optimizing the overall cultivation process. This technology involves the application of automated systems and machinery to various preharvest activities, including irrigation, fertilization, and crop monitoring in several countries as mentioned in Table 14.1. The horticultural practices function in project mission mode. Making judgments about the projects requires tracking and monitoring the projects. There are numerous methods for gathering and storing data from sites (Giran et al., 2023). Data is fed in a system to create various models (Jarial, 2022) for prognosis.

TABLE 14.1 List of AI Models Used in Horticultural Crops

No.	Crop Name	AI Model	Objective	Accuracy (%)	Reference
1	Banana	Deep convolutional neural networks (DCNN)	Innovative and quick techniques for the prompt identification of illnesses and pests	>90	Selvaraj et al. (2019)
2	Apple	YOLOv3-tiny algorithm	To detect and estimate the apple yield	83.64	Mazzia et al. (2020)
3	Grapes	MobileNet-V1	Grape bunch detection in vineyards taking into account the early and medium growth stages	66.96	Aguiar et al. (2021)
4	Papaya	Convolutional neural network (CNN)	Classification and prediction of diseases	91.00	Hossen et al. (2020)
5	Tomato	Backpropagation neural network (BPNN)	To determine the maturity of tomato	99.31	Lyu et al. (2022)
6	Strawberry	UAV and RGB cameras	To calculate the strawberry plant's potential fruit and bloom yield	84.10	Chen et al., 2019
7	Guava	Convolution neural network (CNN)	Estimate the maturity-indexed and differentiate the damaged fruits	99.92	Pathmanaban et al. (2023)
8	Pineapple	YOLOv3	To detect and the pineapple fruits harvest	95.55	Anh et al. (2020)
9	Peach	Convolutional neural network and convolutional autoencoder (CAE) network	To detect the plant diseases of peach	99.35	Bedi et al. (2021)
10	Litchi	YOLOv4 and YOLOv5-l	To detect the male and female flowers in litchi	94.54	Lyu et al. (2023)
11	Cherry	YOLOv5	Identifying the quality of the cheery fruit according to pH value, vitamin C content, sugar–acid ratio	99.60	Han et al. (2022)

TABLE 14.1 *(Continued)*

No.	Crop Name	AI Model	Objective	Accuracy (%)	Reference
12	Coconut	–	To forecast the amount of fertilizer needed based on soil analysis in order to boost production	93.50	Lekshmi et al. (2022)
13	Kiwi	YOLOv3-tiny	To accurate detection of Kiwi fruit in the orchard	–	Fu et al. (2021)
14	Plum	Convolutional neural network (CNN)	To analyze the ripeness of plum	94.00	Miragaia et al. (2021)
15	Citrus	YOLOv5-CS (citrus sort)	Identifying objects and precisely counting green citrus in their natural environment	97.66	Lyu et al. (2022)
16	Custard apple	SIANN-shift invariant artificial neural networks	Early prediction of the disease in leaves for increasing the production of the plant	85.00	Renugadevi et al. (2021)

14.3.1.1 AUTOMATIONS IN GREENHOUSE

Greenhouses are a well-established method to produce fruits and vegetables. Here, wireless technology and the IoT come into play, and we may implement the monitoring and control the weather parameters without the presence of humans on the farm by employing novel communication protocols and sensors (Table 14.2). Establishing plants in a greenhouse uses many types of machinery, sensors, and multiple automation devices that can adjust multiple growing conditions of plants under the greenhouse. The first milestone study on the remote controlled greenhouse of cucumber with the algorithm of artificial intelligence (AI) has been held out successfully by Hemming et al. (2019). Gonzalez-Briones et al. (2019) used a multiple-agent system that enables managing all the operations such as the collection of data, monitoring, and user interaction. This research paves the way for future smart applications in AI, where the proposed system can be employed to enable intelligent and autonomous functionalities. The technologies enable precise monitoring, automated decision-making, and optimization of growing conditions, ultimately leading to enhanced yields and more efficient resource utilization.

TABLE 14.2 List of the AI Models for Greenhouse and Irrigation Management of Horticultural Crops

S. No.	AI Model	Descriptions	References
1	RealSense camera	Shows a reduction in water use of 15.06%, is clearly indicative of the system's performance	Afonso et al. (2019)
2	Sensors	Effectively controlled irrigation and climate, establishing optimal climate setpoints and formulating crop management strategies	Hemming et al. (2019)
3	Artificial neural networks (ANN) and fuzzy control	Efficiently ensure temperature and management of the irrigation and effective prevention of frost disasters	Castañeda-Miranda et al. (2020)
4	Artificial neural networks (ANN)	Effectively manage irrigation and temperature conditions, thus enhancing crop growth and productivity	Chen et al. (2020)
5	Convolutional neural network (CNN)	Impressive performance is shown, with an average intersection over union score of 83% on the test dataset and a validation dataset seed recognition accuracy of 97%	Shadrin et al. (2019)
6	USP-based neuronal network (NN)	Consists of soil moisture, humidity, and temperature sensors that provide informational data to the server via a protocol named ×802.11	Nawandar et al. (2019)
7	Cyber-physical irrigation system	Enables the collection of field data through sensors, makes water application decisions based on a decision system, and activates solenoid valves accordingly. The system allows for monitoring of the irrigation process both on-site and remotely, utilizing a central station and its internet connectivity	Jimenez et al. (2020)
8	K-nearest neighbor (KNN)	The system is fully automated, with devices communicating with each other and applying intelligent decision-making in the process of irrigation	Shekhar et al. (2014)
9	Smart irrigation system (SIS)	Implemented by leveraging open-source IoT cloud computing platform and wireless sensor networks. This platform was utilized for various purposes, including data collection, storage, data analytics, and visualization, enabling efficient and effective management of the irrigation process	Ndunagu et al. (2022)
10	Global system for mobile communication (GSM)	Platform enables the reception and sending of SMS messages to and from the mobile device of the farmer or homeowner. This allows for efficient irrigation control based on soil moisture levels or user instructions	Yasin et al. (2019)

14.3.1.2 PROGNOSIS OF INSECT PEST AND DISEASE

The plant pathology domain has witnessed significant progress in using the capabilities of machine learning and AI. Numerous technologies, including intelligent monitoring systems, drones, sensors, and robots, have found extensive applications in this domain. Specifically, computer vision-based phenotyping has emerged as a promising approach for detecting and assessing plant diseases, as well as evaluating the severity of plant stress. These advancements have gained substantial momentum in the horticultural crops (Prabha et al., 2021). Deep learning models, a subclass of AI, offer significant potential in identifying plant diseases through the analysis of a plant's appearance and visual symptoms. Image classification extensively utilizes machine learning techniques, including artificial neural networks (ANN), support vector machines (SVM), random forest, probabilistic neural networks (PNN), and convolutional neural networks (CNN). For instance, a new field robot named "Robhortic" has been developed by Cubero et al. (2020) which can be operated remotely and equipped with various sensors to examine the horticultural crops. This robot works in environments with controlled lighting. The multispectral sensors on board have a high resolution, ranging from 1 to 2.5 mm per pixel. This resolution surpasses that of sensors on drones, allowing for in-depth examination at the leaf level. Through the utilization of the spectral data collected by these sensors, maps representing various spectral indices can be produced. These advancements highlight the potential of AI and machine learning in transforming disease and insect detection in horticultural crops (Table 14.3), empowering farmers with early detection and targeted treatments for protecting their crops.

14.3.1.3 ORCHARD YIELD ESTIMATION AND HARVESTING

Harvesting and orchard yield estimating robots have gained significant attention and development in recent years as a potential solution for automating the labor-intensive task of harvesting horticultural crops (Khadatkar et al., 2022). Harvesting robots offer a practical solution to labor shortages and rising labor costs, while also enabling the implementation of precision agriculture. Specifically, these robots can ensure that the harvesting process becomes manageable, trackable, and adaptable. Additionally, they have the capability to collect, process, and analyze various types of data related to the targeted fruits, such as harvest time,

TABLE 14.3 List of Algorithms Used for Disease and Insect Pest Detection in Horticultural Crops

No.	Application	Method	Algorithms	Accuracy	Reference
1	Detection of banana disease	Via digital imaging	Deep convolutional neural network (DCNN)	–	Selvaraj et al. (2019)
2	Detection of citrus greening disease	Via visible spectrum image processing	Cost-support vector classification (C-SVC)	91.93%	Deng et al. (2016)
3	Detect and differentiate of Asian citrus psyllid pest	Via machine vision	Convolutional neural networks (CNN)	Precision rate of 80%	Partel et al. (2019)
4	Strawberry powdery mildew detection	Image texture analysis	Support vector machines (SVM) and K-nearest neighbors (kNN)	91.86%	Mahmud et al. (2019)
5	Identify apple foliar diseases	Dataset consisting of image processing	Deep convolutional neural network (DCNN)	97.60%	Liu et al. (2018)
6	Detect the disease of tomato plant and fruits	Via image processing	Convolutional neural networks (CNNs)	95.31%	Khatoon et al. (2021)
7	Identify diseases in peach plants and fruits	Using the VGG-19 architecture	Convolutional neural network (CNN)	94%	Alosaimi et al. (2021)

fruit position and orientation, ripeness level, and defects (Zhou et al., 2022). Accurately and speedily identifying densely clustered orchards and assessing their level of maturity holds immense practical importance for estimating yields in orchards and enabling robotic harvesting. Robotic platforms for assisting farmers via machine vision have proven valuable in various applications, including the nondestructive assessment of production yield before harvest. By providing accurate information, farmers can make informed decisions regarding resources for people, harvesting machinery, storage, transit, and product promotion (Table 14.4). Kolhalkar et al. (2021) developed a unique end-effector, combining both gripping and cutting functionalities, which has been created, tested extensively, and intended for usage with table grapes. The performance and effectiveness of this innovative tool have been validated through rigorous experimentation. Moreover, Park et al. (2021) presented the development and implementation of a comprehensive system that combines IoT, edge computing, AI, and cloud technologies to monitor environmental conditions in hydroponic strawberry cultivation. The integrated system not only collects and stores data regarding the growing environment of strawberries but also provides visualizations of this data. Additionally, a harvest decision system is proposed, which employs a deep learning algorithm to classify the degree of maturity of strawberries according to image analysis. These algorithms are integrated with a knowledge base and a digital twin of the cover crops such as oats, radish, and vetch. Together, these components enable the team of robots to effectively carry out the harvesting process.

TABLE 14.4 Orchard Yield Estimation and Harvesting of Horticultural Crops

No.	Application	Algorithm Used	Results	Reference
1	Detecting litchi fruit for yield estimation	You Only Look Once version 5 (YOLOv5) model	It functions as a valuable resource for offering direction and recommendations for estimating litchi yields and managing subsequent harvesting procedures	Xiong et al. (2023)
2	Access the productivity of apple orchards	Inertial navigation system (INS), multiscale multilayer perceptron (MLP), convolutional neural network (CNN)	The most accurate count estimates for estimating apple orchard yield were obtained using CNN and WS	Bargoti et al. (2014)
3	Estimate yields in kiwi orchards	Support vector machine (SVM)	This approach proves to be efficient for accurately estimating kiwifruit quantities in the captured images	Massah et al. (2022)
4	Identify and harvest tomato crop	Deep-ToMaToS (a deep learning network)	Integrates 6D posture assessment of the fruit and side-stems of the tomatoes that are ready to be harvested, as well as maturity recognition for every tomato object	Kim et al. (2022)
5	Harvesting of watermelon	You Only Look Once version 5—Convolutional block attention module (YOLOv5s–CBAM)	Obtaining a remarkable 89.8% accuracy rate on the test dataset. Furthermore, the watermelon point cloud is better segmented inside the designated region of interest by utilizing the K-means algorithm	Rong et al. (2022)
6	Strawberry-harvesting robot	Artificial neural networks (ANN)	Implemented a unique soft gripper specifically designed for the delicate task of harvesting soft strawberries	Parsa et al. (2022)

14.3.1.4 FRUIT RIPENESS AND IDENTIFYING QUALITY

In recent years, massive research has been conducted in the domain of fruit quality identification using computer vision technology, leading to notable findings. For instance, based on the quality ratio, the mango fruits are classified into two categories. If the quality ratio exceeds the predetermined threshold, the fruit is considered rotten. On the contrary, if the quality ratio is inferior to the threshold, the fruit is deemed good. Consequently, by employing AI algorithm, it becomes possible to sort mango fruits according to their quality, a crucial step in adding value to the fruits. Begum et al. (2022) used transfer learning through Visual Geometry Group (VGG), Inception and Residual Network (ResNet) for the purpose of identifying tomato fruit maturity. To address the classification task, various pretrained models of transfer learning such as VGG16, VGG19, Inception V3, ResNet101, and ResNet152 are utilized. These models have already been trained on large datasets and possess the ability to extract meaningful features from tomato images, enabling accurate classification. The primary focus of this research is to achieve nondestructive tomato ripeness detection. To accomplish this, this study utilizes a newly developed method in computer vision in deep transfer learning. By employing this technique, tomatoes are automatically divided into three groups according to maturity: immature, moderately mature, and mature. Choi et al. (2018) successfully implemented an image processing method by utilizing a charge-coupled device (CCD) camera to analyze the color of fruit images. The color information from the fruit images is utilized for preprocessing and separation, specifically for detecting imperfections. In addition to color analysis, the researchers employed Near InfraRed Spectroscopy (NIRS) to measure the internal qualities of the fruits. As a result of their comprehensive approach, an impressive accuracy rate of 97.4% was achieved through this methodology. Hasanzadeh et al. (2022) utilized hyperspectral imaging with the objective of assessing the characteristics of Red Delicious and Golden Delicious apples' nondestructive quality. Titratable acid (TA), total phenol (TP), soluble solids content (SSC), and pH were among the quality criteria assessed. The researchers used a variety of preprocessing methods in conjunction with the partial least squares (PLS) method to estimate the quality features of the apples. Zhou et al. (2021) created a framework for automatic category of strawberry maturity stages to enable rapid and precise categorization. The system utilized an advanced deep learning technique called You Only Look Once (YOLOv3), known for its proficiency in detecting small objects. The system was trained using this method to detect both strawberry flowers and fruits at various stages of maturity. By leveraging YOLOv3, the system achieved accurate identification and classification of different strawberry maturity stages. MacEachern et al. (2023) created six deep learning ANN models to identify the maturity stage of wild blueberries. Additionally, models were developed to estimate the yield of the blueberries. YOLOv3, YOLOv3-SPP, YOLOv3-Tiny, YOLOv4, YOLOv4-Small, and YOLOv4-Tiny were the six networks that were used. Growers can raise yields and eventually profit margins by learning more about the ripening traits of their farms.

14.3.1.5 AUTOMATION IN ON-FARM SORTING OF FRUITS AND VEGETABLE

Traditional methods of fruit sorting primarily rely on visual evaluation, considering elements such as quality, ripeness, deterioration, illness, and damage (Gabriëls et al., 2020; Kang & Gwak, 2022). With the introduction of hyperspectral imaging, it has become possible to not only identify defects on the fruit's skin but also establish connections between these defects and specific diseases that hold significant significance. This advancement has expanded the capabilities of fruit inspection, enabling more comprehensive analysis and assessment. Nevertheless, the packing house environment provides a favorable setting for effectively addressing these issues and facilitating various postharvest processing tasks (Cubero et al., 2020). On farm sorting of fruits and vegetables can heal these challenges and prevent the infection from spreading. Implementing on-farm sorting for fruits and vegetables can address these challenges and minimize the spread of infections.

A novel approach was introduced by Kang & Gwak (2022), involving a method of ensemble that merges the characteristics of two distinct pretrained deep CNN, namely ResNet-50 and ResNet-101, to classify fruit freshness. Additionally, a multitask learning framework was presented, utilizing fruit type information to improve freshness classification performance. The findings showed that the suggested model performed exceptionally well, with average accuracy for fruit categorization being 97.43% and freshness classification being 98.50%. Gabriëls et al. (2020) employed visible and near-infrared spectroscopy (VNIRS) to predict when the "Keitt" mango halves will experience interior browning. ANNs and PLS, two different analysis methods, were used. When VNIRS is merged with ANN, it can successfully categorize mangoes as either healthy or impacted by interior browning with the accuracy of 80%. Implementing a resilient and dependable categorization system has the potential to enhance quality assessments across the fruit supply chain, consequently minimizing postharvest losses. Another research effort focused on mangoes aimed to visualize the spongy tissue and internal defects, which ultimately contribute to a decline in postharvest quality. Raghavendra et al. (2021) introduced methods of wavelength selection to verify the optimal range of wavelengths for arranging the defected and healthy mango fruits. The experimental results demonstrated that the wavelength of lesser range (673–1100 nm) proved to be effective in identifying mango's interior defects. The research highlights the potential of near-infrared (NIR) systems as a valuable technology for automated assessment of mango fruits, specifically in identifying internal defects during on-farm sorting. Another novel approach was introduced by Wendel et al. (2018) employing a combination of a ground vehicle-mounted navigation system, a hyperspectral camera, and a light detection and ranging (LIDAR) sensor to individually forecast fruit dry matter for several trees in a block of mango orchards. This method enables efficient and accurate assessment of dry matter for many trees, offering valuable insights into the orchard's overall fruit quality. The novel approaches and technologies for on-farm sorting have been introduced by much research in different fruit crops like in apple. Hyperspectral imaging was used, which are nondestructive and novel methods for visualizing the internal defects in apple. Hyperspectral images of Pink Lady apples were examined by Çetin et al. (2022) at various harvest stages to forecast internal characteristics such as the amount of soluble solids and hardness. The prediction accuracy of different models including k-nearest neighbors (KNN), decision trees (DT), partial least squares regression (PLSR), ANN, and multiple linear regression (MLR) was assessed. Further, Wang et al. (2022) reported using nondestructive hyperspectral imaging to access the quality and maturity in apple fruit. The study highlighted the potential of the Normalization Different Spectral Index (NDSI)-Stability Competitive Adaptive Reweighted Sampling PLSR model, utilizing hyperspectral imaging, for tracking the development of apples' quality and maturity on-site. This model demonstrated the ability to generate the mapping of the regional distribution of apple ripeness and quality, offering beneficial insights into ripening patterns, regional growth dynamics, and harvest planning.

14.3.2 AUTOMATION FOR POSTHARVEST PRACTICES IN HORTICULTURE

14.3.2.1 Automation in Grading of Fruits and Vegetables

The postharvest process includes a crucial step known as grading, which holds significant importance. Automation in the grading of fruits has developed as a transformative resolution in the industry of agriculture (Ghazal et al., 2021). Automated grading systems utilize various technologies such as computer vision, machine learning (Behera et al., 2020; Zhang et al., 2018), and robotics to assess fruits based on predetermined quality criteria. Rizwan & Hakim (2022) utilized deep learning to revolutionize and classify the different varieties of mango and assessed for quality. By utilizing CNNs, crucial image characteristics like shape, color, texture defects, and size are automatically learned, resulting in a more accurate and efficient classification and grading process. An intelligent automation system was proposed and developed by Chopra et al. (2021) who presented an innovative system that utilizes computer vision and spectrophotometry to automate fruit isolation based on their grade. By utilizing Microsoft Azure's cloud computing infrastructure, the system can identify fruit with 95% accuracy when it is entered. Subsequently, the system employs spectroscopy and collects the machine learning techniques to forecast the grade of the identified fruit. Mesa et al. (2021) devised an automated system for banana grading that is noninvasive and employs deep learning technology, hyperspectral imaging, and red-green-blue (RGB). Another automated grading technique in banana was introduced by Ganguli et al. (2022), which combines a multiinput deep learning model by fusing deep learning methods, RGB imagery, and hyperspectral imaging. The technique of RGB enables the quantification of external aspects of bananas, including size, color, and texture.

14.3.2.2 Automated On-Farm Transportation

Automated on-farm transportation systems offer significant potential to enhance the productivity sustainability of operations in agriculture by leveraging the capabilities of autonomous vehicles and robotics (Zhou et al. 2023). The system aids to play a vital role in assisting with fruit-harvesting tasks, including the picking of apples (Lu et al., 2022), strawberries (Peng et al., 2021), and grapes (Faheem et al., 2021). Previous investigations in the realm of on-farm transportation have primarily centered around corobot scheduling, maneuvering strategies, and the reduction of product damage during transit. The objective has been to create a collaborative robotic system capable of efficiently carrying trays and transporting fruit within the harvest field (Rysz & Mehta, 2021). Consequently, research efforts have delved into exploring human–robot collaboration models that encompass activities such as picking and walking. These investigations aim to establish effective corobot scheduling and dispatching policies that optimize the overall performance of the system. A modeling framework was created to facilitate all robot-aided and manual harvesting, serving as both an offline simulation tool for system designers and a model that represents for real-time control by the robot. To fulfil these dual purposes, the framework models the actions of both robots and human pickers with accuracy by using hybrid systems. This comprehensive framework enables designers to simulate and evaluate system performance in advance, while also providing a reliable model for controlling robots during actual operations (Seyyedhasani et al., 2020). The growing need for adaptable automation in warehouse has led to a significant increase in the use of automated guided vehicles (AGVs) within automated warehouses. López et al. (2022) introduced a versatile framework for simulating and controlling autonomous transport systems using AGVs. Usually, two stages of simulation are used in the creation of these applications. Dharmasiri et al. (2020) introduced an algorithm for traffic control in a dynamic environment with multiple AGVs. It models the map using the graph theory and uses the Dijkstra path planning method to design paths efficiently. Additionally, a time-based technique is used to simulate and compute optimized velocity profiles for every AGV, guaranteeing that they operate without colliding. Another study was conducted by Chun-Fu et al. (2014). The authors put forth a conceptual framework for the visual navigation's mechanical and control system architecture in AGVs designed specifically for small, agile warehouses. Overall, the visual navigation for AGV proved capable of meeting the demand for adaptable storage in warehouse settings. Witczak et al. (2020) developed multiple AGV models and fault-tolerant control. However, AVG management becomes a difficult issue, especially when bidirectional routes are used to increase the warehouse's flexibility and efficiency (Zhang et al., 2018).

14.3.2.3 AUTOMATED STORAGE AND RETRIEVAL SYSTEM

Automation in storing houses or warehouses concerns to the utilization of technology, machines to streamline and optimize various processes involved in storing and managing inventory (Borisoglebskaya et al., 2019). Automation in warehouses enhances operational efficiency, accuracy, inventory control, and customer satisfaction while reducing costs and improving safety. The various AI models and retrieval systems are listed in Table 14.5.

It enables warehouses to meet the demands of a rapidly evolving supply chain landscape and stay competitive in today's market (Bertolini et al., 2023). Numerous studies have been carried out in the realm of automated warehouses, aiming to enhance the product's quality throughout its journey until it reaches the consumer's doorstep such as queuing model was developed by Chen et al. (2022) which estimates the system response time and throughput capacity by utilizing pre-emptive-resume interrupted service. This model considers random storage, two input/output point locations, and a policy for assigning cranes that divides up requests among the cranes. The validity of the analytical models is verified through simulations using data obtained from real-world scenarios. A software solution based on an open queuing network was presented by Ekren et al. (2021). It can quickly estimate key performance parameters for a preset warehouse design. The tool allows users to input the necessary data values, enabling efficient analysis. To optimize warehousing in a large-scale automated high-rise warehouse product-service system, Leng et al. (2021) suggested a digital twin technique. Using this method, data is mapped from the physical warehouse system in real time to a cyber model. Within this system, a joint optimization model is integrated to facilitate timely optimization of storage assignment and stacked packing for the warehouse product-service system. An increasing number of online merchants are looking at cutting-edge technological ways to expedite the processing of returned goods in their warehouses. Lane-guided transport (LGT) vehicle deployment is one example of such a system. These little electric cars can pick up boxes of returned goods from a depot and deliver them to assigned workstations. By utilizing LGT vehicles, logistics workers can devote their attention to processing the items rather than spending time manually transporting them throughout the warehouse (Emde et al., 2021).

TABLE 14.5 List of AI Models for Storage and Retrieval System

No.	AI Model	Objective	Reference
1.	Nondominated sorting genetic algorithm II (NSGA II)	The first objective of our model is to assess and determine the system's performance	Rajković et al. (2019)
2.	Radio-frequency identification technology	Significant improvements in warehouse space usage and data-handling efficiency. It also enhances warehouse flexibility and capability	Wang et al. (2015)
3.	Warehouse management systems (WMS)	Opportunities for reducing the carbon footprint in warehousing logistics are explored in relation to digitalization	Minashkina et al. (2020)
4.	ThinkSpeak cloud	To collect data on the incoming and outgoing quantities of items within the warehouse	Pandian et al. (2019)
5.	Ant colony optimization (ACO), adaptive large neighborhood search (ALNS)	–	Kazemi et al. (2021)
6.	Assignment-adaptive large neighborhood search (ALNS)	To enhance the selection of the roadway stacker's departure, point from the warehouse	Yang et al. (2021)

14.4 CHALLENGES AND OPPORTUNITIES

Digital methodologies are indispensable in the field of horticultural production, even though their evolution, application, and adoption are still in the initial phases of development (da Silveira et al., 2023b). It is anticipated that horticulture 4.0 would contribute to process improvement and introduce fresh approaches to both persistent and emerging issues (da Silveira et al., 2023a). However, the integration of 4.0 technologies into horticulture has yet to progress in the realm of sensor technologies. This technological advancement possesses the capacity to improve the sustainability and competitiveness (Sharma et al., 2023). Though, horticulture Industry 4.0 seems plausible. However, this does not resolve the issue of lowering carbon dioxide emissions (Tuysuz & Trestian, 2020). Paradoxically, carbon emissions have increased because of increased technosphere interaction with biosphere. Separation of technology from biosphere is required for sustainable life. Building trust in new technologies, data security, safety, and privacy requires proactive steps and ethical frameworks (Jarial, 2022).

The adoption of horticulture 4.0 technologies involves significant upfront costs, in the purchase of equipment like sensors, drones, and automated systems, as well as the implementation of software solutions, big data storage. Infrastructure is also a challenge (da Silveira et al., 2023a). People who bought information and communications technology (ICTs) have a big and positive effect on how technologically advanced a company is in areas like planning production, managing inventory, and analyzing data. There is, however, no proof that customers who invested in ICTs have helped businesses reach the highest levels of technology use. Also, managing and analyzing the big data generated by IoT devices can be daunting, necessitating infrastructure and expertise. Governments should work with experts and offer guarantees to investors in public–private partnership (PPP) projects due to the risks brought on by the size of the investments and the lengthy duration of the projects (Xu et al., 2023). Overcoming these challenges will require collaboration, new organizations, and structures, between stakeholders, including researchers, technology providers, policymakers, and farmers (Jarial, 2022; Sharma et al., 2023).

Power disparities in food systems are expected to worsen because of digitization because transparency can turn into tyranny when it encourages the growth of audits, assessments, and assessment methods (Commandré et al., 2021). When so few powerful tech companies like Amazon, Google, own the technology, it concentrates power, limits competition, and poses risks to equity and accountability in the digital sphere.

Another challenge is the demographic aging of the rural population, which poses threats to the sustainability of smallholder farming. Farmers in rural areas have lower average levels of education and smaller farms because of the ageing population due to land transfer-out and desertion. These changes have a direct impact on agricultural inputs, such as fertilizers and machinery, resulting in decreased agricultural output and labor productivity (da Silveira et al., 2023b). For instance, most of the phosphorus used in fertilizer comes from mines worldwide. All forms of cultivation that do not return minerals to the soil require the mineral. However, because it is more cost-effective to mine new sources of phosphorus than to recycle existing ones, significant amounts of phosphorus are released into the sea, where they are essentially irretrievable (Demay et al., 2023). Future horticulturists require reskilling of competence in Industry 4.0, sustainable regenerative organic horticultural practices with reduced pesticides use coupled with data-driven technologies (Newton et al., 2020).

Biomimicry and Industry 4.0 intersect seamlessly, creating ecosystems where innovation flourishes (Pathak, 2019). Biomimicry informs and enhances the advancement of Industry 4.0 technologies, offering sustainable and efficient solutions inspired by nature (Stevens et al., 2022). Horticulture 4.0 can create such spaces.

14.5 CONCLUSION

The chapter presented the utilization of Industry 4.0 in horticultural crop management for operational excellence, its challenges, and opportunities. Emerging technologies are increasingly utilized to automate various operations such as irrigation, fruit quality and maturity detection, pest, and disease identification, harvest, reduce and tackle labor shortage, insect pest attack, wastage of produce, and so forth to bring operational excellence in the horticulture industry. Sustainable development is actually very simple. Behavioral modification is a big part of it. A different perspective on combining and multi using is necessary. Most individuals can understand it with no trouble at all. This synergy yields advanced, resilient, and adaptable systems, reflecting the intricate and optimized designs found in the natural world, thereby pushing the boundaries of technology and operational excellence.

KEYWORDS

- **horticulture 4.0**
- **operational excellence**
- **postharvest management**
- **sustainable practices**

REFERENCES

Afonso, M., Mencarelli, A., Polder, G., Wehrens, R., Lensink, D., & Faber, N. (2019). Detection of tomato flowers from greenhouse images using color space transformations. In: *EPIA Conference on Artificial Intelligence*. Springer, pp. 146–155.

Aguiar, A.S., Magalhães, S.A., Dos Santos, F.N., Castro, L., Pinho, T., Valente, J., Martins, R., & Boaventura-Cunha, J. (2021). Grape bunch detection at different growth stages using deep learning quantized models. *Agronomy*, 11(9), 1890.

Al Kurdi, B., Alzoubi, H.M., Turki Alshurideh, M., Khalil Alquqa, E., & Hamadneh, S. (2023). Impact of supply chain 4.0 and supply chain risk on organizational performance: An empirical evidence from the UAE food manufacturing industry. *Uncertain Supply Chain Management*, 11, 111–118. https://doi.org/10.5267/j.uscm.2022.11.004

Alosaimi, W., Alyami, H., & Uddin, M.I. (2021). PeachNet: Peach diseases detection for automatic harvesting. *Computers, Materials & Continua*, 67, 1665–1677.

Anh, N.P.T., Hoang, S., Van Tai, D., & Quoc, B.L.C. (2020). Developing robotic system for harvesting pineapples. In: *2020 International Conference on Advanced Mechatronic Systems (ICAMechS)*. IEEE, pp. 39–44.

Bargoti, S., & Underwood, J.P. (2014). Image segmentation for fruit detection and yield estimation in apple orchards. *Journal of Field Robotics*, 34(6), 1039–1060.

Bedi, P., & Gole, P. (2021). Plant disease detection using hybrid model based on convolutional autoencoder and convolutional neural network. *Artificial Intelligence in Agriculture*, 5, 90–101.

Begum, N., & Hazarika, M.K. (2022). Maturity detection of tomatoes using transfer learning. *Measurement: Food*, 7, 100038.

Behera, S.K., Rath, A.K., Mahapatra, A., & Sethy, P.K. (2020). Identification, classification & grading of fruits using machine learning & computer intelligence: A review. *Journal of Ambient Intelligence and Humanized Computing*, 15(1), 117–117.

Bertolini, M., Neroni, M., & Uckelmann, D. (2023). A survey of literature on automated storage and retrieval systems from 2009 to 2019. *International Journal of Logistics Systems and Management*, 44(4), 514–552.

Borisoglebskaya, L.N., Provotorova, E.N., Sergeev, S.M., & Khudyakov, A.P. (2019). Automated storage and retrieval system for Industry 4.0 concept. *IOP Conference Series: Materials Science and Engineering*, 537(3), 032036.

Brini, M. (2023). Digital Agriculture https://www.researchgate.net/publication/371598930_DIGITAL_AGRICULTURE_AN_UPDATED_COMPREHENSIVE_OVERVIEW

Cappelli, I., Fort, A., Pozzebon, A., Tani, M., Trivellin, N., Vignoli, V., & Bruzzi, M. (2022). Autonomous IoT monitoring matching spectral artificial light manipulation for horticulture. *Sensors*, 22(11), 4046.

Castañeda-Miranda, A., & Castaño-Meneses, V.M. (2020). Smart frost measurement for anti-disaster intelligent control in greenhouses via embedding IoT and hybrid AI methods. *Measurement*, 164, 108043.

Çetin, N., Karaman, K., Kavuncuoğlu, E., Yıldırım, B., & Jahanbakhshi, A. (2022). Using hyperspectral imaging technology and machine learning algorithms for assessing internal quality parameters of apple fruits. *Chemometrics and Intelligent Laboratory Systems*, 230, 104650.

Chen, J.I.Z., & Yeh, L.T. (2020). Greenhouse protection against frost conditions in smart farming using IoT enabled artificial neural networks. *Journal of Electronics*, 2(04), 228–232.

Chen, W., Gong, Y., & de Koster, R.B. (2022). Performance estimation of a passing-crane automated storage and retrieval system. *International Journal of Production Research,* 60(4), 1210–1230.

Chen, Y., Lee, W.S., Gan, H., Peres, N., Fraisse, C., Zhang, Y., & He, Y. (2019). Strawberry yield prediction based on a deep neural network using high-resolution aerial orthoimages. *Remote Sensing*, 11(13), 1584.

Choi, H.S., Cho, J.B., Kim, S.G., & Choi, H.S. (2018). A real-time smart fruit quality grading system classifying by external appearance and internal flavor factors. In: *International Conference on Industrial Technology (ICIT)*. IEEE, pp. 2081–2086.

Chopra, H., Singh, H., Bamrah, M.S., Mahbubani, F., Verma, A., Hooda, N., Rana, P.S., Singla, R.K., & Singh, A.K. (2021). Efficient fruit grading system using spectrophotometry and machine learning approaches. *IEEE Sensors Journal*, 21(14), 16162–16169.

Christensen, C.M. (1997). *The innovator's dilemma: When New Technologies Cause Great Firms to Fail*. Harvard Business School. https://www.hbs.edu/faculty/Pages/item.aspx?num=46

Chun-Fu, W., Xiao-Long, W., Qing-Xie, C., Xiao-Wei, C., & Guo-Dong, L. (2014). Research on visual navigation algorithm of AGV used in the small agile warehouse. In: *2014 Chinese Automation Congress (CAC)*. IEEE, pp. 214–222.

Commandré, Y., Macombe, C., & Mignon, S. (2021). Implications for agricultural producers of using blockchain for food transparency, study of 4 food chains by cumulative approach. *Sustainability*, 13(14), 9843. https://doi.org/10.3390/SU13149843

Cubero, S., Marco-Noales, E., Aleixos, N., Barbé, S., & Blasco, J. (2020). Robhortic: A field robot to detect pests and diseases in horticultural crops by proximal sensing. *Agriculture*, 10(7), 276.

da Silveira, F., Barbedo, J.G.A., da Silva, S.L.C., & Amaral, F.G. (2023a). Proposal for a framework to manage the barriers that hinder the development of agriculture 4.0 in the agricultural production chain. *Computers and Electronics in Agriculture*, 214, 108281. https://doi.org/10.1016/J.COMPAG.2023.108281

da Silveira, F., Barbedo, J.G.A., da Silva, S.L.C., & Amaral, F.G. (2023b). Proposal for a framework to manage the barriers that hinder the development of agriculture 4.0 in the agricultural production chain. *Computers and Electronics in Agriculture*, 214, 108281. https://doi.org/10.1016/J.COMPAG.2023.108281

Demay, J., Ringeval, B., Pellerin, S., & Nesme, T. (2023). Half of global agricultural soil phosphorus fertility derived from anthropogenic sources. *Nature Geoscience*, 16(1), 69–74. https://doi.org/10.1038/s41561–022–01092–0

Deng, X., Lan, Y., Hong, T., & Chen, J. (2016). Citrus greening detection using visible spectrum imaging and C-SVC. *Computers and Electronics in Agriculture*, 130, 147–183.

Dharmasiri, P., Kavalchuk, I., & Akbari, M. (2020). Novel implementation of multiple automated ground vehicles traffic real time control algorithm for warehouse operations: Djikstra approach. *Operations and Supply Chain Management: An International Journal*, 13(4), 396–405.

Ekren, B.Y., & Akpunar, A. (2021). An open queuing network-based tool for performance estimations in a shuttle-based storage and retrieval system. *Applied Mathematical Modelling*, 89, 1678–1695.

Emde, S., Tahirov, N., Gendreau, M., & Glock, C.H. (2021). Routing automated lane-guided transport vehicles in a warehouse handling returns. *European Journal of Operational Research*, 292(3), 1085–1098.

Faheem, M., Liu, J., Chang, G., Ahmad, I., & Peng Y. (2021). Hanging force analysis for realizing low vibration of grape clusters during speedy robotic postharvest handling. *International Journal of Agricultural and Biological Engineering*, 14(4), 62–71.

Found, P., Lahy, A., Williams, S., Hu, Q., & Mason, R. (2018). Towards a theory of operational excellence. *Total Quality Management & Business Excellence*, 29(9–10), 1012–1024. https://doi.org/10.1080/14783363.2018.1486544

Fu, L., Feng, Y., Wu, J., Liu, Z., Gao, F., Majeed, Y., Al-Mallahi, A., Zhang, Q., Li, R., & Cui, Y. (2021). Fast and accurate detection of kiwifruit in orchard using improved YOLOv3-tiny model. *Precision Agriculture*, 22, 754–776.

Gabriëls S.H.E.J., Mishra P., Mensink M.G.J., Spoelstra P., & Woltering E.J. (2020). Non-destructive measurement of internal browning in mangoes using visible and near-infrared spectroscopy supported by artificial neural network analysis. *Postharvest Biology and Technology*, 166, 111206.

Ganguli, S., Selvan, P.T., Nayak, M.M., Chaudhury, S., Espina, R.U., & Ofori, I. (2022). Deep learning based dual channel banana grading system using convolution neural network. *Journal of Food Quality*, 9. https://doi.org/10.1155/2022/6050284

Ghazal, S., Qureshi, W.S., Khan, U.S., Iqbal, J., Rashid, N., & Tiwana, M.I. (2021). Analysis of visual features and classifiers for Fruit classification problem. *Computers and Electronics in Agriculture*, 187, 106267.

Giran, O., Türkakın, O.H., & Anbarcı, M. (2023). Location-based tracking and monitoring infrastructural construction works by using business intelligence tool. *Engineering Management Journal*, 36, 244–255. https://doi.org/10.1080/10429247.2023.2253697

González-Briones, A., Mezquita, Y., Castellanos-Garzón, J.A., Prieto, J., & Corchado, J.M. (2019). Intelligent multi-agent system for water reduction in automotive irrigation processes. *Procedia Computer Science*, 151, 971–976.

Han, W., Jiang, F., & Zhu, Z. (2022). Detection of cherry quality using YOLOV5 model based on flood filling algorithm. *Foods*, 11(8), 1127.

Hasanzadeh, B., Abbaspour-Gilandeh, Y., Soltani-Nazarloo, A., Hernández-Hernández, M., Gallardo-Bernal, I., & Hernández-Hernández, J.L. (2022). Non-destructive detection of fruit quality parameters using hyperspectral imaging, multiple regression analysis and artificial intelligence. *Horticulturae*, 8(7), 598.

Hemming, S., de Zwart, F., Elings, A., Righini, I., & Petropoulou, A. (2019). Remote control of greenhouse vegetable production with artificial intelligence—Greenhouse climate, irrigation, and crop production. *Sensors*, 19(8), 1807.

Hossen, M.S., Haque, I., Islam, M.S., Ahmed, M.T., Nime, M.J., & Islam, M.A. (2020). Deep learning-based classification of papaya disease recognition. In: *3rd International Conference on Intelligent Sustainable Systems (ICISS)*. IEEE, pp. 945–951.

Howard, D.A., Ma, Z., Veje, C., Clausen, A., Aaslyng, J.M., & Jørgensen, B.N. (2021). Greenhouse industry 4.0—Digital twin technology for commercial greenhouses. *Energy Informatics*, 4(2), 1–13. https://doi.org/10.1186/S42162-021-00161-9/TABLES/1

Jarial, S. (2022). Internet of Things application in Indian agriculture, challenges and effect on the extension advisory services—A review. *Journal of Agribusiness in Developing and Emerging Economies*. https://doi.org/10.1108/JADEE-05-2021-0121

Jayashree, S., Malarvizhi, C.A., Nurul, M., & Reza, H. (2020). The challenges and opportunities of industry 4.0—A review. *Asia Proceedings of Social Sciences*, 5(2), 143–148. https://doi.org/10.31580/APSS.V5I2.1152

Jimenez, A.F., Cardenas, P.F., Jimenez, F., Ruiz-Canales, A., & López, A. (2020). A cyber-physical intelligent agent for irrigation scheduling in horticultural crops. *Computers and Electronics in Agriculture*, 148, 105777.

Kalla, A., Prombage, P., & Liyanage, M. (2020). Introduction to IoT. In: *IoT Security*. Wiley, pp. 1–25. https://doi.org/10.1002/9781119527978.ch1

Kang, J., & Gwak, J. (2022). Ensemble of multi-task deep convolutional neural networks using transfer learning for fruit freshness classification. *Multimedia Tools and Applications*. 81, 22355–22377.

Kazemi, M., Asef-Vaziri, A., Shojaei, T., & Amiri, S.A.H.S. (2021). A math-heuristic algorithm for concurrent assignment and sequence scheduling in multi-shuttle shared location automated storage and retrieval systems. *Journal of Supply Chain and Operations Management*, 19(1), 1.

Khadatkar, A., Mehta, C.R., & Sawant, C.P. (2022). Application of robotics in changing the future of agriculture. *Journal of Eco-friendly Agriculture*, 14(1), 48–51.

Khatoon, S., Hasan, M.M., Asif, A., Alshmari, M., & Yap, Y. (2021). Image-based automatic diagnostic system for tomato plants using deep learning. *Computers, Materials & Continua*, 67(1), 595–612.

Kim, J., Pyo, H., Jang, I., Kang, J., Ju, B., & Ko, K. (2022). Tomato harvesting robotic system based on Deep-ToMaToS: Deep learning network using transformation loss for 6D pose estimation of maturity classified tomatoes with side-stem. *Computers and Electronics in Agriculture*, 201, 107300.

Kolhalkar, N.R., Krishnan, V.L., Pandit, A.A., Somkuwar, R.G., & Shaaikh, J.A. (2021). Design and performance evaluation of a novel end-effector with integrated gripper cum cutter for harvesting greenhouse produce. *International Journal of Advanced Technology and Engineering Exploration*, 8, 1479–1481.

Lamm, K., Powell, A., & Lombardini, L. (2021). Identifying critical issues in the horticulture industry: A Delphi analysis during the COVID-19 pandemic. *Horticulturae*, 7(11), 416. https://doi.org/10.3390/HORTICULTURAE7110416

Lekshmi, G., & Rekha, P. (2022). AI based IoT framework for soil analysis and fertilization recommendation for smart coconut farming. In: *2022 8th International Conference on Signal Processing and Communication (ICSC)*. IEEE, pp. 26–31.

Leng, J., Yan, D., Liu, Q., Zhang, H., Zhao, G., Wei, L., Zhang, D., Yu, A., & Chen, X. (2021). Digital twin-driven joint optimisation of packing and storage assignment in large-scale automated high-rise warehouse product-service system. *International Journal of Computer Integrated Manufacturing*, 34(7–8), 783–800.

Lin, Y., Zhao, X., & Chen, L. (2023). A multi-case study of business innovation models for manufacturing capacity sharing platforms, based on a resource orchestration perspective. *Engineering Management Journal*, 36(1), 92–102. https://doi.org/10.1080/10429247.2023.2193127

Liu, B, Zhang, Y, He, D, & Li, Y. (2018) Identification of apple leaf diseases based on deep convolutional neural networks. *Symmetry*, 10(1), 11.

López, J., Zalama, E., & Gómez-García-Bermejo, J. (2022). A simulation and control framework for AGV based transport systems. *Simulation Modelling Practice and Theory*, 116, 102430.

Lu, Y., Lu, R., & Zhang, Z. (2022). Development and preliminary evaluation of a new apple harvest assist and in-field sorting machine. *Applied Engineering in Agriculture*, 38(1), 23–35.

Luo, H., Sheng, D., Zhong, B., Chen, K., Sepasgozar, S.M.E., & Xing, X. (2022). Conceptual framework for the service-oriented management of construction labor resource. *Engineering Management Journal*, 34(4), 543–558. https://doi.org/10.1080/10429247.2021.1954819

Lyu, S., Li, R., Zhao, Y., Li, Z., Fan, R., & Liu, S. (2022). Green citrus detection and counting in orchards based on YOLOv5-CS and AI edge system. *Sensors*, 22(2), 576.

Lyu, S., Zhao, Y., Liu, X., Li, Z., Wang, C., & Shen, J. (2023). Detection of male and female litchi flowers using YOLO-HPFD multi-teacher feature distillation and FPGA-embedded platform. *Agronomy*, 13(4), 987.

MacEachern, C.B., Esau, T.J., Schumann, A.W., Hennessy, P.J., & Zaman, Q.U. (2023). Detection of fruit maturity stage and yield estimation in wild blueberry using deep learning convolutional neural networks. *Smart Agricultural Technology*, 3, 100099.

Mahmud, M.S., Chang, Y.K., Zaman, Q.U., Esau, T.J., Price, G.W., & Prithiviraj, B. (2019) Development of an artificial cloud lighting condition system using machine vision for strawberry powdery mildew disease detection. *Computers and Electronics in Agriculture*, 158, 219–225.

Massah, J., Vakilian, K.A., Shabanian, M., & Shariatmadari, S.M. (2021). Design, development, and performance evaluation of a robot for yield estimation of kiwifruit. *Computers and Electronics in Agriculture*, 185, 106132.

Mazzia, V., Khaliq, A., Salvetti, F., & Chiaberge, M. (2020). Real-time apple detection system using embedded systems with hardware accelerators: An edge AI application. *IEEE Access*, 8, 9102–9114.

Mesa, A.R., & Chiang, J.Y. (2021). Multi-input deep learning model with RGB and hyperspectral imaging for banana grading. *Agriculture*, 11(8), 687.

Minashkina, D., & Happonen, A. (2020). Decarbonizing warehousing activities through digitalization and automatization with WMS integration for sustainability supporting operations. In: *E3S Web of Conferences*. EDP Sciences, Vol. 158, p. 03002.

Miragaia, R., Chávez, F., Díaz, J., Vivas, A., Prieto, M.H., & Moñino, M.J. (2021). Plum ripeness analysis in real environments using deep learning with convolutional neural networks. *Agronomy*, 11(11), 2353.

Nawandar, N.K., & Satpute, V.R. (2019). IoT based low cost and intelligent module for smart irrigation system. *Computers and Electronics in Agriculture*, 162, 979–990.

Ndunagu, J.N., Ukhurebor, K.E., Akaaza, M., & Onyancha, R.B. (2022). Development of a wireless sensor network and IoT-based smart irrigation system. In: *Applied and Environmental Soil Science*, p. 13.

Newton, P., Civita, N., Frankel-Goldwater, L., Bartel, K., & Johns, C. (2020). What is regenerative agriculture? A review of scholar and practitioner definitions based on processes and outcomes. *Frontiers in Sustainable Food Systems*, 4, 194. https://doi.org/10.3389/FSUFS.2020.577723/BIBTEX

Pandian, A.P. (2019). Artificial intelligence application in smart warehousing environment for automated logistics. *Journal of Artificial Intelligence*, 1(02), 63–72.

Park, S., & Kim, J. (2021). Design and implementation of a hydroponic strawberry monitoring and harvesting timing information supporting system based on nano AI-cloud and IoT-edge. *Electronics*, 10(12), 1400.

Parsa, S., Parsons, S., & Ghalamzan, A. (2022). Peduncle gripping and cutting force for strawberry harvesting robotic end-effector design. In: *2022 4th International Conference on Control and Robotics (ICCR)*. IEEE, pp. 59–64.

Partel, V., Nunes, L., Stansley, P., & Ampatzidis, Y. (2019) Automated visionbased system for monitoring Asian citrus Psyllid in orchards utilizing artificial intelligence. *Computers and Electronics in Agriculture*, 162, 328–336.

Pathak, S. (2019). Biomimicry: (innovation inspired by nature). *International Journal of New Technology and Research*, 5(6). https://doi.org/10.31871/IJNTR.5.6.14

Pathmanaban, P., Gnanavel, B.K., & Anandan, S.S. (2023). Guava fruit (*Psidium guajava*) damage and disease detection using deep convolutional neural networks and thermal imaging. *The Imaging Science Journal*, 70(2), 102–116.

Peng, C., Vougioukas, S., Slaughter, D., Fei, Z., & Arikapudi, R. (2021). A strawberry harvest-aiding system with crop-transport harvesting-aid robots: Design, development, and field evaluation. *arXiv preprint arXiv:2107.13063*.

Perçin, S. (2023). An integrated interval type-2 fuzzy set model for evaluating circular low carbon suppliers in a developing country. *Engineering Management Journal*, 36(2), 221–243. https://doi.org/10.1080/10429247.2023.2253691

Prabha, K. (2021). Disease sniffing robots to apps fixing plant diseases: Applications of artificial intelligence in plant pathology—A mini review. *Indian Phytopathology*, 74(1), 13–20.

Raghavendra, A., Guru, D.S., & Rao, M.K. (2021). Mango internal defect detection based on optimal wavelength selection method using NIR spectroscopy. *Artificial Intelligence in Agriculture*, 5, 43–51.

Rahutomo, R., Perbangsa, A.S., Lie, Y., Cenggoro, T.W., & Pardamean, B. (2019). Artificial intelligence model implementation in web-based application for pineapple object counting. In: *2019 International Conference on Information Management and Technology (ICIMTech)*. IEEE, Vol. 1, pp. 525–530.

Rajković, M., Zrnić, N., Kosanić, N., Borovinšek, M., & Lerher, T. (2019). A multi-objective optimization model for minimizing investment expenses, cycle times and CO2 footprint of an automated storage and retrieval systems. *Transport*, 34(2), 275–286.

Rathore, N.S. (2021). Application of *Artificial Intelligence in Agriculture* including horticulture. *International Journal of Innovative Horticulture*, 10(2), 138–141.

Renugadevi, A.S. (2021). Classification of custard apple leaves using deep convolutional networks. *Turkish Journal of Computer and Mathematics Education (TURCOMAT)*, 12(6), 3288–3292.

Rizwan Iqbal, H.M., & Hakim, A. (2022). Classification and grading of harvested mangoes using convolutional neural network. *International Journal of Fruit Science*, 22(1), 95–109.

Rong, J., Fu, J., Zhang, Z., Yin, J., Tan, Y., Yuan, T., & Wang, P. (2022). Development and evaluation of a watermelon-harvesting robot prototype: Vision system and end-effector. *Agronomy*, 12(11), 2836.

Rysz, M.W., & Mehta S.S. (2021). A risk-averse optimization approach to human-robot collaboration in robotic fruit harvesting. *Computers and Electronics in Agriculture*, 182, 106018.

Selvaraj, M.G., Vergara, A., Ruiz, H., Safari, N., Elayabalan, S., Ocimati, W., & Blomme, G. (2019). AI-powered banana diseases and pest detection. *Plant Methods*, 15, 1–11.

Seyyedhasani, H., Peng, C., Jang, W.J., & Vougioukas, S.G. (2020). Collaboration of human pickers and crop-transporting robots during harvesting—Part I: Model and simulator development. *Computers and Electronics in Agriculture*, 142, 105324.

Shadrin, D., Menshchikov, A., Ermilov, D., & Somov, A. (2019). Designing future precision agriculture: Detection of seeds germination using artificial intelligence on a low-power embedded system. *IEEE Sensors Journal*, 19(23), 11573–11582.

Sharma, S., Malik, A., Sharma, C., Batra, I., Kaswan, M.S., & Garza-Reyes, J.A. (2023). Adoption of industry 4.0 in different sectors: A structural review using natural language processing. *International Journal on Interactive Design and Manufacturing (IJIDeM)*, 18, 6069–6091. https://doi.org/10.1007/S12008-023-01550-Y

Shekhar, Y., Dagur, E., Mishra, S., & Sankaranarayanan, S. (2014). Intelligent IoT based automated irrigation system. *International Journal of Applied Engineering Research*, 12(18), 7306–7320.

Siddiqui, R., Azmat, M., Ahmed, S., & Kummer, S. (2022). A hybrid demand forecasting model for greater forecasting accuracy: The case of the pharmaceutical industry. *Supply Chain Forum: An International Journal*, 23(2), 124–134. https://doi.org/10.1080/16258312.2021.1967081

Sivagami, A., Hareeshvare, U., Maheshwar, S., & Venkatachalapathy, V.S.K. (2018). Automated irrigation system for greenhouse monitoring. *Journal of the Institution of Engineers (India): Series A*, 99(2), 183–191.

Stevens, L.L., Whitehead, C., & Singhal, A. (2022). Cultivating cooperative relationships: Identifying learning gaps when teaching students systems thinking biomimicry. *Biomimetics (Basel, Switzerland)*, 7(4). https://doi.org/10.3390/BIOMIMETICS7040184

Sushma, K., & Santhosha, B. (2022). *Chapter—5 Post-Harvest Technology in Vegetables*. In: R.K. Naresh, Ed., Vol. 83, p. 69.

Tripathi, V., Chattopadhyaya, S., Mukhopadhyay, A.K., Sharma, S., Kumar, V., Li, C., & Singh, S. (2023). Lean, green, and smart manufacturing: An ingenious framework for enhancing the sustainability of operations management on the shop floor in industry 4.0. *Proceedings of the Institution of Mechanical Engineers, Part E: Journal of Process Mechanical Engineering*, 238(4), 1976–90. https://doi.org/10.1147/09544089231159834.

Tuysuz, M.F., & Trestian, R. (2020). From serendipity to sustainable green IoT: Technical, industrial and political perspective. *Computer Networks*, 182, 107469. https://doi.org/10.1016/j.comnet.2020.107469

Wang, F., Zhao, C., Yang, H., Jiang, H., Li, L., & Yang, G. (2022). Non-destructive and in-site estimation of apple quality and maturity by hyperspectral imaging. *Computers and Electronics in Agriculture*, 195, 106843.

Wang, M., Guo, L., Zhang, Y., & Liang, X. (2015). A method of mobile robotic drilling trajectory planning for large-scale components. In: *ASME International Mechanical Engineering Congress and Exposition* (57366, V02BT02A064).

Wendel, A., Underwood, J., & Walsh, K. (2018). Maturity estimation of mangoes using hyperspectral imaging from a ground based mobile platform. *Computers and Electronics in Agriculture*, 155, 298–313.

Witczak, M., Majdzik, P., Stetter, R., & Lipiec, B. (2020). A fault-tolerant control strategy for multiple automated guided vehicles. *Journal of Manufacturing Systems*, 55, 56–68.

Xiong, Z., Wang, L., Zhao, Y., & Lan, Y. (2023). Precision detection of Dense litchi fruit in UAV images based on improved YOLOv5 model. *Remote Sensing*, 15(16), 4014.

Xu, L.P., Liu, N., Zhou, L., & Lu, L.Q. (2023). Fiscal risk management of public–private partnership projects: A multidimensional identification and assessment framework. *Engineering Management Journal*, 36(2), 166–192. https://doi.org/10.1080/10429247.2023.2231490

Yang, D., Wu, Y., & Ma, W. (2021). Optimization of storage location assignment in automated warehouse. *Microprocessors and Microsystems*, 80, 103356.

Yang, M., Kumar, P., Bhola, J., & Shabaz, M. (2021). Development of image recognition software based on artificial intelligence algorithm for the efficient sorting of apple fruit. *International Journal of System Assurance Engineering and Management*, 13(Suppl 1), 322–330.

Yasin, H.M., Zeebaree, S.R., & Zebari, I.M. (2019). Arduino based automatic irrigation system: Monitoring and SMS controlling. In: *2019 4th Scientific International Conference Najaf (SICN)*. IEEE, 109–114.

Zhang, Z., Guo, Q., Chen, J., & Yuan, P. (2018) Collision-free route planning for multiple agvs in an automated warehouse based on collision classification. *IEEE Access*, 6, 26022–26035.

Zhang, L., Jia, J., Gui, G., Hao, X., Gao, W., & Wang, M. (2018). Deep learning based improved classification system for designing tomato harvesting robot. *IEEE Access*, 6, 67940–67950.

Zhang, X., Qiao, Y., Meng, F., Fan, C., & Zhang, M. (2018). Identification of maize leaf diseases using improved deep convolutional neural networks. *IEEE Access*, 6, 30370–30377.

Zhou, Z., Zahid, U., Majeed, Y., Nisha, Mustafa, S, Sajjad, M.M., Butt, H.D., & Fu, L. (2023). Advancement in artificial intelligence for on-farm fruit sorting and transportation. *Frontiers in Plant Science*, 6(14), 1082860.

Zhou, H., Wang, X., Au, W., Kang, H., & Chen, C. (2022). Intelligent robots for fruit harvesting: Recent developments and future challenges. *Precision Agriculture*, 23(5), 1856–1907.

Zhou, X., Lee, W.S., Ampatzidis, Y., Chen, Y., Peres, N., & Fraisse, C. (2021). Strawberry maturity classification from UAV and near-ground imaging using deep learning. *Smart Agricultural Technology*, 1, 100001.

Index

D

I

J

K

L

M

N

R

T